This **Student's Solutions Manu**
and Lerner's ALGEBRA AND TR
This source, together with the tex
your own pace, and can help you to improve your understanding and skills. Through these efforts you will gain confidence and your performance will improve.

Complete detailed solutions are provided for every fourth exercise in each section, beginning with Exercise 1. Worked out solutions for each sample test question (at the end of every chapter) are also included.

This *Student's Solutions Manual* aids you in three very distinct ways:

- to grasp the difficult concepts that may have slowed you down in class
- to recognize and apply correct mathematical usage
- to arrive at the correct answers *at your own pace* and *on your own time*

—to gain strength and confidence in the subject matter.

Student's Solutions Manual

Third Edition

Algebra and Trigonometry

A Pre-Calculus Approach

Max A. Sobel Norbert Lerner

Margaret Hall Babcock

Montclair State College

PRENTICE HALL, ENGLEWOOD CLIFFS, NEW JERSEY 07632

Editorial/production: Stefanie Siegel & Jean Hunter
Cover design: Meryl Poweski
Manufacturing buyer: Paula Massenaro

A Division of Simon & Schuster
Englewood Cliffs, New Jersey 07632

Printed in the United States of America

10 9 8 7 6 5 4 3 2 1

ISBN 0-13-021627-5 01

PRENTICE-HALL INTERNATIONAL (UK) LIMITED, *London*
PRENTICE-HALL OF AUSTRALIA PTY. LIMITED, *Sydney*
PRENTICE-HALL CANADA INC., *Toronto*
PRENTICE-HALL HISPANOAMERICANA, S.A., *Mexico*
PRENTICE-HALL OF INDIA PRIVATE LIMITED, *New Delhi*
PRENTICE-HALL OF JAPAN, INC., *Tokyo*
SIMON & SCHUSTER ASIA PTE. LTD., *Singapore*
EDITORA PRENTICE-HALL DO BRASIL, LTDA., *Rio de Janeiro*

CONTENTS

PREFACE

This solutions manual is a supplement to the third edition of Algebra and Trigonometry, A Pre-Calculus Approach, by Max A. Sobel and Norbert Lerner. It contains worked-out solutions to every fourth exercise in the book, beginning with number 1. Thus, you will find solutions for numbers 1, 5, 9, 13, ..., for each section. Since the complexity of the exercises usually progresses gradually, the solutions provided in this manual serve as good illustrations for many of the remaining exercises. The manual also contains worked-out solutions to every sample test question for each chapter.

In using this manual, we advise that you make substantial efforts to solve a problem before referring to the solution given here. And when you do consult a given solution, do not just copy the work presented. Rather, try to understand the given solution, and then try to solve the same problem on your own with minimal reference. Also, keep in mind that there are usually a variety of ways in which a problem can be solved correctly, so that your solution does not necessarily have to be the same as the one given.

We wish to thank Margaret Babcock for her careful attention to detail, and all her hard work in preparing the manuscript for this solutions manual. Also, we extend our thanks to Betty Thomas for her excellent typing of the manuscript.

M. A. S.

N. L.

Student's Solutions Manual

Algebra and Trigonometry

CHAPTER 1

SETS OF NUMBERS, EQUATIONS, AND INEQUALITIES

Section 1.1 Sets of Numbers

1) {1, 2, 3, 4}

5) {-2, -1}

9) ∅

13) True

17) False. Some points on the number line are associated with irrational numbers.

21) c, d, f

25) a, b, c, d, f

29) e, f

33)

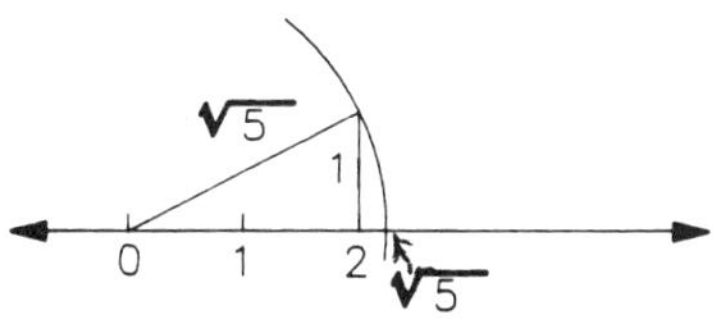

By Pythagorean theorem:

$$x^2 = 1^2 + 2^2$$
$$x^2 = 5$$
$$x = \sqrt{5}$$

Set a compass to the length of the hypotenuse. Using 0 as the center, swing an arc to intersect the number line. The point of intersection is $\sqrt{5}$ units from 0.

37) $5\overline{)4.0}$ = 0.8 $\qquad \frac{4}{5} = 0.8$

41) $13\overline{)45.000000}$ = $3.\overline{461538}$

45) a)
$$\begin{aligned} 100n &= 45.4545\ldots \\ n &= 0.4545\ldots \\ \hline 99n &= 45 \\ n &= \frac{45}{99} = \frac{5}{11} \end{aligned}$$

b)
$$\begin{aligned} 100n &= 37.373737\ldots \\ n &= 0.373737\ldots \\ \hline 99n &= 37 \\ n &= \frac{37}{99} \end{aligned}$$

c)
$$\begin{aligned} 1000n &= 234.234234\ldots \\ n &= 0.234234\ldots \\ \hline 999n &= 234 \\ n &= \frac{234}{999} = \frac{26}{111} \end{aligned}$$

Section 1.2 Properties of the Real Numbers

1) closure property for addition

5) commutative property for multiplication

9) Identity property for addition

13) Multiplication property of zero

17) $7 + 3 = 3 + 7$

21) $5(8 + 7) = (5 \times 8) + (5 \times 7)$

25) (i) $a(-b) = -(ab)$ (ii) Definition of division

(iii) Associative property for multiplication

29) No. $2^3 \neq 3^2$ is a counter example.

33) False. $7 \div 3 = 2.\overline{3}$ and $2.\overline{3}$ is not an integer.

37) True

41) If $10 - x = 7$, then $7 + x = 10$.
If $7 + x = 10$, then $10 - x = 7$.

45) If triangle ABC is congruent to triangle DEF, then the three sides of one triangle are congruent to the three sides of the other triangle.
If three sides of triangle ABC are congruent to three sides of triangle DEF, then triangle ABC is congruent to triangle DEF.

49) Let $\frac{0}{0} = x$, where x is some number. By the definition of division, $0 \cdot x = 0$. This equation is true for any number x, so there is no unique value for $\frac{0}{0}$. For this reason $\frac{0}{0}$ is not defined.

Section 1.3 Introduction to Equations and Problem Solving

1)
$$3x - 2 = 10$$
$$3x - 2 + 2 = 10 + 2$$
$$3x = 12$$
$$\frac{1}{3}(3x) = \frac{1}{3}(12)$$
$$x = 4$$

5)
$$-3x - 5 = 7$$
$$-3x - 5 + 5 = 7 + 5$$
$$-3x = 12$$
$$-\frac{1}{3}(-3x) = -\frac{1}{3}(12)$$
$$x = -4$$

9)
$$2(x + 1) = 11$$
$$2x + 2 = 11$$
$$2x + 2 - 2 = 11 - 2$$
$$2x = 9$$
$$\frac{1}{2}(2x) = \frac{1}{2}(9)$$
$$x = \frac{9}{2}$$

13)
$$\frac{1}{2}x + 7 = 2x - 3$$
$$\frac{1}{2}x + 7 - 7 = 2x - 3 - 7$$
$$\frac{1}{2}x = 2x - 10$$
$$\frac{1}{2}x - 2x = 2x - 10 - 2x$$
$$-\frac{3}{2}x = -10$$
$$-\frac{2}{3}\quad -\frac{3}{2}x = -\frac{2}{3}(-10)$$
$$x = \frac{20}{3}$$

17)
$$\frac{3}{5}(x - 5) = x + 1$$
$$\frac{3}{5}x - 3 = x + 1$$
$$\frac{3}{5}x - 3 + 3 = x + 1 + 3$$
$$\frac{3}{5}x = x + 4$$
$$\frac{3}{5}x - x = x + 4 - x$$
$$-\frac{2}{5}x = 4$$
$$-\frac{5}{2}\quad -\frac{2}{5}x = -\frac{5}{2}(4)$$
$$x = -10$$

21)
$$-3(x + 2) + 1 = x - 25$$
$$-3x - 6 + 1 = x - 25$$
$$-3x - 5 = x - 25$$
$$-3x - 5 + 5 = x - 25 + 5$$
$$-3x = x - 20$$
$$-3x - x = x - 20 - x$$
$$-4x = -20$$
$$-\frac{1}{4}(-4x) = -\frac{1}{4}(-20)$$
$$x = 5$$

25) $x + 2(1/6x + 2) = 6/5x + 16$

$x + 1/3x + 4 = 6/5x + 16$

$4/3x + 4 = 6/5x + 16$

$4/3x + 4 - 4 = 6/5x + 16 - 4$

$4/3x = 6/5x + 12$

$4/3x - 6/5x = 6/5x + 12 - 6/5x$

$2/15x = 12$

$15/2(2/15x) = 15/2(12)$

$x = 90$

29) $N = 10t + u$

$N - u = 10t + u - u$

$N - u = 10t$

$1/10(N - u) = 1/10(10t)$

$\frac{N - u}{10} = t$

33) $6 + 4v = w - 1$

$6 + 4v - 6 = w - 1 - 6$

$4v = w - 7$

$1/4(4v) = 1/4(w - 7)$

$v = \frac{w - 7}{4}$

37) $\frac{5F - 160}{9} = C$

$\frac{5(32) - 160}{9} = C$

$\frac{160 - 160}{9} = C$

$0 = C$

$32^oF = 0^o C$

41)

Value of each stamp	Number of stamps	Value of stamps
10¢	x	$10x$
20¢	$x + 3$	$20(x + 3)$
22¢	$\frac{x}{2}$	$11x$

The value of the stamps is in cents, so change $4.70 to 470 cents.

$10x + 20(x + 3) + 11x = 470$

$10x + 20x + 60 + 11x = 470$

$41x = 410$

$x = 10$

If $x = 10$, then $x + 3 = 13$ and $\frac{x}{2} = 5$. Carlos bought ten 10¢ stamps, thirteen 20¢ stamps, and five 22¢ stamps.

45) Let x represent the first odd integer. $(x + 2)$ and $(x + 4)$ represent the next consecutive odd integers. Their sum is 180.

$$x + (x + 2) + (x + 4) = 180$$
$$3x + 6 = 180$$
$$3x = 174$$
$$x = 58$$

58 is the only root of this equation, but it is even. Therefore there is no odd solution and the measures of the angles of a triangle cannot be represented by consecutive odd integers.

49) Let x represent the first odd integer. $(x + 2)$ and $(x + 4)$ represent the next consecutive odd integers. Their sum is 237.

$$x + (x + 2) + (x + 4) = 237$$
$$3x + 6 = 237$$
$$3x = 231$$
$$x = 77$$

If $x = 77$, then $x + 2 = 79$ and $x + 4 = 81$. The three integers are 77, 79, 81.

53)

Amount invested	Interest Rate	Interest for 1 yr. I = Prt
x	0.09	$0.09x$
$x + 2700$	0.12	$0.12(x + 2700)$

The sum of the two interest amounts is \$1794.

$$0.09x + 0.12(x + 2700) = 1794$$
$$0.09x + 0.12x + 324 = 1794$$
$$0.21x + 324 = 1794$$
$$0.21x = 1470$$
$$21x = 147000$$
$$x = \$7000$$

If $x = \$7000$, then $x + 2700 = \$9700$. The advisor invested \$7000 at 9% and \$9700 at 12%.

Section 1.4 Introduction to Statements of Inequality

1) $-100 < 2$

5) $0.7 < \frac{7}{9}$

9) $-1/3 < -0.33$

13) True

17) True

21)
$$x - 7 \geq -3$$
$$x - 7 + 7 \geq -3 + 7$$
$$x \geq 4$$
$$\{x \mid x \geq 4\}$$

25)
$$3x > -21$$
$$1/3(3x) > 1/3(-21)$$
$$x > -7$$
$$\{x \mid x > -7\}$$

29)
$$5x - 3 \leq 22$$
$$5x - 3 + 3 \leq 22 + 3$$
$$5x \leq 25$$
$$1/5(5x) \leq 1/5(25)$$
$$x \leq 5$$
$$\{x \mid x \leq 5\}$$

33)
$$-5x + 5 < -3x + 1$$
$$-5x + 5 - 5 < -3x + 1 - 5$$
$$-5x < -3x - 4$$
$$-5x + 3x < -3x - 4 + 3x$$
$$-2x < -4$$
$$1/2(-2x) > -1/2(-4)$$
$$x > 2$$
$$\{x \mid x > 2\}$$

37)
$$1/2x - 5 > 1/4x + 3$$
$$1/2x - 5 + 5 > 1/4x + 3 + 5$$
$$1/2x > 1/4x + 8$$
$$1/2x - 1/4x > 1/4x + 8 - 1/4x$$
$$1/4x > 8$$
$$4(1/4x) > 4(8)$$
$$x > 32$$
$$\{x \mid x > 32\}$$

41) $\frac{1}{x} < 0$

The fraction is negative when numerator and denominator have opposite signs.

$1 > 0$, so $x < 0$.

$$\{x \mid x < 0\}$$

45) $3x - 2 \not< 1$

$3x - 2 \geq 1$

$3x - 2 + 2 \geq 1 + 2$

$3x \geq 3$

$1/3(3x) \geq 1/3(3)$

$x \geq 1$

$\{x \mid x \geq 1\}$

49) $5/3x \not< 2x - 1$

$5/3x \geq 2x - 1$

$5/3x - 2x \geq 2x - 1 - 2x$

$-1/3x \geq -1$

$-3(-1/3x) \leq -3(-1)$

$\{x \mid x \leq 3\}$

53) Prove: If $0 < a < 1$, then $a^2 < a$

$a < 1$

$a \cdot a < a \cdot 1$ Because $a > 0$

$a^2 < a$

Section 1.5 Graphs of Inequalities on the Number Line

1) 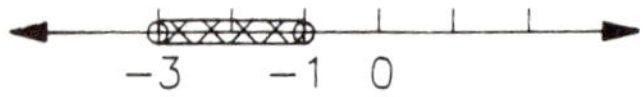

5) 

9) [-5, 2]

13) (-10, 10)

17) [-2, ∞)

21) $-1 \le x < 3$; [-1, 3)

25) $x < 1$; (- ∞, 1)

29) $\frac{1}{2}x + 2 \le 1$

$\frac{1}{2}x \le -1$

$x \le -2$

$\{x \mid x \le -2\}$

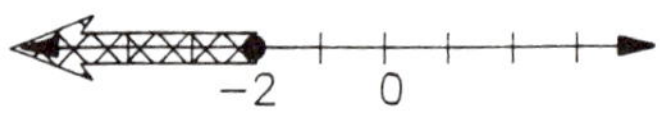

33) a) $-2 < x$

$-2 + 2 < x + 2$

$0 < x + 2$

$x + 2$ is positive

b) $-2 < x$

$-2 + 4 < x + 4$

$2 < x + 4$

$x + 4$ is positive

37) $(x + 1)(x - 3) > 0$

Interval	$x < -1$	$-1 < x < 3$	$3 < x$
Sign of $x + 1$	-	+	+
Sign of $x - 3$	-	-	+
Sign of $(x + 1)(x - 3)$	+	-	+

Solution set: $\{x \mid x < -1 \text{ or } x > 3\}$

41) $(x + 5)(2x - 1) \le 0$

$x + 5 = 0 \qquad 2x - 1 = 0$

$x = -5 \qquad 2x = 1$

$x = 1/2$

Interval	$x < -5$	$-5 < x < 1/2$	$1/2 < x$
Sign of $x + 5$	-	+	+
Sign of $2x - 1$	-	-	+
Sign of $(x + 5)(2x - 1)$	+	-	+

Solution set: $\{x \mid -5 \le x \le 1/2\}$

45) $\dfrac{x + 5}{2 - x} < 0$

$x + 5 = 0 \qquad 2 - x = 0$

$x = -5 \qquad 2 = x$

(But note $x \neq 2$ because $2 - x$ cannot be 0.)

Interval	$x < -5$	$-5 < x < 2$	$2 < x$
Sign of $x + 5$	-	+	+
Sign of $2 - x$	+	+	-
Sign of $\dfrac{x + 5}{2 - x}$	-	+	-

Solution set: $\{x \mid x < -5 \text{ or } x > 2\}$

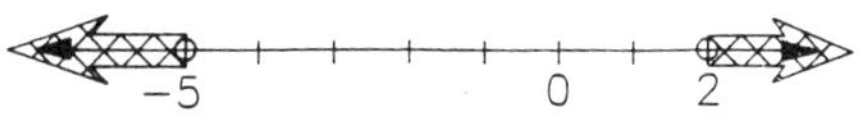

49) $(x + 3)(x - 2)^2 > 0$

Because any square must be positive, $(x - 2)^2$ is always positive. $(x + 3)(x - 2)^2$ will be positive when $(x + 3)$ is positive.

$x + 3 > 0$

$x > -3$

Solution: $\{x \mid x > -3\}$

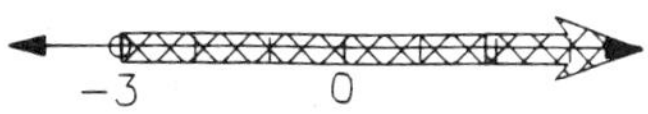

53) $(x - 2)(x - 5) > 0$ when $(x - 2)$ and $(x - 5)$ have the same signs. From the graph, both expressions are negative when $x < 2$, both positive when $x > 5$. The solution set is $\{x \mid x < 2 \text{ or } x > 5\}$.

57) Let x represent the grade on the fourth test. $\frac{2x + 85 + 86 + 93}{5}$ represents the test average.

$$86 < \frac{2x + 85 + 86 + 93}{5} < 90$$

$$430 < 2x + 264 < 450$$
$$166 < 2x < 186$$
$$83 < x < 93$$

The student must score at least 83 on the fourth test but less than 93.

61) a) The maximum girth is determined by the 110-inch limit for length plus girth.

$$\ell + g \leq 110$$
$$42 + g \leq 110$$
$$g \leq 68$$

The minimum girth is determined by the 2-inch minimum on height and width.

$$g = 2w + 2h$$
$$g \geq 2 \cdot 2 + 2 \cdot 2$$
$$g \geq 8$$

Solution: $8 \leq g \leq 68$

b) $g = 2w + 2h$

$g = 2 \cdot 18 + 2h$

$\ell + g \leq 110$

$42 + (2 \cdot 18 + 2h) \leq 110$

$42 + 36 + 2h \leq 110$

$78 + 2h \leq 110$

$2h \leq 32$

$h \leq 16$

Including the 2" minimum, the solution is $2 \leq h \leq 16$.

Section 1.6 Absolute Value

1) False

5) True

9) False

13) $\frac{-12 + 0}{2} = -6$

17) $|x| = \frac{1}{2}$ $\quad x = \frac{1}{2}$ $\quad$ or $-x = \frac{1}{2}$

$x = -\frac{1}{2}$

$x = \pm\frac{1}{2}$

21) $|2x - 3| = 7$

$2x - 3 = 7$ $\quad$ or $\quad$ $-(2x - 3) = 7$

$2x = 10$ $\quad\quad$ $-2x + 3 = 7$

$x = 5$ $\quad\quad$ $-2x = 4$

$x = -2$

$x = 5$ or $x = -2$

25) $|3x + 4| = 16$

$3x + 4 = 16$ $\quad$ or $\quad$ $-(3x + 4) = 16$

$3x = 12$ $\quad\quad$ $3x - 4 = 16$

$x = 4$ $\quad\quad$ $-3x = 20$

$x = -\frac{20}{3}$

$x = 4$ or $x = -\frac{20}{3}$

29) $|x + 1| = 3$

$x + 1 = 3$ $\quad$ or $\quad$ $-(x + 1) = 3$

$x = 2$ $\quad\quad$ $-x - 1 = 3$

$-x = 4$

$x = -4$

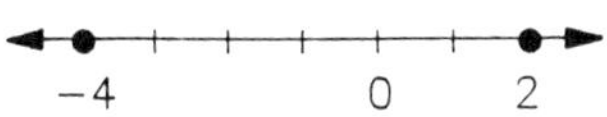

$x = 2$ or $x = -4$

33) $|x + 2| \leq 3$

$-3 \leq x + 2 \leq 3$

$-5 \leq x \leq 1$

37) $|x| \geq 5$

$x \leq -5$ or $x \geq 5$

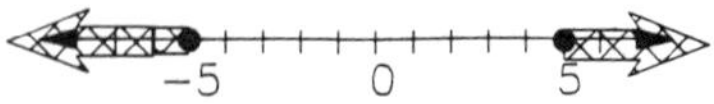

41) $|x - 3| < 0.1$

$-0.1 < x - 3 < 0.1$

$2.9 < x < 3.1$

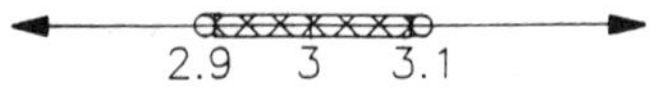

45) $|4 - x| < 2$

$-2 < 4 - x < 2$

$-6 < -x < -2$

$6 > x > 2$

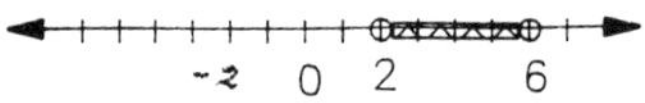

49) $\dfrac{1}{|x - 3|} > 0$

$|x - 3| > 0$

$x \neq 3$

Numerator and denominator must have the same signs. $1 > 0$, so denominator is positive.

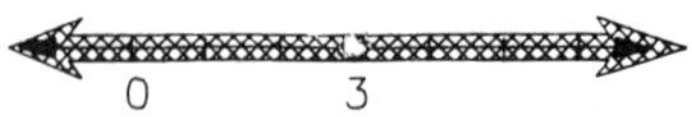

53) If $x < 0$ and $y > 0$, then $\frac{x}{y} < 0$. Therefore $\left|\frac{x}{y}\right| = -\frac{x}{y}$ by the definition of absolute value.

$$= \frac{-x}{y}$$

$$= \frac{|x|}{|y|} \text{ because}$$

$x < 0$ and $y > 0$

57) a) $\frac{9}{10} < x < \frac{11}{10}$

b) $\frac{27}{10} < 3x < \frac{33}{10}$

c) $-\frac{3}{10} < 3x - 3 < \frac{3}{10}$

d) $-\frac{3}{10} < y - 8 < \frac{3}{10}$

e) $|y - 8| < \frac{3}{10}$

Sample Test Questions for Chapter 1

Part I

1)

	Whole Numbers	Integers	Rational numbers	Irrational numbers	Negative Integers
$\frac{-6}{3}$		✓	✓		✓
0.231			✓		
$\sqrt{12}$				✓	
$\sqrt{\frac{9}{4}}$			✓		
1987	✓	✓	✓		

2) Construct a perpendicular of length 1 at the point with coordinate 3. The hypotenuse of this right triangle can be found using the Pythagorean Theorem.

$$1^2 + 3^2 = h^2$$
$$10 = h^2$$
$$\sqrt{10} = h$$

Use a compass to mark off an arc with center at origin and radius equal to the hypotenuse. The point where the arc intersects the number line has coordinate $\sqrt{10}$.

3) a) False. All irrational numbers are real numbers.

b) True.

c) False. The integers are rational numbers so no irrational can can be an integer.

d) True.

e) False. $x < y$
$x - 5 < y - 5$ (by the addition property of order)

f) False. The absolute value of a sum is sometimes less than the sum of the absolute values (See property 8).

g) True.

h) False. It means x is within 3 units of 2 on the number line.

4) a) Distributive property

b) Identity property for addition

c) Identity property for multiplication

d) Definition of subtraction

e) Associative property for multiplication

f) Definition of $a < b$

5) a) If $a = b$, then $a + c = b + c$.

b) If $a = b$, then $ac = bc$.

6) $\frac{-4 + 6}{2} = 1$

7) Let x represent the unknown coordinate.

$$\frac{-7 + x}{2} = -2$$
$$-7 + x = -4$$
$$x = 3$$

8)
$$\frac{3}{4}x - 1 = 11$$
$$\frac{3}{4}x = 12$$
$$\frac{4}{3}\left(\frac{3}{4}x\right) = \frac{4}{3}(12)$$
$$x = 16$$

9)
$$4x + 20 = 2x - 2$$
$$2x + 20 = -2$$
$$2x = -22$$
$$x = -11$$

10)
$$3(x + 2) = x - 3$$
$$3x + 6 = x - 3$$
$$2x + 6 = -3$$
$$2x = -9$$
$$x = \frac{-9}{2}$$

11)
$$\frac{2}{3}(x - 3) + 1 = 2x + 3$$
$$\frac{2}{3}x - 2 + 1 = 2x + 3$$
$$\frac{2}{3}x - 1 = 2x + 3$$
$$-\frac{4}{3}x - 1 = 3$$
$$-\frac{4}{3}x = 4$$
$$-\frac{3}{4}\left(-\frac{4}{3}x\right) = -\frac{3}{4}(4)$$
$$x = -3$$

12) Let w represent the width,

$2w + 5$ represents the length.

$$P = 2w + 2\ell$$
$$52 = 2w + 2(2w + 5)$$
$$52 = 2w + 4w + 10$$
$$52 = 6w + 10$$
$$42 = 6w$$
$$7 = w$$

If $w = 7$, then $2w + 5 = 19$.

The rectangle has width 7 inches and length 19 inches.

13) First make a sketch

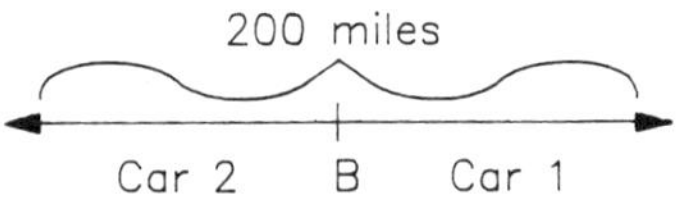

Let t represent the time the first car travels.

$t - 1$ represents the time the second car travels.

Using the formula distance = rate x time, represent the distance each car travels.

$55t$ is the distance the first car travels.

$45(t - 1)$ is the distance the second car travels.

The sum of their distances is 200 miles.

$$55t + 45(t - 1) = 200$$
$$55t + 45t - 45 = 200$$
$$100t = 245$$
$$t = 2.45 \text{ hours}$$

This is <u>not</u> 2 hours and 45 minutes. It is $2\,\frac{45}{100}$ hr. $= 2\,\frac{9}{20}$ hr. = 2 hours and 27 minutes. This is the travel time of the first car, which started at noon.

Answer: At 2:27 pm the cars will be 200 miles apart.

14) 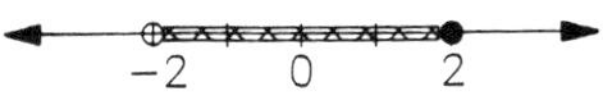

15) 

16) 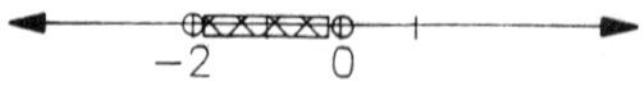

17) 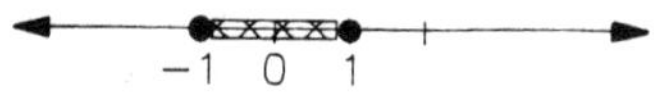

18) $|3 - 4x| = 2$

$$3 - 4x = 2 \quad \text{or} \quad -(3 - 4x) = 2$$
$$-4x = -1 \qquad -3 + 4x = 2$$
$$x = \frac{1}{4} \qquad 4x = 5$$
$$x = \frac{5}{4}$$

$x = 1/4$ or $x = 5/4$

19) $\dfrac{|x + 2|}{x + 2} = -1$

$x + 2 = -|x + 2|$

$x + 2 < 0$

$x < -2$

20) $\dfrac{|x - 2|}{x - 2} = 1$

$x - 2 = |x - 2|$

$x - 2 > 0$

$x > 2$

21) $|x + 2| < 1$

$-1 < x + 2 < 1$

$-3 < x < -1$

22) $2(5x - 1) < x$

$10x - 2 < x$

$9x < 2$

$x < \frac{2}{9}$

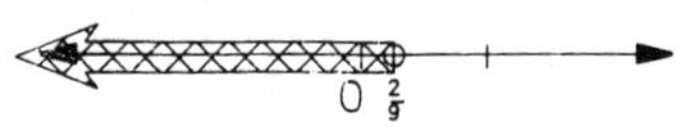

23) $|2x - 1| \geq 3$

$2x - 1 \leq -3$ or $2x - 1 \geq 3$

$2x \leq -2 \qquad 2x \geq 4$

$x \leq -1 \qquad x \geq 2$

$x \leq -1$ or $x \geq 2$

24) $\dfrac{3}{x - 2} > 0$

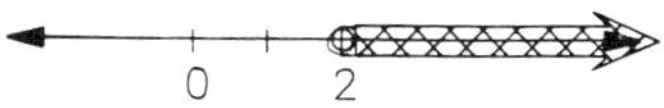

$x - 2 > 0$

$x > 2$

25) $(2x - 1)(x + 5) < 0$

$2x - 1 = 0$ $\quad$ $x + 5 = 0$

$2x = 1$ $\quad$ $x = -5$

$x = 1/2$

Interval	$x < -5$	$-5 < x < 1/2$	$1/2 < x$
Sign of $2x - 1$	-	-	+
Sign of $x + 5$	-	+	+
Sign of $(2x - 1)(x + 5)$	+	-	+

Solution: $-5 < x < 1/2$

Part II

1) (d)

2) (c)

3) (b)

4) (e)

5) (a)

$A = 1/2(b + h)$

$A = 1/2b + 1/2h$

$A - 1/2b = 1/2h$

$2(A - 1/2b) = h$

$2A - b = h$

6) (b)

$-\dfrac{2}{x - 3} < 0$

$\dfrac{2}{x - 3} > 0$

$x - 3 > 0$

$x > 3$

7) (c)

8) (a)

Interval	$x < 1$	$1 < x < 3$	$3 < x$
Sign of $1 - x$	+	-	-
Sign of $x - 3$	-	-	+
Sign of $\frac{1 - x}{x - 3}$	-	+	-

9) (d)

10) (b)

11) (e) $|x - 1| < 2$

$-2 < x - 1 < 2$

$-1 < x < 3$

12) (a)

13) (d)

14) (b) $-3(x + 1) < 2x + 2$

$-3x - 3 < 2x + 2$

$-5x - 3 < 2$

$-5x < 5$

$x > -1$

$(-1, \infty)$

15) (c)

Let x represent the original width.

$3x - 2$ represents the original length.

$x - 1$ is the new width.

$(3x - 2) + 5$ is the new length.

The new length is 5 times the new width.

$(3x - 2) + 5 = 5(x - 1)$

$3x + 3 = 5(x - 1)$

CHAPTER 2
FUNDAMENTALS OF ALGEBRA

Section 2.1 Integral Exponents

1) False $3^4 \cdot 3^2 = 3^6$

5) True

9) False $3^4 + 3^4 = 2 \cdot 3^4$

13) False $\frac{1}{2^{-3}} = 2^3$

17) $2^0 + 2^1 + 2^2 = 1 + 2 + 4 = 7$

21) $\frac{1}{2}^4 (-2)^4 = \frac{1}{16} \cdot 16 = 1$

25) $\frac{2^3 \cdot 3^4 \cdot 4^5}{2^2 \cdot 3^3 \cdot 4^4} = 2^{3-2} \cdot 3^{4-3} \cdot 4^{5-4} = 2 \cdot 3 \cdot 4 = 24$

29) $\frac{2}{3}^{-2} + \frac{2}{3}^{-1} = \frac{3}{2}^2 + \frac{3}{2}^1 = \frac{9}{4} + \frac{3}{2} = \frac{15}{4}$

33) $x^3 \cdot x^9 = x^{12}$

37) $(-2a^2b^0)^4 = (-2)^4(a^2)^4(b^0)^4 = 16 \cdot a^8 \cdot 1^4 = 16a^8$

41) $\frac{x^3}{y^2}^4 \frac{-y}{x^2}^2 = \frac{x^{12}}{y^8} \cdot \frac{(-y)^2}{x^4} = \frac{x^{12}y^2}{x^4y^8} = \frac{x^{12-4}}{y^{8-2}} = \frac{x^8}{y^6}$

45) $\frac{5x^0y^{-2}}{x^{-1}y^{-2}} = 5x^{0-(-1)}y^{-2-(-2)} = 5x^1y^0 = 5x$

49) $\frac{(a + b)^{-2}}{(a + b)^{-8}} = (a + b)^{-2-(-8)} = (a + b)^6$

53) $\frac{(a + 3b)^{-12}}{(a + 3b)^{10}} = (a + 3b)^{-12-10} = (a + 3b)^{-22} = \frac{1}{(a + 3b)^{22}}$

57) $\left(\frac{x^{-2}}{y^3}\right)^{-1} = \frac{(x^{-2})^{-1}}{(y^3)^{-1}} = \frac{x^2}{x^{-3}} = x^2y^3$

61) $2^x \cdot 2^3 = 2^{x+3} = 2^{12}$

$x + 3 = 12$

$x = 9$

65) $\frac{2^x}{2^2} = 2^{x-2} = 2^{-5}$

$x - 2 = -5$

$x = -3$

69) If 1/3 is discarded after each cut, 2/3 will remain.

After 0 cuts $243 \cdot \left(\frac{2}{3}\right)^0 = 243.$

After 5 cuts $243 \cdot \left(\frac{2}{3}\right)^5 = 243 \cdot \frac{32}{243} = 32$ feet.

After n cuts $243 \cdot \left(\frac{2}{3}\right)^n$ feet.

73) If raising to a power were commutative, 2^3 would equal 3^2, but $2^3 = 8$ while $3^2 = 9$.

If the operation were associative,

$(2^3)^4$ would equal $2^{(3^4)}$. However, $(2^3)^4 = 8^4 = 4096$ while

$2^{(3^4)} = 2^{81} \approx 2.4178 \times 10^{24}.$

Section 2.2 Radicals and Rational Exponents

1) $11^{1/2}$

5) $6^{2/3}$

9) $\sqrt{3}$

13) $\frac{1}{\sqrt{2}}$

17) True

21) False

$$\left(-\frac{1}{8}\right)^{-1/3} = \frac{1}{\left(-\frac{1}{8}\right)^{\frac{1}{3}}} = \frac{1}{\sqrt[3]{-\frac{1}{8}}} = \frac{1}{-\frac{1}{2}} = -2$$

25) True

29) $81^{-1/2} = \frac{1}{81^{1/2}} = \frac{1}{\sqrt{81}} = \frac{1}{9}$

33) $(-125)^{2/3} = \left(\sqrt[3]{-125}\right)^2 = (-5)^2 = 25$

37) $\dfrac{\sqrt[3]{-3}}{\sqrt[3]{-24}} = \sqrt[3]{\dfrac{-3}{-24}} = \sqrt[3]{\dfrac{1}{8}} = \dfrac{1}{2}$

41) $\sqrt[3]{(-125)(-1000)} = \sqrt[3]{-125} \cdot \sqrt[3]{-1000} = (-5)(-10) = 50$

45) $\sqrt{\dfrac{2}{3}} \cdot \sqrt{\dfrac{75}{98}} = \sqrt{\dfrac{\cancel{2}^{1}}{\cancel{3}_{1}} \cdot \dfrac{\cancel{75}^{25}}{\cancel{98}_{49}}} = \sqrt{\dfrac{25}{49}} = \dfrac{5}{7}$

49) $\sqrt{144 + 25} = \sqrt{169} = 13$

53) $\left(\dfrac{16}{81}\right)^{3/4} + \left(\dfrac{256}{625}\right)^{1/4} = \dfrac{16^{3/4}}{81^{3/4}} + \dfrac{256^{1/4}}{625^{1/4}}$

$$= \frac{(\sqrt[4]{16}\,)^3}{(\sqrt[4]{81}\,)^3} + \frac{\sqrt[4]{256}}{\sqrt[4]{625}}$$

$$= \frac{2^3}{3^3} + \frac{4}{5}$$

$$= \frac{8}{27} + \frac{4}{5}$$

$$= \frac{148}{135}$$

57) $\left(\dfrac{38}{49} - \dfrac{1}{7}\right)^{1/2} = \left(\dfrac{31}{49}\right)^{1/2} = \sqrt{\dfrac{31}{49}} = \dfrac{\sqrt{31}}{\sqrt{49}} = \dfrac{\sqrt{31}}{7}$

61) $(a^{-4}b^{-8})^{3/4} = \left(\dfrac{1}{a^4b^8}\right)^{3/4} = \dfrac{1^{3/4}}{a^{4\cdot 3/4}b^{8\cdot 3/4}} = \dfrac{1}{a^3b^6}$

65) $\left(\dfrac{64a^6}{b^{-9}}\right)^{2/3}$

$$= (64a^6b^9)^{2/3}$$

$$= (64)^{2/3}a^{6\cdot 2/3}b^{9\cdot 2/3}$$

$$= (\sqrt[3]{64}\,)^2\ a^4b^6$$

$$= 4^2a^4b^6$$

$$= 16a^4b^6$$

69) $\frac{1}{2}(3x^2 + 2)^{-1/2} \cdot 6x = \frac{1}{2} \cdot \frac{1}{(3x^2 + 2)^{1/2}} \cdot 6x$

$= \frac{6x}{2(3x^2 + 2)^{1/2}}$

$= \frac{3x}{(3x^2 + 2)^{1/2}}$

73) a) $d = \sqrt{20^2 + 15^2} = \sqrt{400 + 225} = \sqrt{625} = 25$ cm

b) $d = \sqrt{16^2 + 10^2} = \sqrt{256 + 100} = \sqrt{356} = 18.9$ cm

77) $\sqrt[3]{\frac{x^{3n+1}y^n}{x^{3n+4}y^{4n}}} = \sqrt[3]{\frac{1}{x^3y^{3n}}} = \frac{1}{xy^n}$

Section 2.3 Simplifying Radicals

1) $\sqrt{2} + \sqrt{18} = \sqrt{2} + \sqrt{9} \cdot \sqrt{2} = \sqrt{2} + 3\sqrt{2} = 4\sqrt{2}$

5) $\sqrt{12} - \sqrt{27} = \sqrt{4} \cdot \sqrt{3} - \sqrt{9} \cdot \sqrt{3} = 2\sqrt{3} - 3\sqrt{3} = -\sqrt{3}$

9) $2\sqrt{5} + 3\sqrt{125} = 2\sqrt{5} + 3\sqrt{25} \cdot \sqrt{5}$

$= 2\sqrt{5} + 3 \cdot 5\sqrt{5}$

$= 2\sqrt{5} + 15\sqrt{5}$

$= 17\sqrt{5}$

13) $\sqrt[3]{128} + \sqrt[3]{16} = \sqrt[3]{64} \cdot \sqrt[3]{2} + \sqrt[3]{8} \cdot \sqrt[3]{2}$

$= 4\sqrt[3]{2} + 2\sqrt[3]{2}$

$= 6\sqrt[3]{2}$

17) $\frac{8}{\sqrt{2}} + 2\sqrt{50} = \frac{8}{\sqrt{2}} \cdot \frac{\sqrt{2}}{\sqrt{2}} + 2\sqrt{25} \cdot \sqrt{2}$

$= \frac{8\sqrt{2}}{2} + 2 \cdot 5\sqrt{2}$

$= 4\sqrt{2} + 10\sqrt{2}$

$= 14\sqrt{2}$

21) $3\sqrt{8x^2} - \sqrt{50x^2} = 3\sqrt{4x^2} \cdot \sqrt{2} - \sqrt{25x^2}\sqrt{2}$

$= 3 \cdot 2|x|\sqrt{2} - 5|x|\sqrt{2}$

$= 6|x|\sqrt{2} - 5|x|\sqrt{2}$

$= |x|\sqrt{2}$

25) $3\sqrt{10} + 4\sqrt{90} - 5\sqrt{40} = 3\sqrt{10} + 4\sqrt{9} \cdot \sqrt{10} - 5\sqrt{4} \cdot \sqrt{10}$

$= 3\sqrt{10} + 4 \cdot 3\sqrt{10} - 5 \cdot 2\sqrt{10}$

$= 3\sqrt{10} + 12\sqrt{10} - 10\sqrt{10}$

$= 5\sqrt{10}$

29) $\frac{10}{\sqrt{5}} + 3\sqrt{45} - 2\sqrt{20} = \frac{10}{\sqrt{5}} \cdot \frac{\sqrt{5}}{\sqrt{5}} + 3\sqrt{9} \cdot \sqrt{5} - 2\sqrt{4} \cdot \sqrt{5}$

$= \frac{10\sqrt{5}}{5} + 3 \cdot 3\sqrt{5} - 2 \cdot 2\sqrt{5}$

$= 2\sqrt{5} + 9\sqrt{5} - 4\sqrt{5}$

$= 7\sqrt{5}$

33) $\sqrt{18x} + \sqrt{50x} - \sqrt{2x} = \sqrt{9} \cdot \sqrt{2x} + \sqrt{25} \cdot \sqrt{2x} - \sqrt{2x}$

$= 3\sqrt{2x} + 5\sqrt{2x} - \sqrt{2x}$

$= 7\sqrt{2x}$

37) $\sqrt{x^2y} + \sqrt{8x^2y} + \sqrt{200x^2y} = \sqrt{x^2} \cdot \sqrt{y} + \sqrt{4x^2} \cdot \sqrt{2y}$

$+ \sqrt{100x^2} \cdot \sqrt{2y}$

$= |x|\sqrt{y} + 2|x|\sqrt{2y} + 10|x|\sqrt{2y}$

$= |x|\sqrt{y} + 12|x|\sqrt{2y}$

41) $x\sqrt[4]{x^{11}} + \sqrt[4]{x^{15}} = x\sqrt[4]{x^8 \cdot x^3} + \sqrt[4]{x^{12} \cdot x^3} = x^3\sqrt[4]{x^3} + x^3\sqrt[4]{x^3}$

$= 2x^3\sqrt[4]{x^3}$

45) $\frac{8x}{\sqrt{2}} \cdot \frac{\sqrt{2}}{\sqrt{2}} = \frac{8x\sqrt{2}}{2} = 4x\sqrt{2}$

49) $\frac{1}{\sqrt{18}} \cdot \frac{\sqrt{18}}{\sqrt{18}} = \frac{\sqrt{18}}{\sqrt{18}} = \frac{\sqrt{9} \cdot \sqrt{2}}{18} = \frac{3\sqrt{2}}{18} = \frac{\sqrt{2}}{6}$

53) $\frac{8}{\sqrt[3]{2}} \cdot \frac{\sqrt[3]{4}}{\sqrt[3]{4}} = \frac{8\sqrt[3]{4}}{\sqrt[3]{8}} = \frac{8\sqrt[3]{4}}{2} = 4\sqrt[3]{4}$

57) $\sqrt{\frac{7}{3}} - \frac{21}{\sqrt{21}} = \frac{2\sqrt{21}}{3} = \frac{\sqrt{7}}{\sqrt{3}} \cdot \frac{\sqrt{3}}{\sqrt{3}} - \frac{21}{\sqrt{21}} \cdot \frac{\sqrt{21}}{\sqrt{21}} + \frac{2\sqrt{21}}{3}$

$= \frac{\sqrt{7} \cdot \sqrt{3}}{3} - \frac{21\sqrt{21}}{21} + \frac{2\sqrt{21}}{3}$

$= \frac{\sqrt{21}}{3} - \sqrt{21} + \frac{2\sqrt{21}}{3}$

$= 0$

61) $|xy| = \sqrt{(xy)^2} = \sqrt{x^2y^2} = \sqrt{x^2}\sqrt{y^2} = |x||y|$

Section 2.4 Fundamental Operations with Polynomials

1) $8x^2 - 2x + 7$

5) $2x^3 + x^2 + 11x - 28$

9) $4x^3 - x^2 - 5x - 22$

13) $(y + 2) + (2y + 1) + (3y + 3) =$
$(y + 2y + 3y) + (2 + 1 + 3) = 6y + 6$

17) $5y - [y - (3y + 8)] = 5y - [y - 3y - 8]$
$= 5y - y + 3y + 8$
$= 7y + 8$

21)
$$\begin{array}{r} 5a^2b - 3ab + 8ab^2 \\ a^2b + ab - 2ab^2 \\ \hline 4a^2b - 4ab + 10ab^2 \end{array}$$

25) $2x^2(2x + 1 - 10x^2) =$

$$2x^2(2x) + 2x^2(1) - 2x^2(10x^2)$$

$$= 4x^3 + 2x^2 - 20x^4 = -20x^4 + 4x^3 + 2x^2$$

29) $(4x - 2)(x + 7) = 4x(x) + 4x(7) - 2(x) - 2(7)$

$$= 4x^2 + 28x - 2x - 14$$

$$= 4x^2 + 26x - 14$$

33) $\left(x + \frac{1}{2}\right)\left(x - \frac{1}{4}\right) = x(x) + x\left(-\frac{1}{4}\right) + \frac{1}{2}(x) + \frac{1}{2}\left(-\frac{1}{4}\right)$

$$= x^2 - \frac{1}{4}x + \frac{1}{2}x - \frac{1}{8}$$

$$= x^2 + \frac{1}{4}x - \frac{1}{8}$$

37) $(-2x - 3)(3x - 6) = -2x(3x) - 2x(-6) - 3(3x) - 3(-6)$

$$= -6x^2 + 12x - 9x + 18$$

$$= -6x^2 + 3x + 18$$

41) $(7 - 3x)(4x - 9) = 7(4x) + 7(-9) - 3x(4x) - 3x(-9)$

$$= 28x - 63 - 12x^2 + 27x$$

$$= -12x^2 + 28x + 27x - 63$$

$$= -12x^2 + 55x - 63$$

45) $(x - 0.1)(x + 0.1) = x^2 - (0.1)^2 = x^2 - 0.01$

49) $(\sqrt{x} - 10)(\sqrt{x} + 10) = (\sqrt{x})^2 - 10^2 = x - 100$

53) $(\sqrt{3} - \sqrt{2})(\sqrt{3} + \sqrt{2}) = (\sqrt{3})^2 - (\sqrt{2})^2 = 3 - 2 = 1$

57) $(2x^2 - 3)(4x^2 + 6x + 9) = (2x^2 - 3)(4x^2) + (2x^2 - 3)(6x) + (2x^2 - 3)9$

$$= 8x^4 + 12x^3 - 12x^2 + 18x^2 - 18x - 27$$

$$= 8x^4 + 12x^3 + 6x^2 - 18x - 27$$

Note: It is very easy to lose a term in the last few steps of such a problem. It is always a good idea to rewrite the polynomial in order so that like terms appear next to each other, which makes combining easier. A shorter method is to combine directly from the mixed-up form, keeping track of the work by crossing out each term in the old line as you copy it or combine it for the new line.

61)

$$\begin{array}{r} x^4 + 2x^3 + 4x^2 + 8x + 16 \\ x - 2 \\ \hline -2x^4 - 4x^3 - 8x^2 - 16x - 32 \\ x^5 + 2x^4 + 4x^3 + 8x^2 + 16x \\ \hline x^5 - 32 \end{array}$$

65) $3x(1 - x)(1 - x) = 3x(1 - 2x + x^2)$

$$= 3x - 6x^2 + 3x^3$$

69) $(x^3 - 2x + 1)(2x) + (x^2 - 2)(3x^2 - 2) =$

$2x^4 - 4x^2 + 2x + 3x^4 - 2x^2 - 6x^2 + 4 =$

$2x^4 + 3x^4 - 4x^2 - 2x^2 - 6x^2 + 2x + 4 =$

$5x^4 - 12x^2 + 2x + 4$

73) $(x - 1)^3 = (x - 1)(x - 1)^2 = (x - 1)(x^2 - 2x + 1) =$

$(x - 1)(x^2) - (x - 1)(2x) + (x - 1)(1) =$

$x^3 - x^2 + 2x^2 + 2x + x - 1 =$

$x^3 - 3x^2 + 3x - 1$

77) Using the formula from Example 7, let

$2x = a$ and $3 = b$.

$$(a + b)^3 = a^3 + 3a^2b + 3ab^2 + b^3$$

$$\begin{aligned}(2x + 3)^3 &= (2x)^3 + 3(2x)^2(3) + 3(2x)(3)^2 + (3)^3 \\ &= 8x^3 + 3(4x^2)(3) + 3(2x)(9) + 27 \\ &= 8x^3 + 36x^2 + 54x + 27\end{aligned}$$

81) $$\frac{12}{\sqrt{5} - \sqrt{3}} \cdot \frac{\sqrt{5} + \sqrt{3}}{\sqrt{5} + \sqrt{3}} = \frac{12\sqrt{5} + 12\sqrt{3}}{5 - 3} = \frac{12\sqrt{5} + 12\sqrt{3}}{2} = 6\sqrt{5} + 6\sqrt{3}$$

85) $$\frac{\sqrt{x} + \sqrt{y}}{\sqrt{x} - \sqrt{y}} \cdot \frac{\sqrt{x} + \sqrt{y}}{\sqrt{x} + \sqrt{y}} = \frac{(\sqrt{x})^2 + 2\sqrt{x}\sqrt{y} + (\sqrt{y})^2}{x - y} = \frac{x + 2\sqrt{xy} + y}{x - y}$$

89) $$\frac{\sqrt{x} + \sqrt{y}}{\sqrt{x} - \sqrt{y}} \cdot \frac{\sqrt{x} - \sqrt{y}}{\sqrt{x} - \sqrt{y}} = \frac{x - y}{(\sqrt{x})^2 - 2\sqrt{x}\sqrt{y} + (\sqrt{y})^2} = \frac{x - y}{x - 2\sqrt{xy} + y}$$

Section 2.5 Factoring Polynomials

1) $5x - 5 = 5(x - 1)$

5) $4a^2b - 6ab = 2ab(2a) - 2ab(3) = 2ab(2a - 3)$

9) $10x(a - b) + 5y(a - b) = (a - b)(10x + 5y) = 5(a - b)(2x + y)$

13) $4x^2 - 9 = (2x)^2 - 3^2 = (2x - 3)(2x + 3)$

17) $x^3 + 64 = x^3 + 4^3 = (x + 4)(x^2 - 4x + 4^2)$

$= (x + 4)(x^2 - 4x + 16)$

21) $8x^3 + 343y^3 = (2x)^3 + (7y)^3$

$$= (2x + 7y)[(2x)^2 - (2x)(7y) + (7y)^2]$$

$$= (2x + 7y)(4x^2 - 14xy + 49y^2)$$

25) $x - 36 = (\sqrt{x})^2 - (6)^2 = (\sqrt{x} - 6)(\sqrt{x} + 6)$

29) $7 + a^3 = (\sqrt[3]{7})^3 + a^3 = (\sqrt[3]{7} + a)[(\sqrt[3]{7})^2 - a\sqrt[3]{7} + a^2]$

$$= (\sqrt[3]{7} + a)(\sqrt[3]{49} - a\sqrt[3]{7} + a^2)$$

33) $3x - 4 = (\sqrt[3]{3x})^3 - (\sqrt[3]{4})^3$

$$= (\sqrt[3]{3x} - \sqrt[3]{4})[(\sqrt[3]{3x})^2 + \sqrt[3]{3x} \cdot \sqrt[3]{4} + (\sqrt[3]{4})^2]$$

$$= (\sqrt[3]{3x} - \sqrt[3]{4})(\sqrt[3]{9x^2} + \sqrt[3]{12x} + \sqrt[3]{16})$$

37) $-y-x + 1 + xy = xy - y - x + 1$

$$= y(x - 1) - (x - 1)$$

$$= (x - 1)(y - 1)$$

41) $7x^3 + 7h^3 = 7(x^3 + h^3) = 7(x + h)(x^2 - hx + h^2)$

45) $x^4 - y^4 = (x^2 - y^2)(x^2 + y^2)$

$$= (x - y)(x + y)(x^2 + y^2)$$

49) $a^5 - 32 = (a - 2)(a^4 + 2a^3 + 2^2a^2 + 2^3a + 2^4)$

$$= (a - 2)(a^4 + 2a^3 + 4a^2 + 8a + 16)$$

53) $3(a + 1)x^3 + 24(a + 1) = 3(a + 1)(x^3 + 8)$

$$= 3(a + 1)(x + 2)(x - 2x + 4)$$

57) $x^5 - 16xy^4 - 2x^4y + 32y^5 = x^5 - 2x^4y - 16xy^4 + 32y^5$

$$= x^4(x - 2y) - 16y^4(x - 2y)$$
$$= (x - 2y)(x^4 - 16y^4)$$
$$= (x - 2y)(x^2 - 4y^2)(x^2 + 4y^2)$$
$$= (x - 2y)(x - 2y)(x + 2y)(x^2 + 4y^2)$$
$$= (x - 2y)^2(x + 2y)(x^2 + 4y^2)$$

61) $(x^3 + 1)^3(2x) + (x^2 - 1)(3)(x^3 + 1)^2(3x^2)$

$$= x(x^3 + 1)^2[2(x^3 + 1) + 9x(x^2 - 1)]$$
$$= x(x^3 + 1)^2[2(x + 1)(x^2 - x + 1) + 9x(x + 1)(x - 1)]$$
$$= x(x + 1)(x^3 + 1)^2[2(x^2 - x + 1) + 9x(x - 1)]$$
$$= x(x + 1)(x^3 + 1)^2(2x^2 - 2x + 2 + 9x^2 - 9x)$$
$$= x(x + 1)(x^3 + 1)^2(11x^2 - 11x + 2)$$
$$= x(x + 1)(x + 1)^2(x^2 - x + 1)^2(11x^2 - 11x + 2)$$
$$= x(x + 1)^3(x^2 - x + 1)^2(11x^2 - 11x + 2)$$

65) $(x - 1)(x + 1)(x^4 + x^2 + 1) = (x - 1)(x^2 + x + 1)(x + 1)(x^2 - x + 1)$

Divide each side by $(x - 1)(x + 1)$.

$$x^4 + x^2 + 1 = (x^2 + x + 1)(x^2 - x + 1)$$

Section 2.6 | Factoring Trinomials

1) $x^2 + 4x + 4 = (x + 2)^2$

5) $1 + 2b + b^2 = (1 + b)^2$

9) $9x^2 - 18xy + 9y^2 = 9(x^2 - 2xy + y^2) = 9(x - y)^2$

13) $x^2 + 5x + 6 = (x + ?)(x + ?)$

List factors of 6: 6 and 1, 3 and 2 and test to arrive at $x^2 + 5x + 6 = (x + 3)(x + 2)$.

17) $20a^2 - 9a + 1 = (?a - 1)(?a - 1)$

Factors of 20: 20 and 1, 10 and 2, 5 and 4

$20a^2 - 9a + 1 = (5a - 1)(4a - 1)$

21) $9x^2 + 6x + 1 = (3x + 1)^2$

25) $3x^2 + 20x + 12 = (3x + ?)(x + ?) = (3x + 2)(x + 6)$

29) $8x^2 - 9x + 1 = (?x - 1)(?x - 1) = (8x - 1)(x - 1)$

33) $b^2 + 18b + 45 = (b + ?)(b + ?) = (b + 15)(b + 3)$

37) $12a^2 - 25a + 12 = (?a - ?)(?a - ?)$

Possibilities for the first blanks are

$(12a - ?)(a - ?)$

$(6a - ?)(2a - ?)$

$(4a - ?)(3a - ?)$

Use trial and error to test factors of 12 in the second blanks. Remember to try each pair <u>both</u> ways. In this case

$(4a - ?)(3a - ?)$ with 4 and 3 gives

$(4a - 4)(3a - 3) = 12a^2 - 24a + 12$

But reversing the second factors gives the correct factorization:

$(4a - 3)(3a - 4) = 12a^2 - 25a + 12$

41) $6a^2 + 5a - 21$

The two binomial factors will have opposite signs which should be put in last. Begin with

$(6a \quad ?)(a \quad ?)$ or $(3a \quad ?)(2a \quad ?)$

When testing, remember to <u>subtract</u> partial products.

That is $(6a \quad 7)(a \quad 3) = 6a^2 \pm (18a - 7a) - 21$

$= 6a^2 \pm (11a) - 21$

When you have arrived at the correct choice of $(3a \quad 7)(2a \quad 3)$, try the + and - signs both ways to find which arrangement makes 5a positive. $(3a + 7)(2a - 3) = 6a^2 + 5a - 21$

45) $24a^2 + 25ab + 6b^2$

$(?a + 6)(?a + 1) \quad \text{or} \quad (?a + 3)(?a + 2)$

$(8a + 3b)(3a + 2b) = 24a^2 + 25ab + 6b^2$

49) $4ax^2 + 4ax + a = a(4x^2 + 4x + 1) = a(2x + 1)^2$

53) $6x^2 + 2x - 20 = 2(3x^2 + x - 10)$

$2(3x \quad ?)(x \quad ?)$

$2(3x - 5)(x + 2) = 6x^2 + 2x - 20$

57) $2x^2 - 2x - 112 = 2(x^2 - 1 - 56) = 2(x - 8)(x + 7)$

61) $a^3b - 2a^2b^2 + ab^3 = ab(a^2 - 2ab + b^2) = ab(a - b)^2$

65) $25a^2 + 50ab + 25b^2 = 25(a^2 + 2ab + b^2) = 25(a + b)^2$

69) $6x^5y - 3x^3y^2 - 30xy^3 = 3xy(2x^4 - x^2y - 10y^2) =$

$3xy(2x^2 - 5y)(x^2 + 2y)$

73) $$\begin{aligned} 5(4x^2 + 4x + 1)^4(8x + 4) &= 5(4x^2 + 4x + 1)^4(4)(2x + 1) \\ &= 20(4x^2 + 4x + 1)^4(2x + 1) \\ &= 20[(2x + 1)^2]^4(2x + 1) \\ &= 20(2x + 1)^8(2x + 1) \\ &= 20(2x + 1)^9 \end{aligned}$$

Section 2.7 Fundamental Operations with Rational Expressions

1) False. $\frac{5}{7} - \frac{2}{3} = \frac{15}{21} - \frac{14}{21} = \frac{1}{21}$ 5) True

9) $\frac{45x^3 + 15x^2}{15x^2} = \frac{\cancel{15x^2}(3x + 1)}{\cancel{15x^2}} = 3x + 1$

13) $\frac{a^2b^2 + ab^2 - a^2b^3}{ab^2} = \frac{ab^2(a + 1 - ab)}{ab^2} = a - ab + 1$

17) $\frac{x^2 - 5x}{5 - x} = \frac{x(x - 5)}{5 - x} = \frac{-x\cancel{(5 - x)}}{\cancel{5 - x}} = -x$

21) $\frac{3x^2 + 3x - 6}{2x^2 + 6x + 4} = \frac{3(x^2 + x - 2)}{2(x^2 + 3x + 2)} = \frac{3\cancel{(x + 2)}(x - 1)}{2\cancel{(x + 2)}(x + 1)} = \frac{3(x - 1)}{2(x + 1)}$

25) $\frac{a^2 - 16b^2}{a^3 + 64b^3} = \frac{\cancel{(a + 4b)}(a - 4b)}{\cancel{(a + 4b)}(a^2 - 4ab + 16b^2)} = \frac{a - 4b}{a^2 - 4ab + 16b^2}$

29) $\frac{2a}{3} \cdot \frac{3}{a^2} \cdot \frac{1}{a} = \frac{6a}{3a^3} = \frac{2}{a^2}$

33) $\frac{a - 2b}{2} - \frac{3a + b}{3} = \frac{3(a - 2b)}{6} - \frac{2(3a + b)}{6}$

$= \frac{3a - 6b - 6a - 2b}{6}$

$= \frac{-3a - 8b}{6}$

37) $\frac{1 - x}{2 + x} \div \frac{x^2 - x}{x^2 + 2x} = \frac{1 - x}{2 + x} \cdot \frac{x^2 + 2x}{x^2 - x}$

$= \frac{1 - x}{2 + x} \cdot \frac{x(x + 2)}{x(x - 1)}$

$= \frac{-1(x - 1)}{x + 2} \cdot \frac{x(x + 2)}{\cancel{x}(x - 1)}$ 1

$= \frac{-1}{1} = -1$

41) $\frac{3y}{y + 1} + \frac{2y}{y - 1} = \frac{3y(y - 1)}{(y + 1)(y - 1)} + \frac{2y(y + 1)}{(y + 1)(y - 1)}$

$= \frac{3y^2 - 3y + 2y^2 + 2y}{(y + 1)(y - 1)}$

$= \frac{5y^2 - y}{y^2 - 1}$

45) $\dfrac{5}{x^2 - 4} - \dfrac{3 - x}{4 - x^2} = \dfrac{5}{x^2 - 4} + \dfrac{3 - x}{x^2 - 4}$

$$= \frac{5 + 3 - x}{x^2 - 4}$$

$$= \frac{8 - x}{x^2 - 4}$$

49) $\dfrac{x + 3}{5 - x} - \dfrac{x - 5}{x + 5} + \dfrac{2x^2 + 30}{x^2 - 25} = \dfrac{-(x + 3)}{x - 5} - \dfrac{x - 5}{x + 5} + \dfrac{2(x^2 + 15)}{(x - 5)(x + 5)}$

$$= \frac{(-x - 3)(x + 5)}{(x - 5)(x + 5)} - \frac{(x - 5)(x - 5)}{(x - 5)(x + 5)} + \frac{2(x^2 + 15)}{(x - 5)(x + 5)}$$

$$= \frac{-x^2 - 8x - 15 - x^2 + 10x - 25 + 2x^2 + 30}{(x - 5)(x + 5)}$$

$$= \frac{2x - 10}{(x - 5)(x + 5)}$$

$$= \frac{2\cancel{(x - 5)}}{\cancel{(x - 5)}(x + 5)}$$

$$= \frac{2}{x + 5}$$

53) $\dfrac{n^3 - 8}{n + 2} \cdot \dfrac{2n^2 + 8}{n^3 - 4n} \cdot \dfrac{n^3 + 2n^2}{n^3 + 2n^2 + 4n}$

$$= \frac{\cancel{(n - 2)}\cancel{(n^2 + 2n + 4)}}{\cancel{n + 2}} \cdot \frac{2(n^2 + 4)}{\cancel{n}(n + 2)\cancel{(n - 2)}} \cdot \frac{\cancel{n^2}\cancel{(n + 2)}}{\cancel{n}\cancel{(n^2 + 2n + 4)}}$$

$$= \frac{2(n^2 + 4)}{n + 2}$$

57) $$\frac{\frac{1}{4+h} - \frac{1}{4}}{h} = \frac{\frac{4}{4(4+h)} - \frac{4+h}{4(4+h)}}{h} = \frac{\frac{4-4-h}{4(4+h)}}{h} = \frac{\frac{-h}{4(4+h)}}{h}$$

$$= \frac{-h}{4(4+h)} \div h = \frac{\overset{-1}{\cancel{-h}}}{4(4+h)} \cdot \frac{1}{\cancel{h}_1} = \frac{-1}{4(4+h)}$$

61) $$\frac{\frac{1}{x^2} - \frac{1}{16}}{x+4} = \frac{\frac{16}{16x^2} - \frac{x^2}{16x^2}}{x+4} = \frac{\frac{16-x^2}{16x^2}}{x+4}$$

$$= \frac{(4-x)(4+x)}{16x^2} \div (x+4) = \frac{(4-x)\cancel{(4+x)}}{16x^2} \cdot \frac{1}{\cancel{x+4}}$$

$$= \frac{4-x}{16x^2}$$

65) $$\frac{(1+x^2)(-2x) - (1-x^2)(2x)}{(1+x^2)^2} = \frac{-2x - 2x^3 - 2x + 2x^3}{(1+x^2)^2}$$

$$= \frac{-4x}{(1+x^2)^2}$$

69) $$\frac{a^{-1} - b^{-1}}{a-b} = \frac{\frac{1}{a} - \frac{1}{b}}{a-b} = \frac{\frac{b-a}{ab}}{a-b}$$

$$= \frac{b-a}{ab} \cdot \frac{1}{a-b} = -\frac{1}{ab}$$

73) It is easier to begin with the expression on the right and simplify to get the expression on the left.

a) $$\frac{\frac{AD}{B} + C}{D} = \frac{\frac{AD}{B}}{D} + \frac{C}{D} = \frac{AD}{BD} + \frac{C}{D} = \frac{A}{B} + \frac{C}{D}$$

b) $$\left[\frac{\left(\frac{AB}{D} + C\right) D}{F} + E\right] \cdot F = \left[\frac{AB + CD}{F} + E\right] \cdot F$$

$$= AB + CD + EF$$

77) For this exercise it is helpful to start with the first equation, work on both sides, and try to "build up" to the second equation.

$$y = \frac{x^2}{8} - \frac{2}{x^2}$$

$$y^2 = \left(\frac{x^2}{8} - \frac{2}{x^2}\right)^2$$

$$y^2 = \frac{x^4}{64} - \frac{4x^2}{8x^2} + \frac{4}{x^4}$$

$$y^2 = \frac{x^4}{64} - \frac{1}{2} + \frac{4}{x^4}$$

$$1 + y^2 = \frac{x^4}{64} - \frac{1}{2} + \frac{4}{x^4} + 1$$

$$1 + y^2 = \frac{x^4}{64} + \frac{1}{2} + \frac{4}{x^4}$$

The right side does not appear to be factorable. But the first and third terms suggest $\left(\frac{x^2}{8} + \frac{2}{x^2}\right)^2$. The middle term of this expansion would be $\frac{4x^2}{8x^2}$, which is another form of $\frac{1}{2}$.

$$1 + y^2 = \frac{x^4}{64} + \frac{4x^2}{8x^2} + \frac{4}{x^4}$$

$$1 + y^2 = \left(\frac{x^2}{8} + \frac{2}{x^2}\right)^2$$

$$\sqrt{1 + y^2} = \frac{x^2}{8} + \frac{2}{x^2}$$

Section 2.8 Introduction to Complex Numbers

1) True

5) False

9) $-5 + 0i$

13) $\sqrt{-144} = \sqrt{-1} \cdot \sqrt{144} - i \cdot 12 = 12i$

17) $-\sqrt{-5} = -\sqrt{-1} \cdot \sqrt{5} = -\, i \cdot \sqrt{5} = -\sqrt{5}\, i$

21) $\sqrt{-3} \cdot \sqrt{-2} = \sqrt{-1} \cdot \sqrt{3} \cdot \sqrt{-1} \cdot \sqrt{2} = \sqrt{3}\, i \cdot \sqrt{2}\, i$

$= \sqrt{6}\, i^2 = (\sqrt{6})(-1) = -\sqrt{6}$

25) $\sqrt{-9} + \sqrt{-81} = \sqrt{-1} \cdot \sqrt{9} + \sqrt{-1} \cdot \sqrt{81} = 3i + 9i = 12i$

29) $\sqrt{-9} - \sqrt{-3} = \sqrt{-1}\sqrt{9} - \sqrt{-1}\sqrt{3} = 3i - \sqrt{3}\, i = (3 - \sqrt{3})i$

33) $(8 + 2i) - (3 + 5i) = (8 - 3) + (2 - 5)i = 5 - 3i$

37) $2i(3 + 5i) = 6i + 10i^2 = 6i + 10(-1) = -10 + 6i$

41) $(5 - 2i)(3 + 4i) = 15 + 20i - 6i - 8i^2 = 15 + 20i - 6i + 8 = 23 + 14i$

45) $\dfrac{5 + 3i}{2 + i} \cdot \dfrac{2 - i}{2 - i} = \dfrac{10 - 5i + 6i - 3i^2}{4 - i^2} = \dfrac{10 - 5i + 6i + 3}{4 + 1}$

$= \dfrac{13 + i}{5} = \dfrac{13}{5} + \dfrac{1}{5}i$

49) $\overline{z + w} = \overline{(-6 + 8i) + \frac{1}{2} + \frac{1}{2}i}$

$= \overline{\left(-6 + \frac{1}{2}\right) + \left(8 + \frac{1}{2}\right)i}$

$= \overline{-5\frac{1}{2} + 8\frac{1}{2}i}$

$= -5\frac{1}{2} - 8\frac{1}{2}i$

$\overline{z} + \overline{w} = (\overline{-6 + 8i}) + \overline{\frac{1}{2} + \frac{1}{2}i}$

$= (-6 - 8i) + \frac{1}{2} - \frac{1}{2}i$

$= -6 + \frac{1}{2} + \left(-8 - \frac{1}{2}\right)i$

$= -5\frac{1}{2} - 8\frac{1}{2}i$

53) $3i^3 = 3(-i) = -3i$

57) $-4i^{18} = -4(i^{16})(i^2) = -4(i^4)^4(i^2) = -4(1)^4 i^2 = -4(-1) = 4$

61) $\sqrt{(-4)(-9)} = \sqrt{36} = 6$ but $\sqrt{-4} \cdot \sqrt{-9} = 2i \cdot 3i = 6i^2 = -6$

65)
$$\begin{aligned}
[(3 + i)(3 - i)](4 + 3i) &= [9 - i^2](4 + 3i) \\
&= (9 + 1)(4 + 3i) \\
&= 10(4 + 3i) \\
&= 40 + 30i
\end{aligned}$$

$$\begin{aligned}
(3 + i)[(3 - i)(4 + 3i)] &= (3 + i)[12 + 9i - 4i - 3i^2] \\
&= (3 + i)(15 + 5i) \\
&= 45 + 15i + 15i + 5i^2 \\
&= 40 + 30i
\end{aligned}$$

69)
$$\begin{aligned}
\frac{(2 + i)^2(3 - i)}{2 + 3i} &= \frac{(4 + 4i + i^2)(3 - i)}{2 + 3i} \\
&= \frac{(3 + 4i)(3 - i)}{2 + 3i} \\
&= \frac{9 - 3i + 12i - 4i^2}{2 + 3i} \\
&= \frac{13 + 9i}{2 + 3i} \cdot \frac{2 - 3i}{2 - 3i} \\
&= \frac{26 - 39i + 18i - 27i^2}{4 - 9i^2} \\
&= \frac{53 - 21i}{13} \\
&= \frac{53}{13} - \frac{21}{13}i
\end{aligned}$$

73) $3x^2 + 75 = 3(x^2 + 25) = 3[x^2 - (5i)^2] = 3(x + 5i)(x - 5i)$

77) $$\overline{\left(\frac{z}{w}\right)} = \overline{\left(\frac{a + bi}{c + di}\right)} = \overline{\left(\frac{(a + bi)(c - di)}{(c + di)(c - di)}\right)} = \overline{\left(\frac{ac - adi + bci - bdi^2}{c^2 - (di)^2}\right)}$$

$$= \overline{\left(\frac{ac + bd + (bc - ad)i}{c^2 + d^2}\right)} = \overline{\left(\frac{ac + bd}{c^2 + d^2} + \frac{bc - ad}{c^2 + d^2}i\right)}$$

$$= \frac{ac + bd}{c^2 + d^2} - \frac{bc - ad}{c^2 + d^2}i$$

$$\frac{\overline{z}}{\overline{w}} = \frac{\overline{a + bi}}{\overline{c + di}} = \frac{a - bi}{c - di} = \frac{(a - bi)(c + di)}{(c - di)(c + di)}$$

$$= \frac{ac + adi - bci - bdi^2}{c^2 - di^2} = \frac{ac + bd + (ad - bc)i}{c^2 + d^2}$$

$$= \frac{ac + bd}{c^2 + d^2} - \frac{bc - ad}{c^2 + d^2}i$$

Sample Test Questions for Chapter 2

PART I

1) False $\quad \dfrac{x^3(-x)^2}{5} = \dfrac{x^5}{5} \neq 5$ $\qquad$ 2) True $\quad \left(\dfrac{3}{2 + a}\right)^{-1} = \dfrac{2 + a}{3} = \dfrac{2}{3} + \dfrac{a}{3}$

3) False $\quad (-27)^{-\frac{1}{3}} = \dfrac{1}{(-27)^{\frac{1}{3}}} = \dfrac{1}{-3} \neq 3$

4) False $\quad (x + y)^{3/5} = (\sqrt[5]{x + y})^3 \neq (\sqrt[3]{x + y})^5$

5) True $\quad \sqrt{9x^2} = \sqrt{9} \cdot \sqrt{x^2} = 3|x|$

6) False $\quad (8 + a^3)^{1/3} = \sqrt[3]{8 + a^3} = \sqrt[3]{(2 + a)(4 + 2a + a^2)}$

7)a) $\dfrac{12^{\frac{11}{3}} \cdot 12^{-\frac{4}{3}}}{12^{1/3}} = 12^{\frac{11}{3} + -\frac{4}{3} - \frac{1}{3}} = 12^{\frac{6}{3}} = 12^2 = 144$

b) $\dfrac{4^2(-3)^3 2^0}{6^3(12^2)} = \dfrac{4^2(-3)(-3)^2(1)}{(3^3)(2^3)(4^2)(3^2)} = \dfrac{-1}{3^2 \cdot 2^3} = \dfrac{-1}{9 \cdot 8} = -\dfrac{1}{72}$ (with -1 above 4^2 and 3^2 below the denominator marked as cancellations)

8)a) $\dfrac{(2x^3y^{-2})^2}{x^{-2}y^3} = \dfrac{2^2x^6y^{-4}}{x^{-2}y^3} = \dfrac{4x^8}{y^7}$

b) $\dfrac{(3x^2y^{-3})^{-1}}{(2x^{-2}y^2)^{-2}} = \dfrac{(2x^{-2}y^2)^2}{3x^2y^{-3}} = \dfrac{2^2x^{-4}y^4}{3x^2y^{-3}} = \dfrac{4y^7}{3x^6}$

9)a) $(a^{1/2}b^{-1/2})(a^{-1/3}b^{1/3}) = \dfrac{a^{1/2}b^{1/3}}{a^{1/3}b^{1/2}} = \dfrac{a^{3/6}b^{2/6}}{a^{2/6}b^{3/6}} = \dfrac{a^{1/6}}{b^{1/6}}$

b) $\dfrac{a^{2/3}b^{-1/2}}{a^{-1/2}b^{1/4}} = \dfrac{a^{2/3} \cdot a^{1/2}}{b^{1/2} \cdot b^{1/4}} = \dfrac{a^{2/3+1/2}}{b^{1/2+1/4}} = \dfrac{a^{7/6}}{b^{3/4}}$

10)a) $\sqrt{8} \cdot \sqrt{6} = \sqrt{48} = \sqrt{16} \cdot \sqrt{3} = 4\sqrt{3}$

b) $\dfrac{\sqrt{360}}{2\sqrt{2}} = \dfrac{1}{2}\sqrt{\dfrac{360}{2}} = \dfrac{1}{2}\sqrt{180} = \dfrac{1}{2}\sqrt{36} \cdot \sqrt{5} = \dfrac{1}{2} \cdot 6\sqrt{5} = 3\sqrt{5}$

10)c) $\dfrac{\sqrt[3]{-243x^8}}{\sqrt[3]{3x^2}} = \sqrt[3]{\dfrac{-243x^8}{3x^2}} = \sqrt[3]{-81x^6} = \sqrt[3]{-27x^6} \cdot \sqrt[3]{3} = -3x^2\sqrt[3]{3}$

11)a) $\sqrt{50} + 3\sqrt{18} - 2\sqrt{8} = \sqrt{25} \cdot \sqrt{2} + 3\sqrt{9} \cdot \sqrt{2} - 2\sqrt{4} \cdot \sqrt{2}$

$= 5\sqrt{2} + 3 \cdot 3\sqrt{2} - 2 \cdot 2\sqrt{2}$

$= 5\sqrt{2} + 9\sqrt{2} - 4\sqrt{2}$

$= 10\sqrt{2}$

b) $\dfrac{12}{\sqrt{3}} + 2\sqrt{3} = \dfrac{12}{\sqrt{3}} \cdot \dfrac{\sqrt{3}}{\sqrt{3}} + 2\sqrt{3}$

$= \dfrac{12\sqrt{3}}{3} + 2\sqrt{3}$

$= 4\sqrt{3} + 2\sqrt{3}$

$= 6\sqrt{3}$

12) $5\sqrt{3x^2} + 2\sqrt{27x^2} - 3\sqrt{48x^2}$

$= 5|x|\sqrt{3} + 2\sqrt{9x^2} \cdot \sqrt{3} - 3\sqrt{16x^2} \cdot \sqrt{3}$

$= 5|x|\sqrt{3} + 2 \cdot 3|x|\sqrt{3} - 3 \cdot 4|x|\sqrt{3}$

$= 5|x|\sqrt{3} + 6|x|\sqrt{3} - 12|x|\sqrt{3}$

$= -|x|\sqrt{3}$

13) $\sqrt[3]{-16x^3} + 2\sqrt[3]{54x^3} - \sqrt[4]{32x^4} = \sqrt[3]{-8x^3} \cdot \sqrt[3]{2} + 2\sqrt[3]{27x^3}\sqrt[3]{2}$

$- \sqrt[4]{16x^4} \cdot \sqrt[4]{2}$

$= -2x\sqrt[3]{2} + 6x\sqrt[3]{2} - 2|x|\sqrt[4]{2}$

$= 4x\sqrt[3]{2} - 2|x|\sqrt[4]{2}$

14) $(ab^2 + 5cd) - (3ab^2 - 2cd + a^2) + (2ab^2 - 5a^2)$

$= ab^2 + 5cd - 3ab^2 + 2cd - a^2 + 2ab^2 - 5a^2$

$= -a^2 - 5a^2 + ab^2 - 3ab^2 + 2ab^2 + 5cd + 2cd$

$= -6a^2 + 7cd$

15) $(5x + 3)(2x - 4) - (3 - x)(3 + x)$
$= 10x^2 - 20x + 6x - 12 - (9 - x^2)$
$= 10x^2 - 14x - 12 - 9 + x^2$
$= 11x^2 - 14x - 21$

16) $(x^2 + 3x)(3x^2) + (x^3 - 1)(2x + 3)$
$= 3x^4 + 9x^3 + 2x^4 + 3x^3 - 2x - 3$
$= 3x^4 + 2x^4 + 9x^3 + 3x^3 - 2x - 3$
$= 5x^4 + 12x^3 - 2x - 3$

17) $64 - 27b^3 = 4^3 - (3b)^3 = (4 - 3b)(16 + 12b + 9b^2)$

18) $3x^5 - 48x = 3x(x^4 - 16)$
$= 3x(x^2 - 4)(x^2 + 4)$
$= 3x(x - 2)(x + 2)(x^2 + 4)$

19) $6x^2 - 7x - 3 = (2x - 3)(3x + 1)$

20) $x^5 - 32 = x^5 - 2^5$
$= (x - 2)(x^4 + 2x^3 + 2^2x^2 + 2^3x + 2^4)$
$= (x - 2)(x^4 + 2x^3 + 4x^2 + 8x + 16)$

21) $2x^2 - 6xy - 3y^3 + xy^2 = 2x(x - 3y) + y^2(-3y + x)$
$= (x - 3y)(2x + y^2)$

22) $ax^2 - 2x + 3ax - 6 = x(ax - 2) + 3(ax - 2)$
$= (x + 3)(ax - 2)$

23) $\dfrac{x^2 - 9}{x^3 + 4x^2 + 4x} \cdot \dfrac{2x^2 + 4x}{x^2 + 2x - 15} = \dfrac{(x+3)(x-3)}{x(x^2 + 4x + 4)} \cdot \dfrac{2x(x + 2)}{(x + 5)(x - 3)}$

$$= \frac{(x + 3)\cancel{(x - 3)}}{\cancel{x}\cancel{(x + 2)^2}\,(x + 2)} \cdot \frac{2\cancel{x}\cancel{(x + 2)}}{(x + 5)\cancel{(x - 3)}}$$

$$= \frac{2(x + 3)}{(x + 2)(x + 5)}$$

24) $$\frac{x^3 + 8}{x^2 - 4x - 12} \div \frac{x^3 - 2x^2 + 4x}{x^3 - 6x^2} = \frac{x^3 + 8}{x^2 - 4x - 12} \cdot \frac{x^3 - 6x^2}{x^3 - 2x^2 + 4x}$$

$$= \frac{\overset{1}{\cancel{(x+2)}}\cancel{(x^2 - 2x + 4)}}{\cancel{(x-6)}\underset{1}{\cancel{(x+2)}}} \cdot \frac{\overset{x}{\cancel{x^2}}\cancel{(x-6)}}{\cancel{x}\underset{1}{\cancel{(x^2 - 2x + 4)}}}$$

$$= x$$

25) $$\frac{\frac{1}{x^2} - \frac{1}{49}}{x - 7} = \frac{\frac{49 - x^2}{49x^2}}{x - 7}$$

$$= \frac{49 - x^2}{49x^2} \cdot \frac{1}{x - 7}$$

$$= \frac{\overset{-1}{\cancel{(7 - x)}}7 + x)}{49x^2} \cdot \frac{1}{\cancel{x - 7}}$$

$$= -\frac{7 + x}{49x^2}$$

26) $$\frac{4x - 1}{2x^2 - x - 3} + \frac{2}{3 - 2x} = \frac{4x - 1}{(2x - 3)(x + 1)} + \frac{-2}{2x - 3}$$

$$= \frac{4x - 1 - 2(x + 1)}{(2x - 3)(x + 1)}$$

$$= \frac{4x - 1 - 2x - 2}{(2x - 3)(x + 1)}$$

$$= \frac{\overset{1}{\cancel{2x - 3}}}{\cancel{(2x - 3)}(x + 1)}$$

$$= \frac{1}{x + 1}$$

27) $\frac{1}{x + 3} - \frac{2}{x^2 - 9} + \frac{x}{2x^2 + x - 15} = \frac{1}{x + 3} - \frac{2}{(x - 3)(x + 3)} + \frac{x}{(2x - 5)(x + 3)}$

$$= \frac{(2x - 5)(x - 3) - 2(2x - 5) + x(x - 3)}{(x - 3)(x + 3)(2x - 5)}$$

$$= \frac{2x^2 - 6x + 5x + 15 - 4x + 10 + 2x - 3x}{(x - 3)(x + 3)(2x - 5)}$$

$$= \frac{3x^2 - 18x + 25}{(x - 3)(x + 3)(2x - 5)}$$

28) $(3 + 7i)(5 - 4i) = 15 - 12i + 35i - 28i^2$

$$= 15 + 28 + 23i$$

$$= 43 + 23i$$

29) $\frac{3 + 7i}{5 - 4i} \cdot \frac{5 + 4i}{5 + 4i} = \frac{15 + 12i + 35i + 28i^2}{25 - 16i^2}$

$$= \frac{15 - 28 + 47i}{25 + 16}$$

$$= - \frac{13 + 47i}{41}$$

$$= - \frac{13}{41} + \frac{47}{41} i$$

30) $(2 - i)^{-2} = \frac{1}{(2 - i)^2}$

$$= \frac{1}{4 - 4i + i^2}$$

$$= \frac{1}{4 - 1 - 4i}$$

$$= \frac{1}{3 - 4i} \cdot \frac{3 + 4i}{3 + 4i}$$

$$= \frac{3 + 4i}{9 - 16i^2}$$

$$= \frac{3 + 4i}{9 + 16}$$

$$= \frac{3 + 4i}{25}$$

$$= \frac{3}{25} + \frac{4}{25} i$$

Part II

1) (c)

I. $(-3)^2 = 9$ but $-3^2 = -9$

II. $\left(\frac{x}{y}\right)^2 = \frac{x^2}{y^2} \neq \frac{x^5}{y^3}$

III. True

2) (e)

I. $x^2 \cdot x^4 = x^6 \neq x^8$

II. $\frac{x^{12}}{x^3} = x^9 \neq x^4$

III. $x^{-5} = \frac{1}{x^5} \neq -\frac{1}{x^5}$

3) (b)

4) (a)

I. True

II. $(-x)^{-1/3} = \frac{1}{(-x)^{1/3}} \neq x^{1/3}$

III. $x^{3/4} = (\sqrt[4]{x})^3 \neq (\sqrt[3]{x})^4$

5) (d)

6) (a)

$(-64)^{2/3} = (\sqrt[3]{-64})^2 = (-4)^2 = 16$

7) (b)

$$\frac{8}{\sqrt[3]{2}} \cdot \frac{\sqrt[3]{4}}{\sqrt[3]{4}} = \frac{8\sqrt[3]{4}}{\sqrt[3]{8}} = \frac{8\sqrt[3]{4}}{2} = 4\sqrt[3]{4}$$

8) (c)

$$\frac{8}{\sqrt{5} - 1} \cdot \frac{\sqrt{5} + 1}{\sqrt{5} + 1} = \frac{8(\sqrt{5} + 1)}{5 - 1} = \frac{8(\sqrt{5} + 1)}{4} = 2(\sqrt{5} + 1)$$

9) (b)

10) (d)

$$\begin{aligned} x^2(x - 1) - 2x(x - 1) + (x - 1) &= (x - 1)(x^2 - 2x + 1) \\ &= (x - 1)(x - 1)^2 \\ &= (x - 1)^3 \end{aligned}$$

11) (e)

I. $x^3 - y^3 = (x - y)(x^2 + xy + y^2) \neq (x - y)(x^2 + y^2)$

II. $(x + y)(x + y) = x^2 + 2xy + y^2 \neq x^2 + y^2$

III. $(2x - 3y)^2 = 4x^2 - 12xy + 9y^2 \neq 4x^2 - 6xy + 9y^2$

12) (c)

$$\frac{6x^2 - 7x - 3}{x^2 - 2x + 1} \cdot \frac{x - 1}{x^2} \div \frac{3x + 1}{x^2 - x} = \frac{6x^2 - 7x - 3}{x^2 - 2x + 1} \cdot \frac{x - 1}{x^2} \cdot \frac{x^2 - x}{3x + 1}$$

$$= \frac{(3x + 1)(2x - 3)}{(x - 1)(x - 1)} \cdot \frac{x - 1}{x^2} \cdot \frac{\not{x}(x - 1)}{3x + 1}$$

$$= \frac{2x - 3}{x}$$

13) (a)

$$\frac{9 - x^3}{2x^2 + 6x} + \frac{x}{2} - \frac{x - 3}{x + 3} = \frac{9 - x^3}{2x(x + 3)} + \frac{x}{2} - \frac{x - 3}{x + 3}$$

$$= \frac{(9 - x^3) + x(x)(x + 3) - (x - 3)(2x)}{2x(x + 3)}$$

$$= \frac{9 - x^3 + x^3 + 3x^2 - 2x^2 + 6x}{2x(x + 3)}$$

$$= \frac{x^2 + 6x + 9}{2x(x + 3)}$$

$$= \frac{(x + 3)(x + 3)}{2x(x + 3)}$$

$$= \frac{x + 3}{2x}$$

14) (d)

15) (b)

$$\frac{2 + i}{3 - i} \cdot \frac{3 + i}{3 + i} = \frac{6 + 2i + 3i + i^2}{9 - i^2}$$

$$= \frac{6 - 1 + 5i}{9 + 1}$$

$$= \frac{5 + 5i}{10} = \frac{1}{2} + \frac{1}{2}i$$

CHAPTER 3
LINEAR FUNCTIONS AND EQUATIONS

Section 3.1 Introduction to the Function Concept

1) Function; each value for x gives only one value for y.
All reals; any real value for x will give a real value for y.

5) Not a function; any positive value for x will give two values for y. Example: $x = 2$, $y = \pm 2$.

9) Not a function; for each non-zero x there will be two y values. Example: $x = 3$, $y = \frac{1}{4}$ and $-\frac{1}{2}$.

13) Not a function; when $x = 3$, y is both 4 and 1.

17) False
$3f(2) = 3[-(2)^2 + 3] = 3(-4 + 3) = 3(-1) = -3.$

21) False

$$\begin{aligned} f(x) - f(4) &= (-x^2 + 3) - [-(4)^2 + 3] \\ &= -x^2 + 3 - (-16 + 3) \\ &= -x^2 + 3 - (-13) \\ &= -x^2 + 3 + 13 \\ &= -x^2 + 16 \end{aligned}$$

25) $f(x) = 2x - 1$
a) $f(-1) = 2(-1) - 1 = -2 - 1 = -3$
b) $f(0) = 2(0) - 1 = 0 - 1 = -1$
c) $f\left(\frac{1}{2}\right) = 2\left(\frac{1}{2}\right) - 1 = 1 - 1 = 0$

29) $f(x) = x^3 - 1$
a) $f(-1) = (-1)^3 - 1 = -1 - 1 = -2$
b) $f(0) = (0)^3 - 1 = -1$
c) $f\left(\frac{1}{2}\right) = \left(\frac{1}{2}\right)^3 - 1 = \frac{1}{8} - 1 = -\frac{7}{8}$

33) $f(x) = \frac{1}{x - 1}$

a) $f(-1) = \frac{1}{-1 - 1} = \frac{1}{-2} = -\frac{1}{2}$

b) $f(0) = \frac{1}{0 - 1} = \frac{1}{-1} = -1$

c) $f\left(\frac{1}{2}\right) = \frac{1}{\frac{1}{2} - 1} = \frac{1}{-\frac{1}{2}} = -2$

37) $g(x) = x^2 - 2x + 1$

a) $g(10) = (10)^2 - 2(10) + 1$
$= 100 - 20 + 1$
$= 81$

b) $5g(2) = 5[(2)^2 - 2(2) + 1]$
$= 5(4 - 4 + 1)$
$= 5 \cdot 1$
$= 5$

c) $g\left(\frac{1}{2}\right) + g\left(\frac{1}{3}\right) = \left[\left(\frac{1}{2}\right)^2 - 2\left(\frac{1}{2}\right) + 1\right] + \left[\left(\frac{1}{3}\right)^2 - 2\left(\frac{1}{3}\right) + 1\right]$
$= \frac{1}{4} - 1 + 1 + \frac{1}{9} - \frac{2}{3} + 1$
$= \frac{25}{36}$

d) $g\left(\frac{1}{2} + \frac{1}{3}\right) = \left(\frac{1}{2} + \frac{1}{3}\right)^2 - 2\left(\frac{1}{2} + \frac{1}{3}\right) + 1$
$= \left(\frac{5}{6}\right)^2 - 2\left(\frac{5}{6}\right) + 1$
$= \frac{25}{36} - \frac{10}{6} + 1$
$= \frac{1}{36}$

41) $y = \begin{cases} 2x - 1 \text{ if } x \le -2 \\ 1 - 2x \text{ if } x > 2 \end{cases}$

a) $x = -5$ $\quad y = 2(-5) - 1 = -11$

b) $x = -2$ $\quad y = 2(-2) - 1 = -5$

c) not defined

d) not defined

e) $x = 5$ $\quad y = 1 - 2(5) = -9$

45) $f(x) = \frac{1}{x}$ $\quad f(3) = \frac{1}{3}$

$$\frac{f(x) - f(3)}{x - 3} = \frac{\frac{1}{x} - \frac{1}{3}}{x - 3} = \frac{\frac{3 - x}{3x}}{x - 3} = \frac{3 - x}{3x} \cdot \frac{1}{x - 3} = -\frac{1}{3x}$$

49) $f(x) = x$ $\quad f(2) = 2$ $\quad f(2 + h) = 2 + h$

$$\frac{f(2 + h) - f(2)}{h} = \frac{2 + h - 2}{h} = \frac{h}{h} = 1$$

53) $f(x) = \frac{1}{x^2}$ $\quad f(2) = \frac{1}{4}$ $\quad f(2 + h) = \frac{1}{4 + 4h + h^2}$

$$\frac{f(2 + h) - f(2)}{h} = \frac{\frac{1}{4 + 4h + h^2} - \frac{1}{4}}{h}$$

$$= \frac{\frac{4 - 4 - 4h - h^2}{4(4 + 4h + h^2)}}{h}$$

$$= \frac{\frac{-4h - h^2}{4(4 + 4h + h^2)}}{h}$$

$$= \frac{\frac{-h(4 + h)}{4(2 + h)^2}}{h}$$

$$= -\frac{4 + h}{4(2 + h)^2}$$

Section 3.2 Graphing Lines in the Rectangular Coordinate System

1) $y = x - 2$

x	-3	-2	-1	0	1	2
y	-5	-4	-3	-2	-1	0

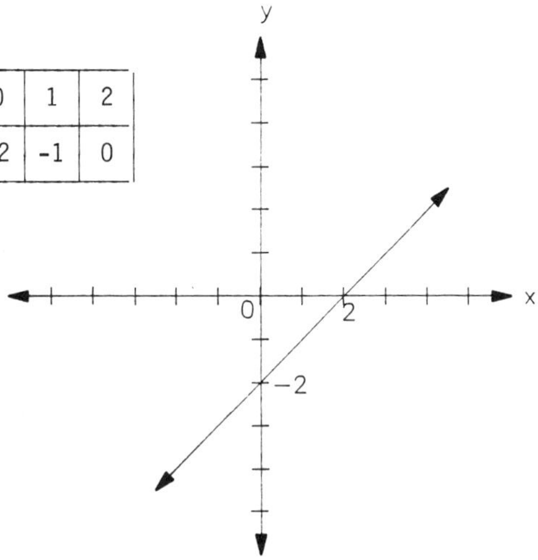

5) $x + 2y = 4$

x-intercept: Let $y = 0$ $\quad x + 2 \cdot 0 = 4$

$x = 4$

y-intercept: Let $x = 0$ $\quad 0 + 2y = 4$

$y = 2$

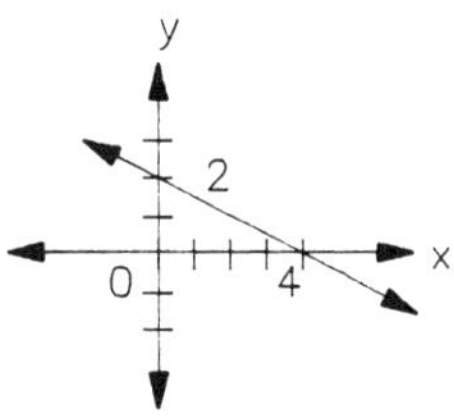

9) $2x - 3y = 6$

x-intercept: Let $y = 0$, $2x - 3 \cdot 0 = 6$

$2x = 6$

$x = 3$

y-intercept: Let $x = 0$. $2 \cdot 0 - 3y = 6$

$y = -2$

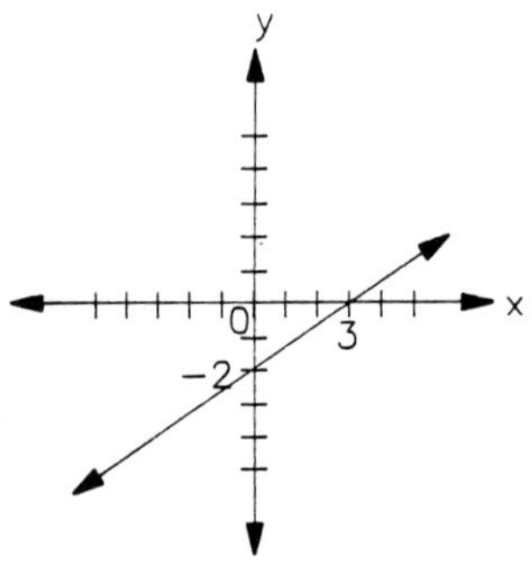

13) $y = 2x - 1$

x-intercept: Let $y = 0$. $0 = 2x - 1$

$1 = 2x$

$\frac{1}{2} = x$

y-intercept: Let $x = 0$. $y = 2 \cdot 0 - 1$

$y = -1$

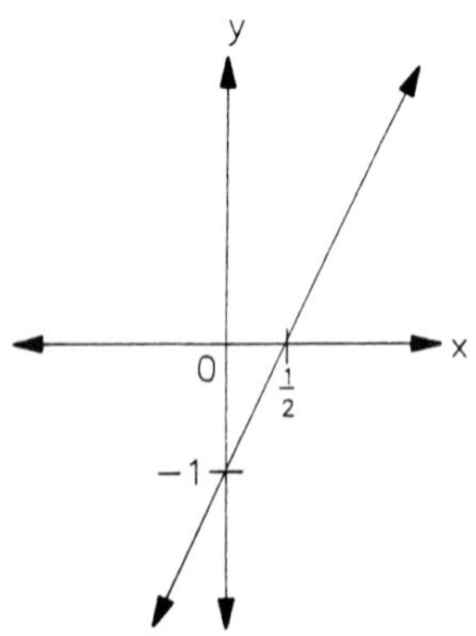

17) $y = \frac{1}{2}x$; $-6 \le x \le 6$

x	-6	0	6
y	-3	0	3

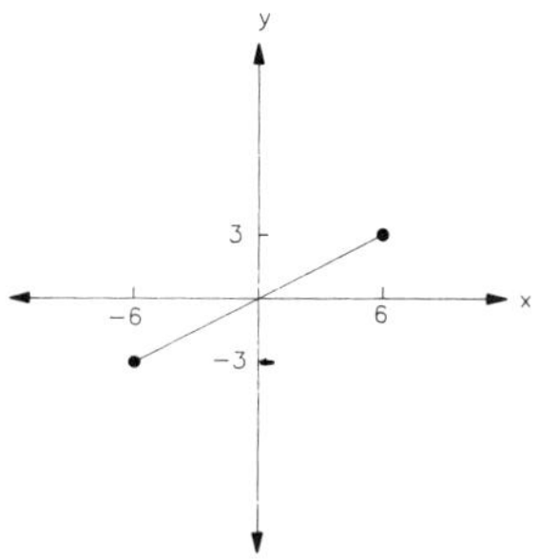

21) $m = \dfrac{y_2 - y_1}{x_2 - x_1}$ A(-2, 3) B(0, 2) C(2, 1) D(4, 0) E(6, -1)

a) $A = (-2, 3) = (x_2, y_2)$ $C = (2, 1) = (x_1, y_1)$

$$m = \frac{3 - 1}{-2 - 2} = \frac{2}{-4} = -\frac{1}{2}$$

b) $B = (0, 2) = (x_2, y_2)$ $D = (4, 0) = (x_1, y_1)$

$$m = \frac{2 - 0}{0 - 4} = \frac{2}{-4} = -\frac{1}{2}$$

c) $C = (2, 1) = (x_2, y_2)$ $D = (4, 0) = (x_1, y_1)$

$$m = \frac{1 - 0}{2 - 4} = \frac{1}{-2}$$

d) $A = (-2, 3) = (x_2, y_2)$ $E = (6, -1) = (x_1, y_1)$

$$m = \frac{3 - (-1)}{-2 - 6} = \frac{4}{-8} = -\frac{1}{2}$$

e) $B = (0, 2) = (x_2, y_2)$ $E = (6, -1) = (x_1, y_1)$

$$m = \frac{2 - (-1)}{0 - 6} = \frac{3}{-6} = -\frac{1}{2}$$

f) $C = (2, 1) = (x_2, y_2)$ $E = (6, -1) = (x_1, y_1)$

$$m = \frac{1 - (-1)}{2 - 6} = \frac{2}{-4} = -\frac{1}{2}$$

25) $m = \frac{-7 - (-7)}{6 - 106} = \frac{0}{-100} = 0$

29) Beginning at (0, 2), go up 3 and right 4 to locate a second point on the line, (4, 5).

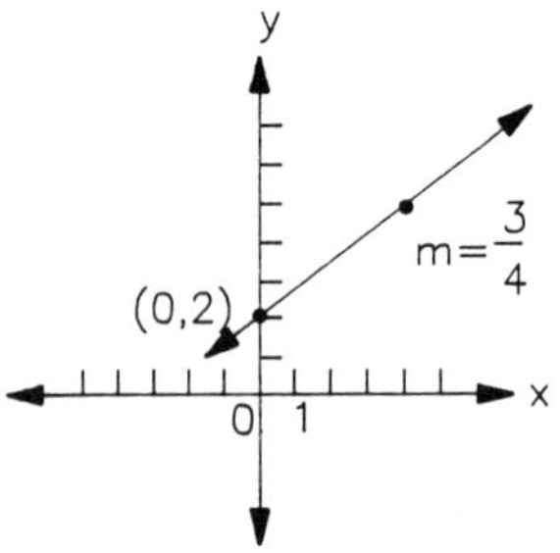

33) The slope of 0 indicates that the line runs parallel to the x-axis.

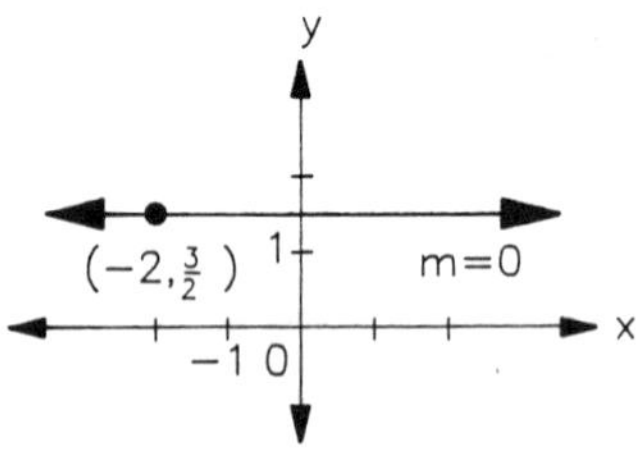

37) $m_1 = \frac{-5 - (-8)}{6 - 8} = \frac{3}{-2} = -\frac{3}{2}$

$m_2 = \frac{12 - 6}{-3 - 1} = \frac{6}{-4} = -\frac{3}{2}$

The two lines are parallel because they have the same slope, $-\frac{3}{2}$.

41) A vertical line has an undefined slope for which no reciprocal exists.
A horizontal line has a slope of 0, which also has no reciprocal.

45) a) For ℓ_1, the slope m_1 is $\frac{\text{change in } y}{\text{change in } x} = \frac{DA}{CD}$

Because CD = 1, $\frac{DA}{CD} = \frac{DA}{1} = DA$

b) For ℓ_2, $m_2 = \frac{\text{change in } y}{\text{change in } x} = \frac{DB}{CD} = \frac{DB}{1} = DB$

Because ℓ_2 "drops off" from left to right, $m_2 < 0$

45) continued

c) By definition of a mean proportional, $\frac{DA}{CD} = \frac{CD}{BD}$

Substitute CD = 1 and DA = m_1, and note that if DB = m_2 then reversing direction gives BD = $-m_2$.

$$\frac{m_1}{1} = \frac{1}{-m_2}$$

Section 3.3 Linear Functions

1) $y = 2x + 3$

5) $y = 0x + 5$
$y = 5$

9) $y = \frac{1}{4}x - 2$

13) $y - 3 = 4[x - (-2)]$
$y - 3 = 4(x + 2)$

17) $y - 1 = \frac{1}{2}(x - 2)$

21) $y - (-\sqrt{2}) = 10(x - \sqrt{2})$
$y + \sqrt{2} = 10(x - \sqrt{2})$

25) $6x - 3y = 1$
$-3y = -6x + 1$
$y = 2x - \frac{1}{3}$
$m = 2;\ b = -\frac{1}{3}$

29) $4x - 3y - 7 = 0$
$-3y = -4x + 7$
$y = \frac{4}{3}x - \frac{7}{3}$
$m = \frac{4}{3},\ b = -\frac{7}{3}$

33) $m = \frac{2 - 3}{3 - 2} = \frac{-1}{1} = -1$
$y - 3 = -1(x - 2)$
$y - 3 = -x + 2$
$x + y - 5 = 0$

37) $m = \frac{1 - (-13)}{-8 - (-1)} = \frac{1 + 13}{-8 + 1} = \frac{14}{-7} = -2$
$y - 1 = -2[x - (-8)]$
$y - 1 = -2(x + 8)$
$y - 1 = -2x - 16$
$2x + y + 15 = 0$

41) $y = -7, \quad x = 5$

45) $y = 3x - 1; \quad (4, 7)$

$m_1 = 3$ Slope of the perpendicular line: $m_2 = -\frac{1}{3}$

$y - 7 = -\frac{1}{3}(x - 4)$

49) (Note: In the second printing of this edition question 49, page 148 will become:

49. In Exercise 48, write the equations of the three altitudes of the triangle.

The solution to this new version is given below.)

The side connecting (1,3) and (4,2) has slope -1/3. The altitude from vertex (-1, -1) to this side is perpendicular to this side and therefore its slope is $-\left(\frac{1}{-1/3}\right) = 3$, and its equation is:

$y + 1 = 3(x+1)$ or $y = 3x + 2$

Similarly, the altitude from vertex (1,3) to the side connecting (-1, -1) and (4,2) has slope $-\left(\frac{1}{3/5}\right) = -\frac{5}{3}$, and its equation is:

$y-3 = -\frac{5}{3}(x-1)$ or $y = -\frac{5}{3}x + \frac{14}{3}$

Also, the altitude from vertex (4,2) to the side connecting (1,3) and (-1, -1) has slope $-\left(\frac{1}{2}\right) = -\frac{1}{2}$ and its equation is

$y - 2 = -\frac{1}{2}(x-4)$ or $y = -\frac{1}{2}x + 4$.

53) $y = f(x) = x + 3$ $f(-2) = -2 + 3 = 1$

$m = 1 \quad b = 3$ $(-2, 1)$

Start at (0, 3) and move up one and right one to locate another point on the line, (1, 4).

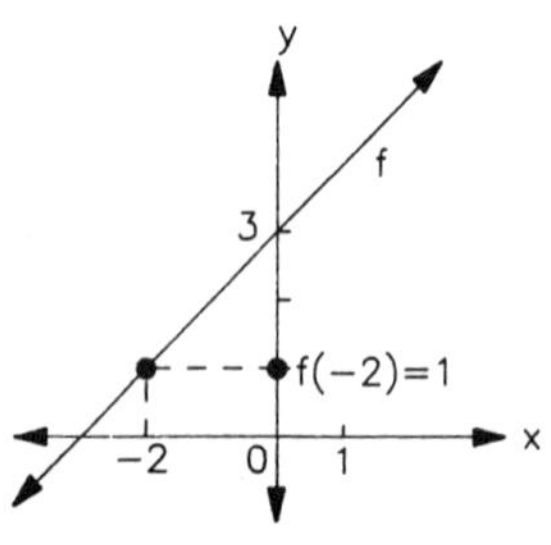

57) $y = 3x - 7 \qquad x \geq 0$

$m = 3 \qquad b = -7$ Start at (0, -7) and use the slope to locate (1, -4).

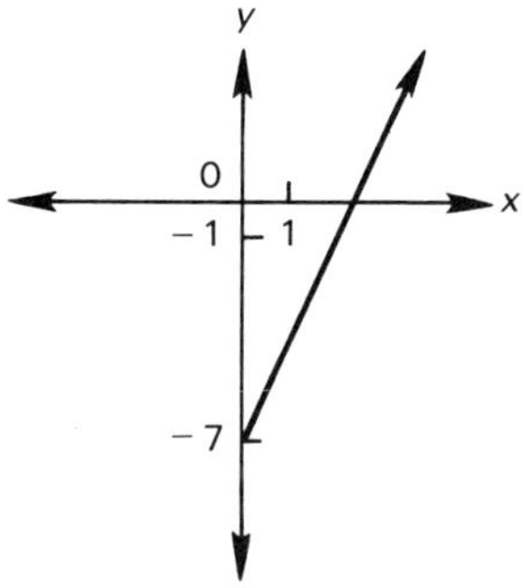

61) $x = -3$; not a function because y can have any value when $x = -3$.

65) $y = \begin{cases} -1 \text{ for } -1 \leq x < 0 \\ 3 \text{ for } 0 \leq x < 3 \\ 1 \text{ for } 3 \leq x \leq 5 \end{cases}$

This is a function because each value of x is associated with only one value for y.

Domain: $-1 \leq x \leq 5$ Range: $\{-1, 1, 3\}$

69) $\dfrac{y - y_1}{y_2 - y_1} = \dfrac{x - x_1}{x_2 - x_1}$ Let $(-2, 3) = (x_1, y_1)$

$(5, -2) = (x_2, y_2)$

$$\frac{y - 3}{-2 - 3} = \frac{x - (-2)}{5 - (-2)}$$

$$\frac{y - 3}{-5} = \frac{x + 2}{5 + 2}$$

$$\frac{y - 3}{-5} = \frac{x + 2}{7}$$

$$y - 3 = -\frac{5}{7}(x + 2)$$

Section 3.4 Special Functions and Linear Inequalities

1) Since $99 \leq 99.1 < 100$, $[99.1] = 99$ 5) Since $-4 \leq -3.5 < -3$, $[-3.5] = -4$

9) $y = |x - 1|$

$x - 1 \geq 0$ when $x \geq 1$

For $x - 1 \geq 0$, $y = x - 1$, $m = 1$.

For $x - 1 < 0$, $y = -(x - 1) = -x + 1$, $m = -1$.

At $x = 1$, $y = 0$.

Domain: all reals.

Range: $y \geq 0$

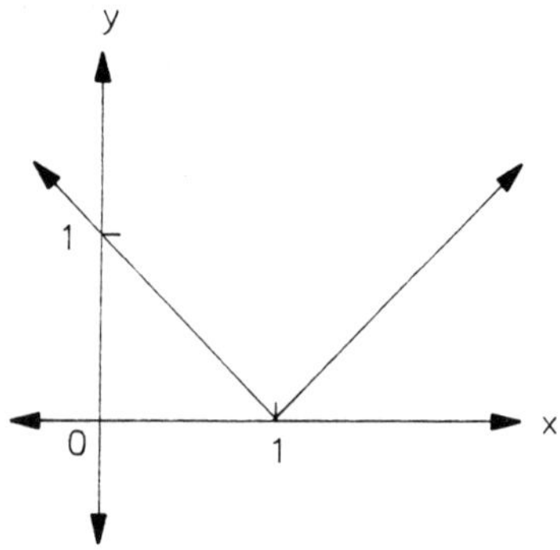

13) $y = |3 - 2x|$

$3 - 2x \geq 0$ when $\frac{3}{2} \geq x$

For $x \leq \frac{3}{2}$, $y = -2x + 3$, $m = -2$.

For $x > \frac{3}{2}$, $y = -(3 - 2x) = 2x - 3$, $m = 2$.

At $x = \frac{3}{2}$, $y = 0$.

Domain: all reals

Range: $y \geq 0$

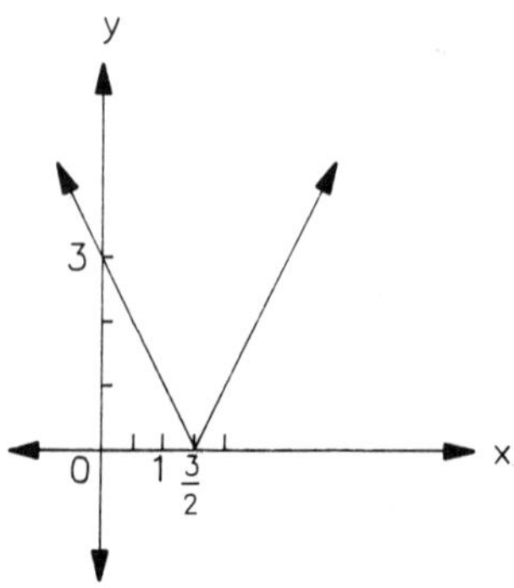

17) $y = \begin{cases} x & \text{if } -2 < x \le 0 \\ 2x & \text{if } 0 < x \le 2 \\ -x+3 & \text{if } 2 < x \le 3 \end{cases}$

This graph is segments of three different lines.
For each segment find the slope and a point.

$y = x$, $m = 1$, $(0, 0)$

$y = 2x$, $m = 2$, $(2, 4)$

$y = -x + 3$, $m = -1$, $(3, 0)$

Domain: $-2 < x \le 3$

Range: $-2 < y \le 4$

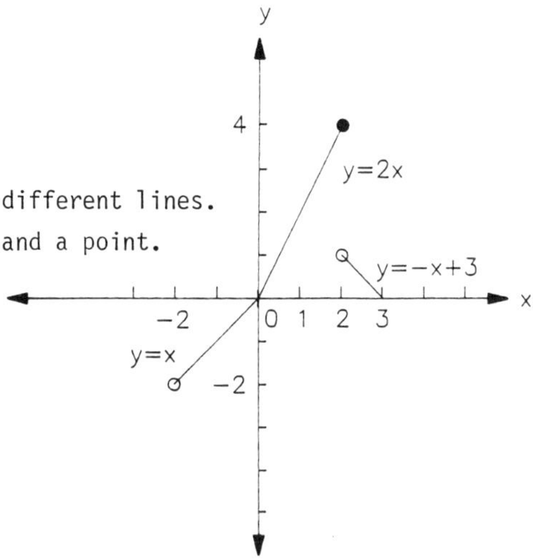

21) $y = \dfrac{x + 3}{|x + 3|}$; all $x \neq -3$

When $x > -3$, $x + 3 > 0$. $|x + 3| = x + 3$ and $\dfrac{x + 3}{|x + 3|} = 1$.

When $x < -3$, $x + 3 < 0$. $|x + 3| = -(x + 3)$ and $\dfrac{x + 3}{|x + 3|} = -1$.

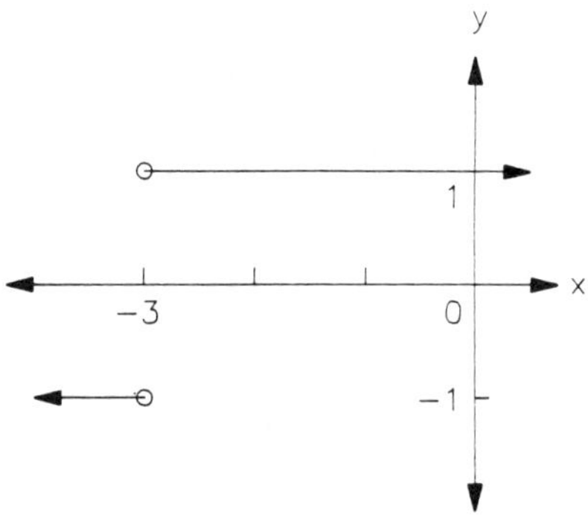

25) $y = 2[x]$; $0 \le x < 2$

For $0 \le x < 1$, $2[x] = 2(0) = 0$.

For $1 \le x < 2$, $2[x] = 2(1) = 2$.

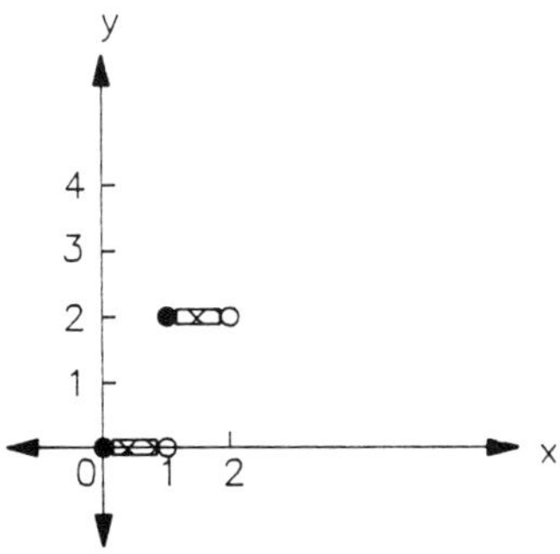

29) For $0 < x < 1$, $[x] = 0$. Thus

$$P(x) = -8$$
$$= .8 + 0$$
$$= \frac{1}{10}(8 + 0)$$
$$= \frac{1}{10}(8 + [x])$$

For $1 \leq x < 2$, $[x] = 1$. Thus

$$P(x) = .9$$
$$= .8 + .1$$
$$= \frac{1}{10}(8 + 1)$$
$$= \frac{1}{10}(8 + [x])$$

By similar argument the same result is obtained for the remaining values of x.

Thus, $P(x) = \frac{1}{10}(8 + [x])$ for $0 < x < 10$.

33) Since the graph consists of all points below the line $y = 2x$ as well as all points on this line the inequality for the graph is $y \leq 2x$.

37) $y > x + 2$

First graph $y = x + 2$ with a broken line.

Since the y values of the inequality must be greater than those on the line, shade the region above the line.

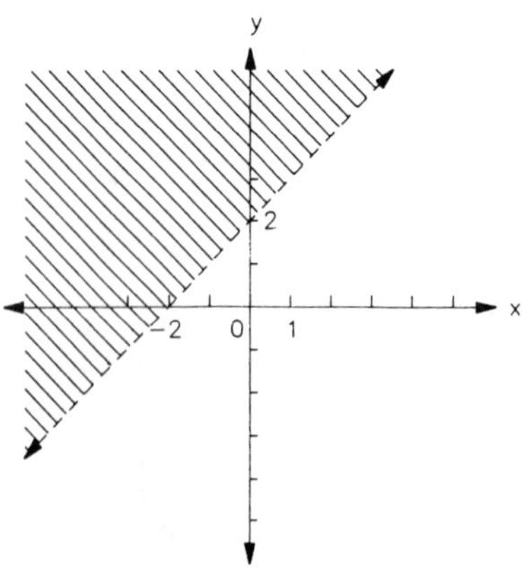

41) $y > x$

First graph $y = x$ with a broken line. The y values of the inequality must be greater than those on this line, so shade the region above the line.

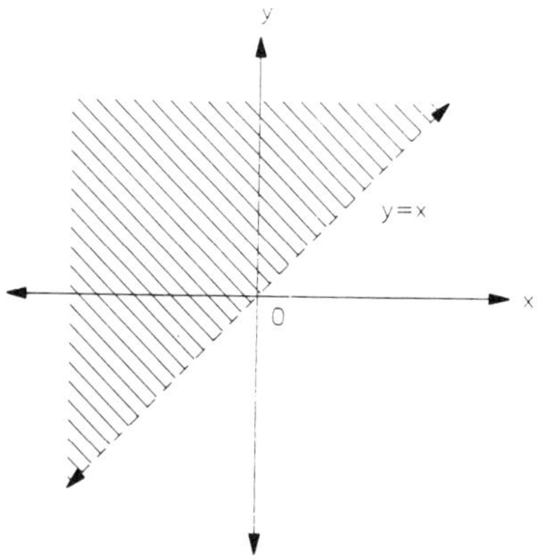

45) $2y + x - 4 < 0$

$2y < -x + 4$

$y < -\frac{1}{2}x + 2$

First graph $-\frac{1}{2}x + 2$ with a broken line. Because y values of the inequality must be less than those on the line, shade the region under the line.

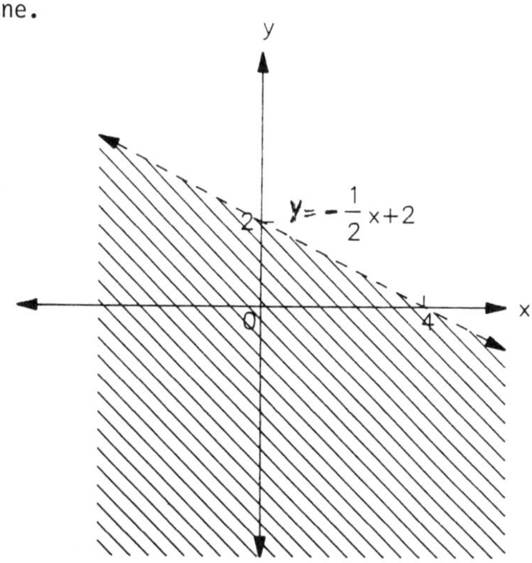

Section 3.5 | Extracting Functions from Geometric Figures

1) $\frac{h}{x} = \frac{5}{2}$ (because the triangles are similar)

$h = \frac{5x}{2}$

$h(x) = \frac{5}{2}x$

5) These two right triangles are similar, and the leg of the larger triangle has length (x + s).

$\frac{s}{x + s} = \frac{6}{24}$

$6(x + s) = 24s$

$6x + 6s = 24s$

$6x = 18s$

$\frac{6}{18}x = s$

$s(x) = \frac{1}{3}x$

9) The formula for volume of a rectangular solid is $V = \ell wh$.
Since the box will have a square bottom, $\ell w = s^2$, so you need to find representations for the length of one side and for the height.
Study the illustration in the text to see that the height of the box will be X.
Because two corners are cut from each side, the sides will measure 50 - 2x.
Substitute these expressions into the volume formula.

$V = s^2h$

$V(x) = (50 - 2x)^2x$

13) A sketch helps to organize the information. It is clear that $V = (x)(2x)(h)$, but this makes V a function of x and h. To eliminate h, make use of the information about the surface area.
120 represents the sum of the areas of all the sides. The sides are rectangles, in three matching pairs because opposite sides are congruent.

13) continued

$120 = 2xh + 2(2x)(x) + 2(h)(2x)$

$120 = 2xh + 4x^2 + 4xh$

$120 = 6xh + 4x^2$

$60 = 3xh + 2x^2$

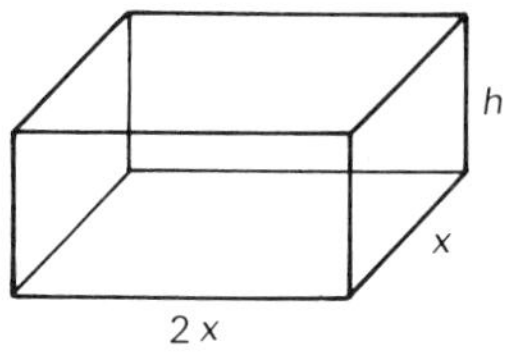

Following the hint, solve for h.

$60 - 2x^2 = 3xh$

$\frac{60}{3x} - \frac{2x^2}{3x} = h$

$\frac{20}{x} - \frac{2}{3}x = h$

Now substitute x, 2x, and $\frac{20}{x} - \frac{2}{3}x$ for w, ℓ, and h in the volume formula.

$$V = (x)(2x)\left(\frac{20}{x} - \frac{2}{3}x\right)$$

$$V = 2x^2\left(\frac{20}{x}\right) - 2x^2\left(\frac{2}{3}x\right)$$

$$V = 40x - \frac{4}{3}x^3$$

$$V(x) = \frac{4}{3}x \cdot (30 - x^2)$$

17) The area of this rectangle can be easily expressed using x and y because the length is 2x and the width is y. A = 2xy. But the function must be in terms of x, so use the additional information to express y in terms of x.

Draw a line from the origin to the point (x, y).

This line forms the hypotenuse of a right triangle with legs measureing x and y. The line is also a radius, so it measures 12.

By the Pythagorean theorem,

$$12^2 = x^2 + y^2$$

$$144 - x^2 = y^2$$

$$\sqrt{144 - x^2} = y$$

Substitute this expression into the area formula.

$$A = 2xy$$

$$A(x) = 2x\sqrt{144 - x^2}$$

Sample Test Questions for Chapter 3

Part I

1) $y = \sqrt{x^2 - 1}$

$y^2 - 1 \geq 0$

$x^2 \geq 1$; Domain: $x \leq -1$ or $x \geq 1$

2) $g(x) = \frac{3}{x}$

a) $g(2 + x) = \frac{3}{2 + x}$

b) $g(2) + g(x) = \frac{3}{2} + \frac{3}{x}$

$= \frac{3x + 6}{2x}$

3) $g(x) = x^2 - x$

$g(9) = 9^2 - 9$

$= 81 - 9 = 72$

$$\frac{g(x) - g(9)}{x - 9} = \frac{x^2 - x - 72}{x - 9} = \frac{(x - 9)(x + 8)}{x - 9} = x + 8$$

4) $g(x) = \frac{2}{x}$ $\quad g(3 + h) = \frac{2}{3 + h}$ $\quad g(3) = \frac{2}{3}$

$$\frac{g(3 + h) - g(3)}{h} = \frac{\frac{2}{3 + h} - \frac{2}{3}}{h}$$

$$= \frac{\frac{6 - 2(3 + h)}{3(3 + h)}}{h}$$

$$= \frac{\frac{6 - 6 - 2h}{3(3 + h)}}{h}$$

$$= \frac{-2h}{3(3 + h)} \cdot \frac{1}{h}$$

$$= -\frac{2}{3(3 + h)}$$

5) $3x - 2y = 6$

x-intercept, let $y = 0$. $\quad 3x - 2 \cdot 0 = 6 \quad\quad x = 2$

y-intercept, let $x = 0$. $\quad 3 \cdot 0 - 2y = 6 \quad\quad y = -3$

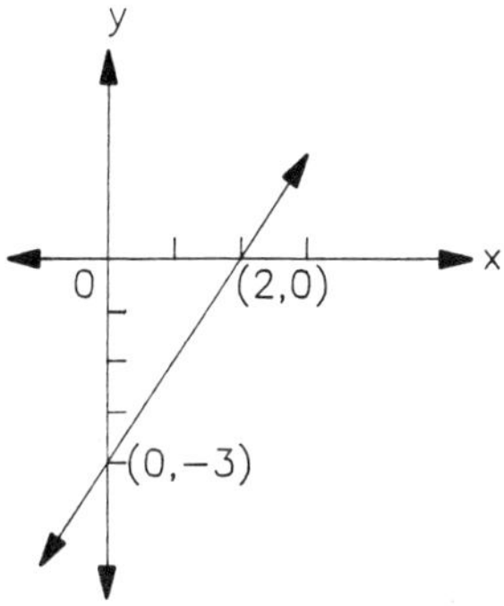

6) Let $(2, -3) = (x_1, y_1)$

$(-1, 4) = (x_2, y_2)$

$$m = \frac{4 - (-3)}{-1 - 2} = \frac{4 + 3}{-3} = -\frac{7}{3}$$

7) Let $(-3, 2) = (x_1, y_1)$

$(4, 2) = (x_2, y_2)$

$$m = \frac{2 - 2}{4 - (-3)} = 0$$

8) $y = 2x + 1 \quad\quad m = 2 \quad\quad b = 1$

Domain: all reals

Range: all reals

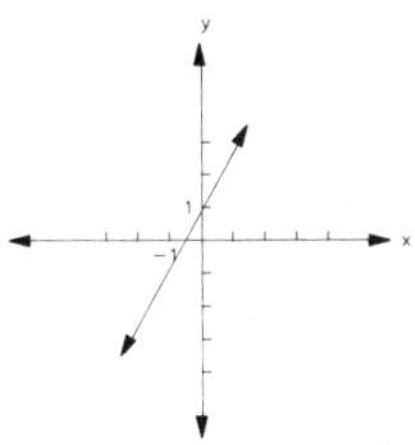

9) $y = -2$

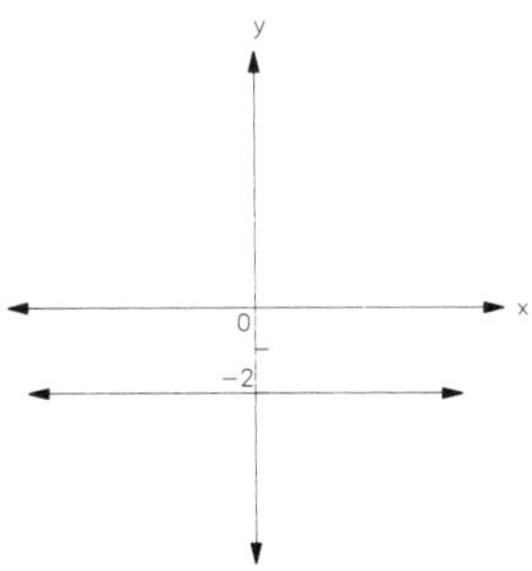

Domain: all reals

Range: $y = -2$

10) a) $x = 3$ b) $y = -2$

11) $y = mx + b$

$y = \frac{1}{2}x - 3$

12) $2x - 3y = 5$

$-3y = -2x + 5$

$y = \frac{2}{3}x - \frac{5}{3}$

$m = \frac{2}{3} \qquad b = -\frac{5}{3}$

13) $y - y_1 = m(x - x_1)$

$y - 3 = -\frac{3}{4}(x + 2)$

14) (3, -5) and (-2, 4)

$m = \frac{-5 - 4}{3 + 2} = -\frac{9}{5}$

$y - 4 = -\frac{9}{5}(x + 2)$

$y - 4 = -\frac{9}{5}x - \frac{18}{5}$

$y = -\frac{9}{5}x + \frac{2}{5}$

15) $-3x + 4y - 2 = 0$

$4y = 3x + 2$

$y = \frac{3}{4}x + \frac{1}{2}$

$m = \frac{3}{4} \qquad b = \frac{1}{2}$

16) First find the slope of the given line.

$3x - y = 6$

$-y = -3x + 6$

$y = 3x - 6$

$m = 3$

Now use point-slope form with the given point (-1, 4).

$y - 4 = 3(x + 1)$

17) The second line will have as its slope the negative reciprocal of the first.

$m_1 = -\frac{2}{5} \qquad m_2 = \frac{5}{2}$

Use point-slope form and the given point (2, 8).

$y - 8 = \frac{5}{2}(x - 2)$

18) a) Because -3.5 is more than -4 but less than -3, $[-3.5] = -4$.

b) Because $\sqrt{3}$ is more than 1 but less than 2, $[\sqrt{3}] = 1$.

19) $y = |x + 2|$

When $x \geq -2$, $x + 2 \geq 0$ and $y = x + 2$.

When $x < -2$, $x + 2 < 0$ and $y = -(x + 2)$.

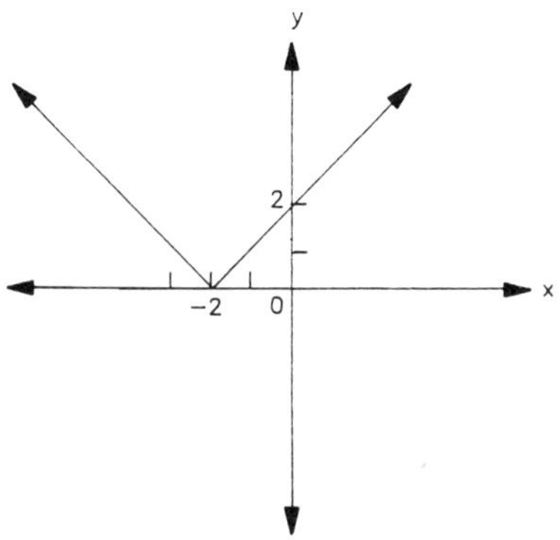

20) $y = \dfrac{|x + 1|}{x + 1}$

Note $x \neq -1$.

When $x > -1$, $x + 1 > 0$, $= \dfrac{x + 1}{x + 1} = 1$.

When $x < -1$, $x + 1 < 0$, $= \dfrac{-(x + 1)}{x + 1} = -1$

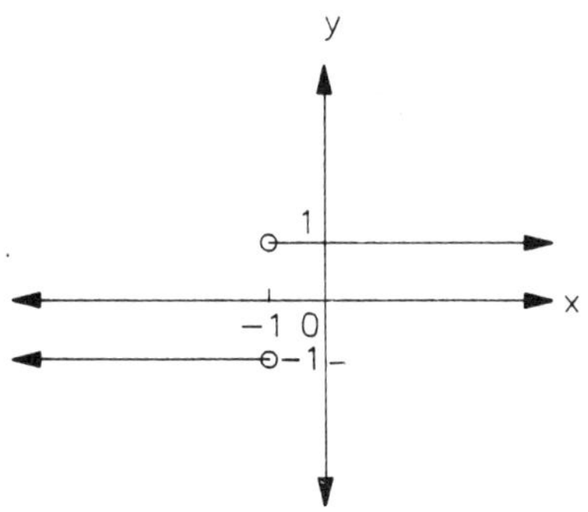

21) $y = 2 - x \qquad -1 \leq x \leq 2$

$m = -1$ | when $x = -1$, $y = 3$ when $x = 2$, $y = 0$

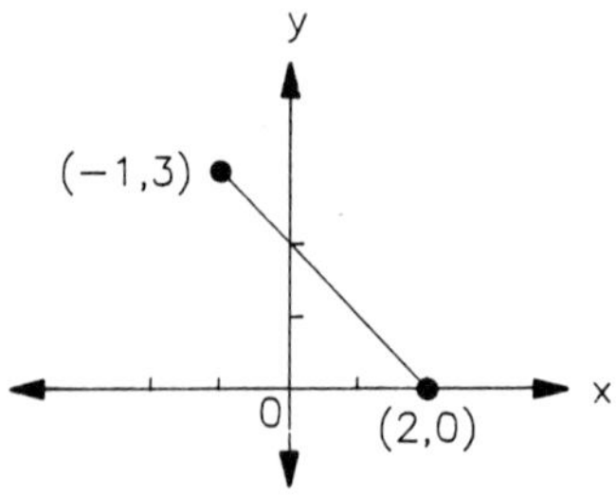

22) $y \geq x - 2$

First graph $y = x - 2$, $m = 1$, $b = -2$.

Because the y values of the inequality must be greater than or equal to those on the line shade the region above the line and make the line solid.

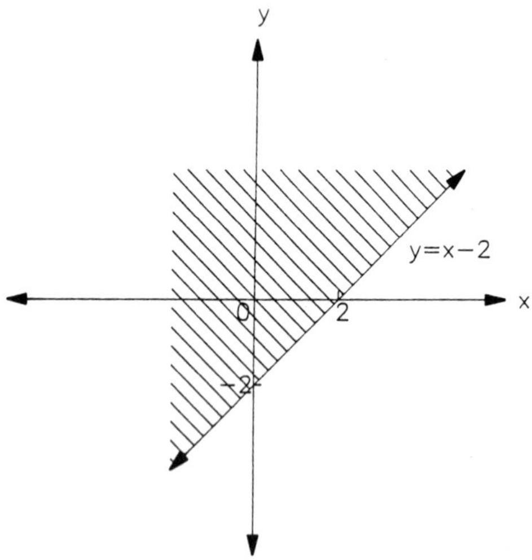

23) $2x + y > 6$

$y > -2x + 6$

First graph $y = -2x + 6$, $m = -2$, $b = 6$, using a broken line. Because the values of y for the inequality must be greater than those on the line, shade the region above the line.

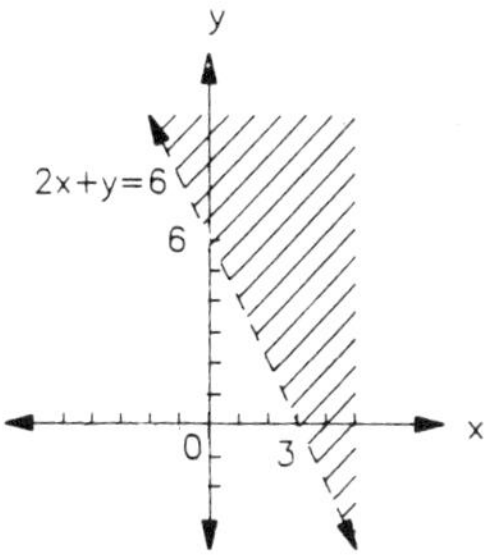

24) First draw a sketch. The volume of a rectangular solid with a square base is $V = s^2h$.

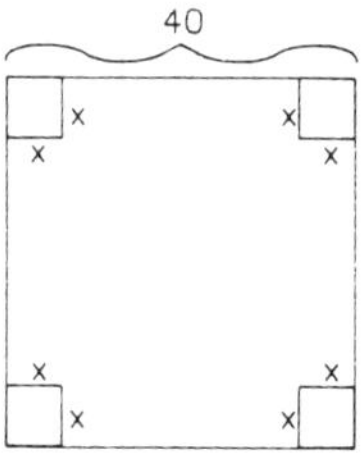

The box formed by folding up the edges will have height x. Each side of the base will have length $40 - 2x$.

$V = s^2h$

$V = (40 - 2x)^2x$

25) First draw a sketch. The area of the rectangle can be expressed as $A = xy$, but A must be a function of x only. To rewrite y in terms of x, consider the little triangle with vertices (x, y), $(4, 0)$, and $(x, 0)$. The base of this triangle measures $(4 - x)$ and the height is y.

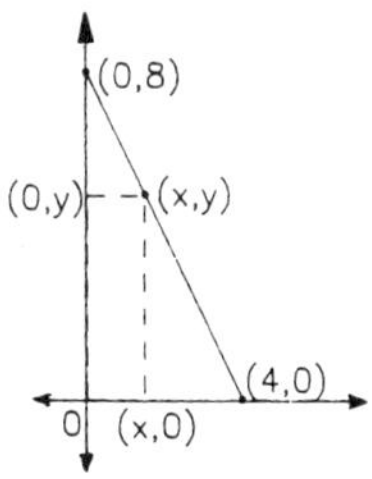

Because the small triangle is similar to the large one:

$$\frac{4 - x}{4} = \frac{y}{8} \longrightarrow 2(4 - x) = y \longrightarrow 8 - 2x = y$$

This expression for y can now be substituted into the area formula for the rectangle. $A = xy = x(8 - 2x) = 8x - 2x^2$

Part II | Multiple Choice

1) (b)

Find the slope of the given line.

$$2x - 3y = 5$$
$$-3y = -2x + 5$$
$$y = \frac{2}{3}x - \frac{5}{3}$$
$$m = \frac{2}{3}$$

Use this slope, the point (-8, 4), and the point-slope form of the equation

$$y - 4 = \frac{2}{3}(x + 8)$$

2) (c)

$$y = \sqrt{4 - x^2}$$

$$4 - x^2 \geq 0$$
$$4 \geq x^2$$
$$-2 \leq x \leq 2$$

3) (e)

$$f(x) = 2x^2 - 3$$

I. $f\left(\frac{1}{3}\right) = 2\left(\frac{1}{3}\right)^2 - 3 = \frac{2}{9} - 3 = -\frac{25}{9}$

$\frac{1}{f(3)} = \frac{1}{2(3)^2 - 3} = \frac{1}{15}$

II. $f(-1) = [2(-1)^2 - 3] = 2 - 3 = -1$

$-f(1) = -[2(1)^2 - 3] = -[-1] = 1$

III. $[f(3)]^2 = [2(3)^2 - 3]^2 = (18 - 3)^2 = (15)^2 = 225$

$f(9) = 2(9)^2 - 3 = 162 - 3 = 159$

4) (a)

$$2x - 3y > 6$$

Consider $2x - 3y = 6$

$$-3y = -2x + 6$$
$$y = \frac{2}{3}x - 2$$

The inequality will be satisfied by all points below this line.

4) continued

III. $2x - 3y = 6$

$-1(2x - 3y) = -1(6)$

$-2x + 3y = -6$

This is the same line, but only the points <u>below</u> it will satisfy the inequality.

5) (d)

$g(x) = (x - 2)^2$ $\qquad g(7) = (7 - 2)^2 = 5^2 = 25$

$$\frac{g(x) - g(7)}{x - 7} = \frac{(x - 2)^2 - 25}{x - 7}$$
$$= \frac{x^2 - 4x + 4 - 25}{x - 7}$$
$$= \frac{x^2 - 4x - 21}{x - 7}$$
$$= \frac{(x - 7)(x + 3)}{(x - 7)}$$
$$= x + 3$$

6) (a) 7) (c)

8) (b)

Find the slope of $2x - 3y = 6$.

$-3y = -2x + 6$

$3y = 2x - 6$

$y = \frac{2}{3}x - 2$ so $m = \frac{2}{3}$

Write a new equation using the negative reciprocal slope, and b for any constant.

$y = -\frac{3}{2}x + b$

$2y = -3x + b$

$3x + 2y = b$ All the values of b give the equations of the lines perpendicular to the first line.

9) (b)

$m = \frac{6 + 3}{-1 - 2} = \frac{9}{-3} = -3$

$y - 6 = -3(x + 1)$

$y - 6 = -3x - 3$

$y = -3x + 3$

10) (d)

First draw a sketch.

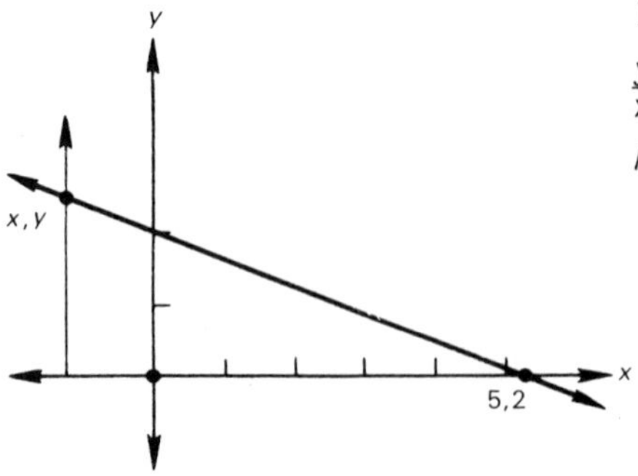

If (5, 2) is on the hypotenuse, then

$\frac{y}{x} = \frac{2}{5},\ y = \left(\frac{2}{5}x\right).$

$$A = \frac{1}{2}bh$$
$$= \frac{1}{2}xy$$
$$= \frac{1}{2}x\ \frac{2}{5}x$$
$$= \frac{1}{5}x^2$$

11) (e)

Test the given values in each equation.

12) (c)

I. $\frac{-2 - 3}{5 + 2} = \frac{-5}{7} \neq \frac{5}{7}$

II. $3x - 2y - 6 = 0$
$$-2y = -3x + 6$$
$$y = \frac{3}{2}x - 3 \qquad m = \frac{3}{2} \neq -\frac{2}{3}$$

III. $y + 3 = 2(x + 2)$
$$y + 3 = 2x + 4$$
$$y = 2x + 1$$

13) (a)

$$f(x) = \begin{cases} x - 2 \text{ if } -3 \leq x \leq 1 \\ -x + 1 \text{ if } 1 < x \end{cases}$$

-3 is in the first part of the domain, so
$$f(x) = x - 2 = -3 - 2 = -5$$
1 is also in this part of the domain,
$$f(1) = x - 2 = 1 - 2 = -1$$
$f(-3) + f(1) = -5 + (-1) = -6$

14) (b)

15) (d)

If the absolute value of x is taken, then the domain for the greatest integer function becomes not $-2 < x < -1$ but $1 < x < 2$. For this domain, the greatest integer is always 1.

CHAPTER 4

QUADRATIC FUNCTIONS AND EQUATIONS

Section 4.1 Graphing Quadratic Functions

1) a) $y = x^2$

b) $y = (x - 1)^2$ Shift $y = x^2$ to the right one unit.

c) $y = (x - 1)^2 + 3$ Shift $y = (x - 1)^2$ up 3 units.

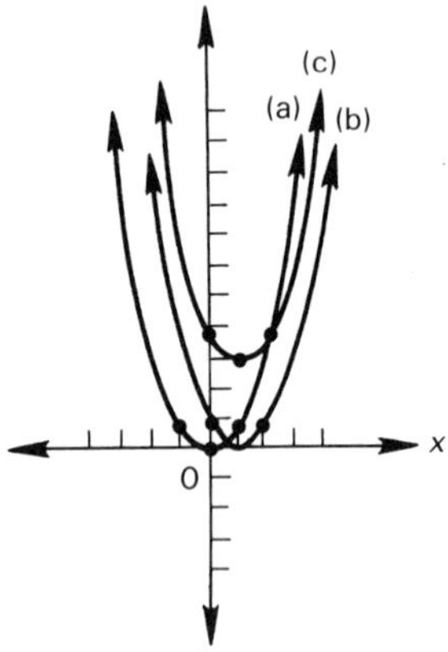

5) a) $y = x^2$

b) $y = 2x^2$
Each y-value is 2 times what it would be for $y = x^2$.

c) $y = 3x^2$
Each y-value is 3 times what it would be for $y = x^2$.

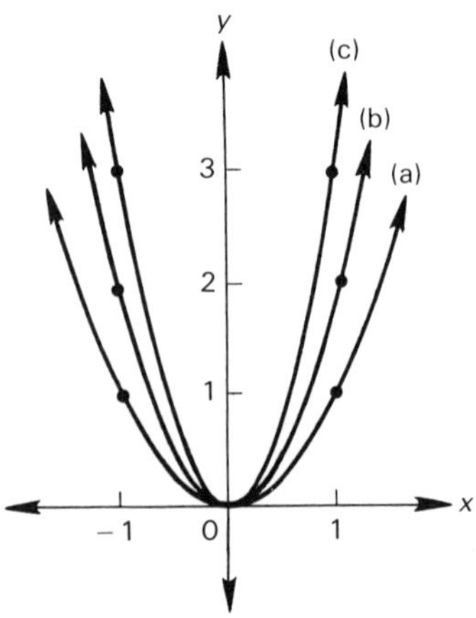

9) $y = -x^2 + 3$

The graph of $y = x^2$ is shifted up three then reflected downward.

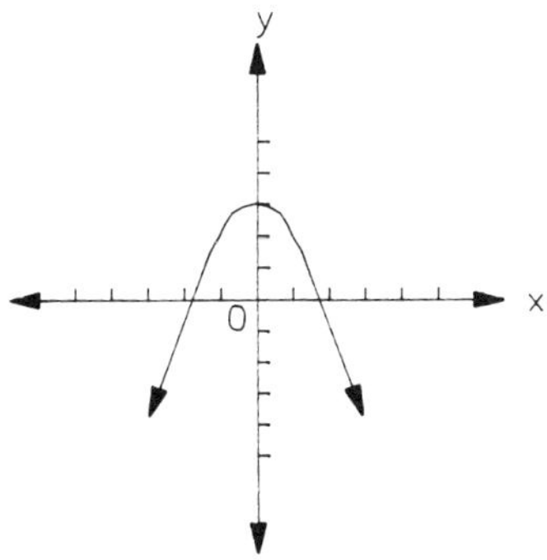

13) $y = \frac{1}{4}x^2$

Each y-value is only $\frac{1}{4}$ of what it would be for $y = x^2$.

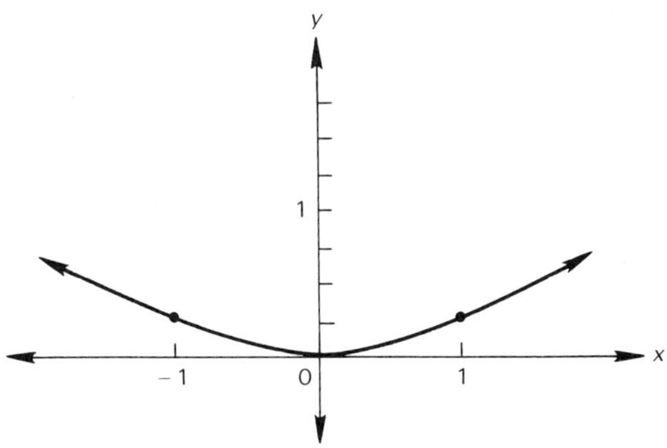

17) $y = -2x^2 + 2$

The y-values are "stretched" to 2 times what they would be for $y = x^2$. The graph of $y = 2x^2$ is reflected through the x-axis and then shifted 2 units up.

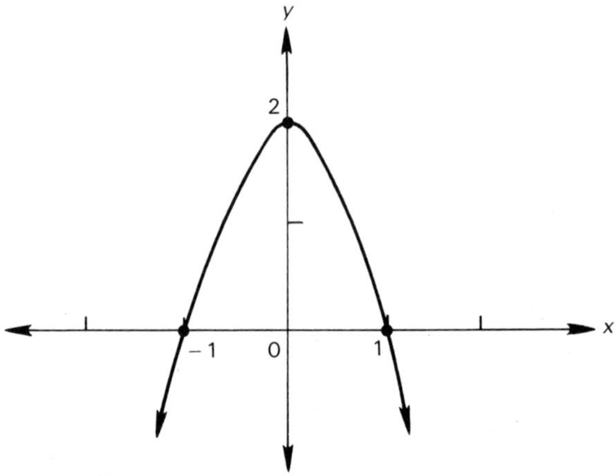

21) $f(x) = -(x + 1)^2 + 2$

The graph of $y = -x^2$ is shifted one unit left and two units up.

It is increasing on $(-\infty, -1]$; decreasing on $[-1, \infty)$; concave down.

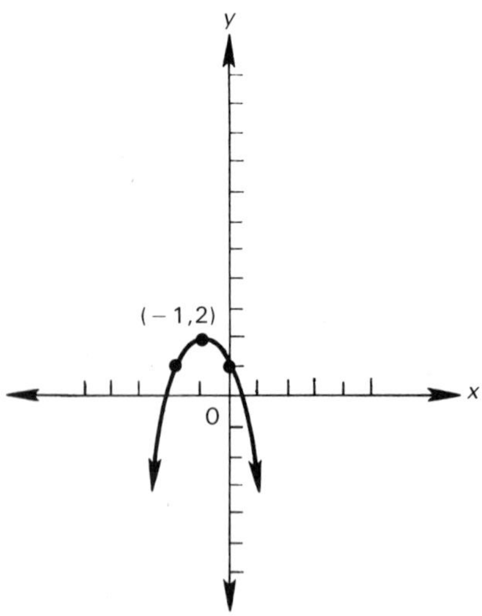

25) $y = f(x) = (y - 3)^2 + 5$

a) vertex: (3, 5)

b) axis of symmetry: $x = 3$

c) domain: all reals

d) (parabola opens up.) Range: $y \geq 5$

29) $y = f(x) = 2(x + 1)^2 - 3$

a) vertex: (-1, -3)

b) axis of symmetry: $x = -1$

c) domain: all reals

d) (Parabola opens up.) Range: $y \geq -3$

33) $y = f(x) = \frac{1}{4}(x + 2)^2 - 4$

a) vertex: (-2, -4)

b) axis of symmetry: $x = -2$

c) domain: all reals

d) (Parabola opens up.) Range: $y \geq -4$

37) $x = y^2$

This graph is congruent to the graph of $y = x^2$, but is symmetric around the y axis and opens right. It is not a function because for each $x > 0$ there are two y-values.

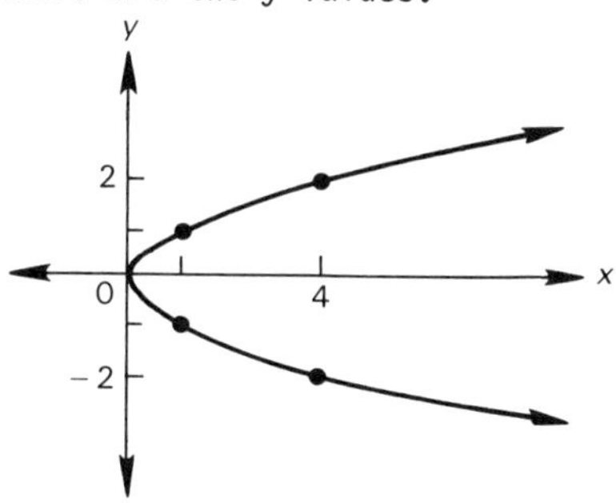

41) $y = ax^2$ If the graph passes through the point (1, -2), then $x = 1$, $y = -2$ satisfies the equation. Put these values into the equation and solve for a.

$$-2 = a(1)^2$$
$$-2 = 1a$$
$$-2 = a$$

45) $x = y^2 - 4$

The graph of $x = y^2$ is shifted 4 units left.

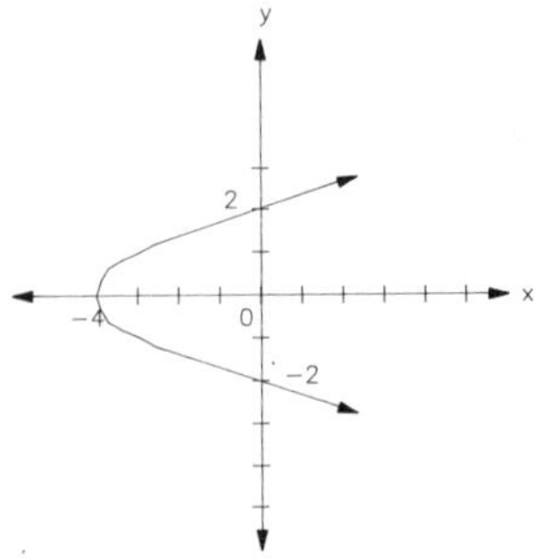

49) $x = 2y^2$ For any given y-value, the x-value is 2 times what it would be for $x = y^2$.

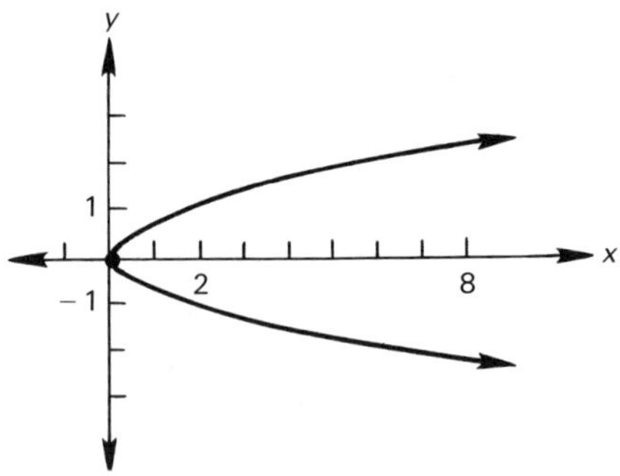

53) Begin with $y = x^2$.

Because the graph is shifted right 4 units, $y = (x - 4)^2$. It is also shifted up two units, which gives $y = (x - 4)^2 + 2$.

57) $y \geq (x + 2)^2$

First graph $y = (x + 2)^2$. Because the y-values for the inequality are greater than or equal to those on the curve shade the region above the curve and make the curve solid.

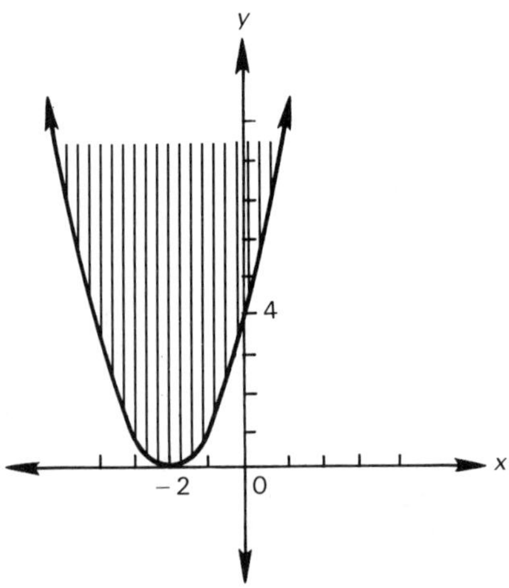

61) $y \geq x^2 - 4$

First graph $y = x^2 - 4$. Because the y-values for the inequality are greater than or equal to those on the curve, shade the region above the curve and make the curve solid.

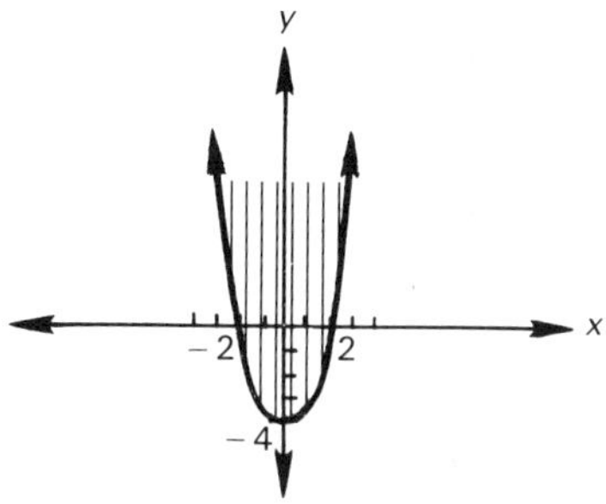

65) $y + (x - 1)^2 > 3$

$y > -(x - 1)^2 + 3$

Graph $y = -(x - 1)^2 + 3$ with a broken curve. Because the values of y for the inequality are greater than those on the curve, shade the region above the curve.

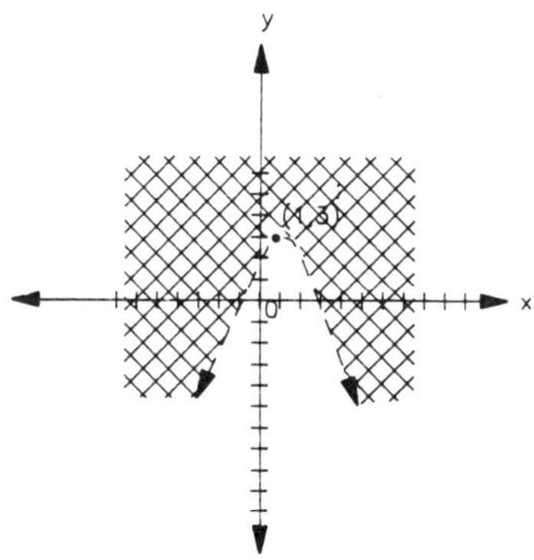

69) Each x-value has only one associated y-value. Yes.

73) Each x-value has only one associated y-value. Yes.

Section 4.2 Completing the Square

1) $y = x^2 + 6x + 9$

5) $y = x^2 - 3x + \frac{9}{4}$ (Take half of 3 and square it: $(\frac{3}{2})^2 = \frac{9}{4}$.)

9) $y = -x^2 - 6x + 2 = -(x^2 + 6x + \quad) + 2 = -(x^2 + 6x + 9) + 2 + 9$

(Because of the negative sign in front of the parentheses, -9 is being added so +9 must be put outside the parentheses.)

$y = -(x + 3)^2 + 11$

13) $y = x^2 + 5x - 2 = (x^2 + 5x + \quad) - 2$

Find half of 5 and square it: $\left(\frac{5}{2}\right)^2 = \frac{25}{4}$

$= \left(x^2 + 5x + \frac{25}{4}\right) - 2 - \frac{25}{4}$

$= \left(x + \frac{5}{2}\right)^2 - \frac{33}{4}$

17) $y = -5 + 6x + 3x^2 = 3x^2 + 6x - 5 = 3(x^2 + 2x + \quad) - 5$

Find half of 2 and square it. $(1)^2 = 1$

$= 3(x^2 + 2x + 1) - 5 - 3 = 3(x + 1)^2 - 8$

21) $y = \frac{3}{4}x^2 - x - \frac{1}{3} = \frac{3}{4}(x^2 - \frac{4}{3}x + \quad) - \frac{1}{3}$

Find half of $\frac{4}{3}$ and square it: $\frac{4}{6} = \frac{2}{3}, \left(\frac{2}{3}\right)^2 = \frac{4}{9}$.

Now consider the $\frac{3}{4}$ in front of the parentheses.

$\frac{3}{4} \cdot \frac{4}{9} = \frac{1}{3}$, so $\frac{1}{3}$ is being added. Subtract $\frac{1}{3}$ outside the parentheses.

$= \frac{3}{4}\left(x^2 - \frac{4}{3}x + \frac{4}{9}\right) - \frac{1}{3} - \frac{1}{3}$

$y = \frac{3}{4}\left(x - \frac{2}{3}\right)^2 - \frac{2}{3}$

25) $y = (x + 1)^2 - 3(x + 1) - \frac{3}{4} = x^2 + 2x + 1 - 3x - 3 - \frac{3}{4}$

$= x^2 - x - \frac{11}{4}$

Take half of 1 and square it: $\left(\frac{1}{2}\right)^2 = \frac{1}{4}$

$= \left(x^2 - x + \frac{1}{4}\right) - \frac{11}{4} - \frac{1}{4} = \left(x - \frac{1}{2}\right)^2 = \frac{12}{4} = \left(x - \frac{1}{2}\right)^2 - 3$

29) $y = x^2 - 4x + 7 = (x^2 - 4x + \quad) + 7 = (x^2 - 4x + 4) + 7 - 4$

$y = (x - 2)^2 + 3$

Vertex: (2, 3)

Axis of symmetry: $x = 2$

y-intercept: $y = (0 - 2)^2 + 3 = 7$

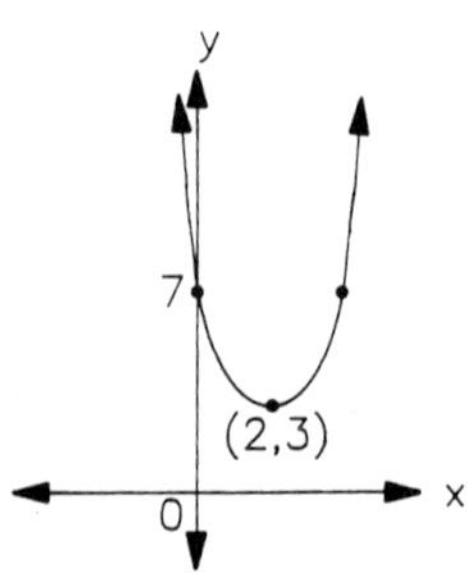

33) $y = 2x^2 - 4x - 4 = 2(x^2 - 2x + \quad) - 4 = 2(x^2 - 2x + 1) - 4 - 2$

$y = 2(x - 1)^2 - 6$

Vertex: (1, -6)

Axis of symmetry: x = 1

y-intercept: $y = 2(0 - 1)^2 - 6 = -4$

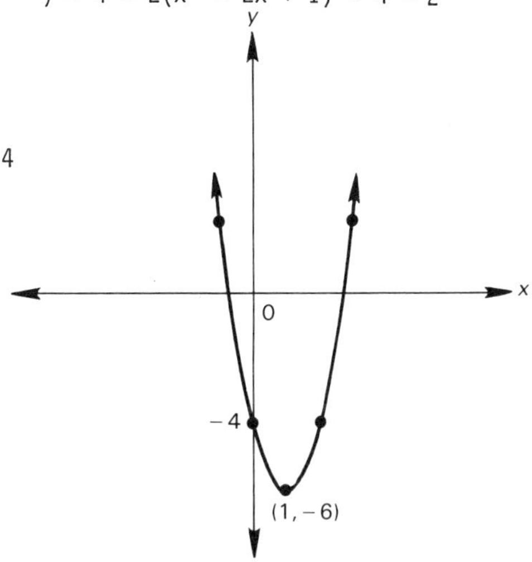

37) $y = \frac{1}{2}x^2 + \frac{5}{2}x + 5$

$= \frac{1}{2}(x^2 + 5x + \quad) + 5$

$= \frac{1}{2}(x^2 + 5x + \frac{25}{4}) + 5 - \frac{25}{8}$

$= \frac{1}{2}(x + \frac{5}{2})^2 + \frac{15}{8}$

y-intercept: $\frac{1}{2}(0 + \frac{5}{2})^2 + \frac{15}{8} = 5$

$(0, c) = (0, 5)$

$(h, k) = \left(-\frac{5}{2}, \frac{15}{8}\right)$

$(2h, c) = (-5, 5)$

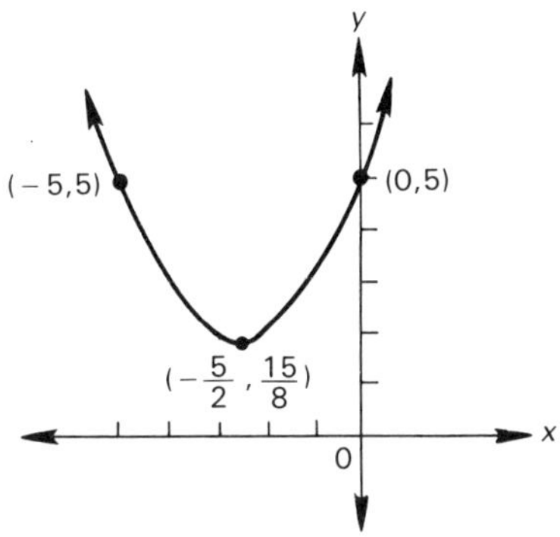

41) $f(x) = |9 - x^2|$

First graph $y = -x^2 + 9$ but use a broken curve for any section below the x-axis because those y-values would be negative. Then reflect that section upward through the x-axis.

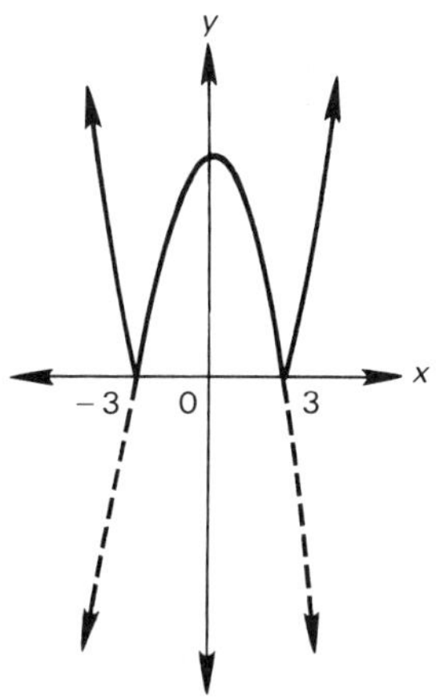

45) $$f(x) = \begin{cases} 1 & \text{if } -3 \le x < 0 \\ x^2 - 4x + 1 & \text{if } 0 \le x < 5 \\ -2x + 16 & \text{if } 5 \le x < 9 \end{cases}$$

Apply the appropriate graphing techniques to each part of the function.

$y = 1$ is a horizontal line.

$y = x^2 - 4x + 1$ is a parabola. Complete the square.

$= (x^2 - 4x + \quad) + 1$

$= (x^2 - 4x + 4) + 1 - 4 = (x - 2)^2 - 3$

$y = -2x + 16$ is a line with $m = -2$, $b = 16$.

Graph each part for the indicated values of x.

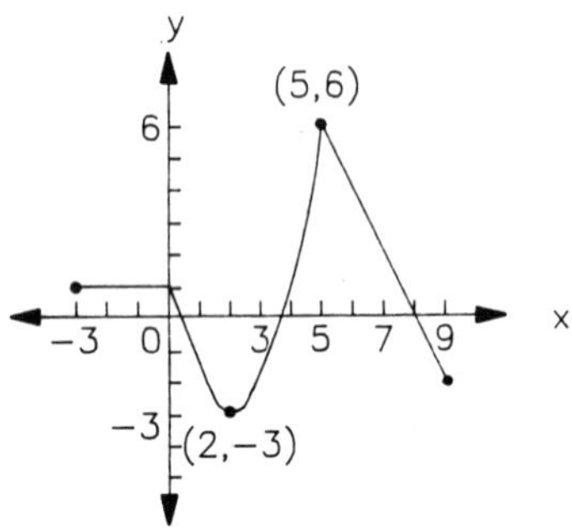

49) $y = a(x - h)^2 + k$ will be concave up when $a > 0$. It will have range $y \ge 2$ when $k = 2$.

Section 4.3 The Quadratic Formula

1) $x^2 - 5x + 6 = 0$

$(x - 3)(x - 2) = 0$

$x - 3 = 0$ or $x - 2 = 0$

$x = 3$ or $x = 2$

5) $10x^2 - 13x - 3 = 0$

$(5x + 1)(2x - 3) = 0$

$5x + 1 = 0$ or $2x - 3 = 0$

$5x = -1$ or $2x = 3$

$x = -\frac{1}{5}$ or $x = \frac{3}{2}$

9) $2x(x + 6) = 22x$

$2x^2 + 12x = 22x$

$2x^2 - 10x = 0$

$2x(x - 5) = 0$

$2x = 0$ or $x - 5 = 0$

$x = 0$ or $x = 5$

13) $x^2 - 3x - 10 = 0$

$$x = \frac{-(-3) \pm \sqrt{(-3)^2 - 4(1)(-10)}}{2(1)}$$

$$x = \frac{3 \pm \sqrt{49}}{2}$$

$$x = \frac{3 \pm 7}{2}$$

$x = \frac{10}{2}$ or $x = -\frac{4}{2}$

$x = 5$ or $x = -2$

17) $3x^2 + 7x + 2 = 0$

$$x = \frac{-(7) \pm \sqrt{(7)^2 - 4(3)(2)}}{2(3)}$$

$$x = \frac{-7 \pm \sqrt{25}}{6}$$

$$x = \frac{-7 \pm 5}{6}$$

$x = -\frac{2}{6}$ or $x = -\frac{12}{6}$

$x = -\frac{1}{3}$ or $x = -2$

21) $-x^2 + 6x - 14 = 0$

$$x = \frac{-(6) \pm \sqrt{(6)^2 - 4(-1)(-14)}}{2(-1)}$$

$$x = \frac{-6 \pm \sqrt{-20}}{-2}$$

$$x = \frac{-6 \pm 2i\sqrt{5}}{-2}$$

$x = 3 \pm i\sqrt{5}$

25) $4 + 3x + x^2 = 0$

$x^2 + 3x + 4 = 0$

$$x = \frac{-(3) \pm \sqrt{(3)^2 - (4)(1)}}{2(1)}$$

$$x = \frac{-3 \pm \sqrt{-7}}{2}$$

$$x = \frac{-3 \pm i\sqrt{7}}{2}$$

29) $y = 2x^2 - 5x - 3$

x-intercepts, if any:

$0 = 2x^2 - 5x - 3$

$$x = \frac{-(-5) \pm \sqrt{(-5)^2 - (4)(2)(-3)}}{2(2)}$$

$$x = \frac{5 \pm \sqrt{49}}{4}$$

$$x = \frac{5 \pm 7}{4}$$

$x = \frac{12}{4}$ or $x = \frac{-2}{4}$

$x = 3$ or $x = -\frac{1}{2}$

33) $y = 2x^2 - 7x + 6$

x-intercepts if any:

$0 = 2x^2 - 7x + 6$

$$x = \frac{-(-7) \pm \sqrt{(-7)^2 - 4(2)(6)}}{2(2)}$$

$$x = \frac{7 \pm \sqrt{1}}{4}$$

$$x = \frac{7 \pm 1}{4}$$

$x = \frac{8}{4}$ or $x = \frac{6}{4}$

$x = 2$ or $x = \frac{3}{2}$

37) (a)

$x^2 - 8x + 16 = 0$

$b^2 - 4ac = (-8)^2 - 4(1)(16)$

$= 0$

41) (d)

$2x^2 - x + 1 = 0$

$b^2 - 4ac = (-1)^2 - 4(2)(1)$

$= -7$

45) (c)

$6x^2 + 7x - 4 = 0$

$b^2 - 4ac = (7)^2 - 4(6)(-4)$

$= 145$

49) $y = x^2 - 4x + 4$

$b^2 - 4ac = (-4)^2 - 4(1)(4)$

$= 0$

Once; the parabola is tangent to the x-axis

a) Vertex:

$0 = x^2 - 4x + 4$

$= (x - 2)^2$

$x = 2$

$(2, 0)$

b) y-intercept

$y = (0)^2 - 4(0) + 4$

$y = 4$

c) x-intercept: 2

53) $y = 2x^2 - 4x + 3$

$b^2 - 4ac = (-4)^2 - 4(2)(3)$

$= -8$

None; the parabola never crosses the x-axis.

a) Vertex:

$y = 2x^2 - 4x + 3$

$= 2(x^2 - 2x + \quad) + 3$

$= 2(x^2 - 2x + 1) + 3 - 2$

$= 2(x - 1)^2 + 1$

$(1, 1)$

b) y-intercept:

$y = 2(0^2) - 4(0) + 3$

$y = 3$

c) None

57) $f(x) = x^2 - bx + 7$

$0 = b^2 - 4ac = (-b)^2 - 4(1)(7)$

$0 = b^2 - 28$

$28 = b^2$

$\pm\sqrt{28} = b$

$\pm 2\sqrt{7} = b$

61) $f(x) = kx^2 - x - 1$

$0 < b^2 - 4ac = (-1)^2 - 4(k)(-1)$

$0 < 1 + 4k$

$-1 < 4k$

$-\frac{1}{4} < k$

65) $f(x) = tx^2 - x - 1$

$0 > b^2 - 4ac = (-1)^2 - 4(t)(-1)$

$0 > 1 + 4t$

$-1 > 4t$

$-\frac{1}{4} > t$

69) $6x^2 + 5x - 4 = 0$

$-\frac{b}{a} = -\frac{5}{6} \qquad \frac{c}{a} = \frac{-4}{6} = -\frac{2}{3}$

$x = \frac{-5 \pm \sqrt{(5)^2 - 4(6)(-4)}}{2(6)}$

$= \frac{-5 \pm \sqrt{121}}{12}$

$= \frac{-5 \pm 11}{12}$

$x = \frac{6}{12} \quad \text{or} \quad x = \frac{-16}{12}$

$x = \frac{1}{2} \qquad x = -\frac{4}{3}$

sum: $\frac{1}{2} + \left(-\frac{4}{3}\right) = \frac{3}{6} - \frac{8}{6} = \frac{-5}{6}$

product: $\frac{1}{2}\left(-\frac{4}{3}\right) = \frac{-4}{6} = -\frac{2}{3}$

73) $2x^2 + 6x - 9 = 0$

$$-\frac{b}{a} = -\frac{6}{2} = -3 \qquad \frac{c}{a} = \frac{-9}{2} = -\frac{9}{2}$$

$$x = \frac{-(6) \pm \sqrt{(6)^2 - 4(2)(-9)}}{2(2)}$$

$$= \frac{-6 \pm \sqrt{108}}{4}$$

$$= \frac{-6 \pm 6\sqrt{3}}{4}$$

$$x = \frac{-3 \pm 3\sqrt{3}}{2}$$

sum: $\dfrac{-3 + 3\sqrt{3}}{2} + \dfrac{-3 - 3\sqrt{3}}{2} = \dfrac{-6}{2} = -3$

product: $\left(\dfrac{-3 + 3\sqrt{3}}{2}\right)\left(\dfrac{-3 - 3\sqrt{3}}{2}\right) = \dfrac{9 - 9 \cdot 3}{4} = \dfrac{-18}{4} = -\dfrac{9}{2}$

77) $x^2 + 3x - 10 \geq 0$

Consider $x^2 + 3x - 10 = 0$

$(x + 5)(x - 2) = 0$

$x + 5 = 0$ or $x - 2 = 0$

$x = -5$ $\quad$ $x = 2$

The parabola opens up and intercepts the x-axis at -5 and 2. Between these values the parabola is below the axis and $y < 0$. So $x^2 + 3x - 10 \geq 0$ when $x \leq -5$ or $x \geq 2$.

81) $f(x) = (x - 3)(x + 1) = x^2 - 2x - 3$

x-intercepts: $0 = (x - 3)(x + 1)$

$x - 3 = 0$ or $x + 1 = 0$

$x = 3$ $\quad$ $x = -1$

y-intercept: $y = (0 - 3)(0 + 1) = -3$

$a > 0$, so parabola opens up.

Find h and k to get the vertex.

$y = (x^2 - 2x + \quad) - 3$

$y = (x^2 - 2x + 1) - 3 - 1$

$y = (x - 1)^2 - 4$

vertex: $(1, -4)$

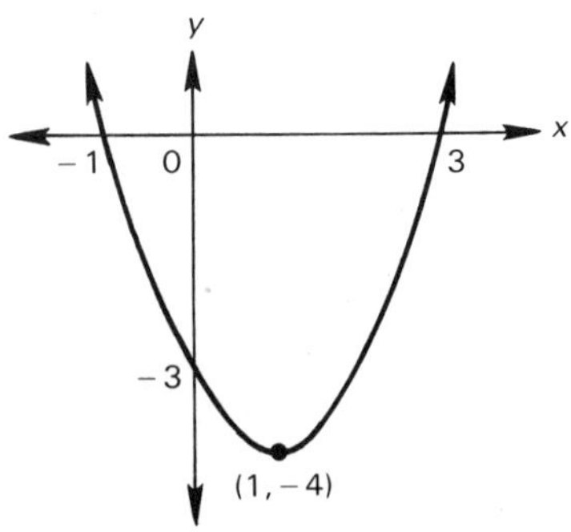

85) $f(x) = 3x^2 - 4x + 1$

$= (3x - 1)(x - 1)$

x-intercepts: $0 = (3x - 1)(x - 1)$

$3x - 1 = 0 \quad \text{or} \quad x - 1 = 0$

$3x = 1$

$x = \frac{1}{3} \quad \text{or} \quad x = 1$

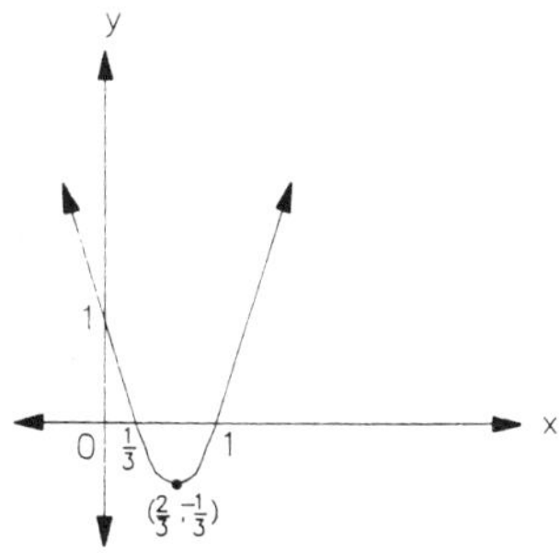

y-intercept: $y = 3(0)^2 - 4(0) + 1 = 1$

Find h and k: $y = 3x^2 - 4x + 1$

$= 3(x^2 - \frac{4}{3}x + \quad) + 1$

$= 3(x^2 - \frac{4}{3}x + \frac{4}{9}) + 1 - \frac{4}{3}$

$= 3\left(x - \frac{2}{3}\right)^2 - \frac{1}{3}$

vertex: $\left(\frac{2}{3}, -\frac{1}{3}\right)$

a > 0, so the parabola opens up.

89) $2x^4 - 13x^2 - 7 = 0$

Let $x^2 = u$

$2u^2 - 13u - 7 = 0$

$(2u + 1)(u - 7) = 0$

$2u + 1 = 0 \quad \text{or} \quad u - 7 = 0$

$2u = -1 \qquad u = 7$

$u = -\frac{1}{2}$

$x^2 = -\frac{1}{2} \quad \text{or} \quad x^2 = 7$

$x = \pm\sqrt{\frac{1}{2}}\, i \qquad x = \pm\sqrt{7}$

$x = \pm\frac{\sqrt{2}}{2}\, i$

93) $\dfrac{4x}{2x - 1} + \dfrac{3}{2x + 1} = 0$

$$\frac{4x}{2x - 1} = \frac{-3}{2x + 1}$$

$$4x(2x + 1) = -3(2x - 1)$$

$$8x^2 + 4x = -6x + 3$$

$$8x^2 + 10x - 3 = 0$$

$$x = \frac{-(10) \pm \sqrt{(10)^2 - 4(8)(-3)}}{2(8)}$$

$$x = \frac{-10 \pm \sqrt{196}}{16}$$

$$x = \frac{-10 \pm 14}{16}$$

$x = \dfrac{-24}{16}$ or $x = \dfrac{4}{16}$

$x = -\dfrac{3}{2}$ or $x = \dfrac{1}{4}$

Section 4.4 Applications of Quadratic Functions

1) Let x represent the first integer.

$x + 1$ represents the next consecutive integer.

$x(x + 1)$ represents their product.

$x(x + 1) = 210$

$x^2 + x = 210$

$x^2 + x - 210 = 0$

$(x - 14)(x + 15) = 0$

$x - 14 = 0$ or $x + 15 = 0$

$x = 14$ or $x = -15$

Reject this solution because the integers must be positive.

If $x = 14$, then $x + 1 = 15$. 14 x 15 is 210.

Answer: The integers are 14 and 15.

5) A sketch will help to solve this problem.

4–20

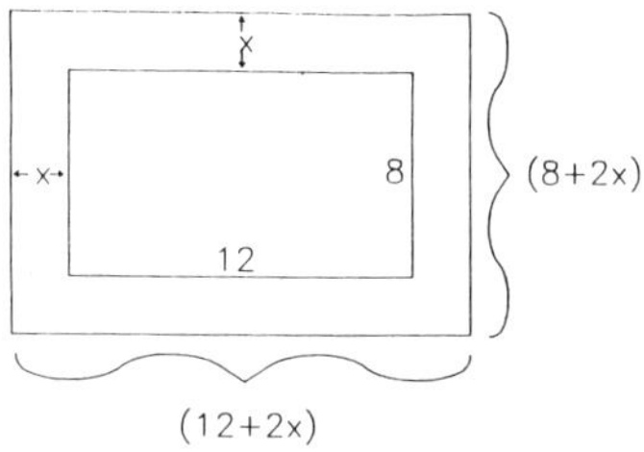

If a border of width x is added on all sides, the expanded rectangle has length $(12 + 2x)$ and width $(8 + 2x)$. The original area is $8 \times 12 = 96$.

The new area is $(12 + 2x)(8 + 2x)$, and is twice the original area.

$$(12 + 2x)(8 + 2x) = 2(96)$$
$$96 + 40x + 4x^2 = 192$$
$$4x^2 + 40x - 96 = 0$$
$$x^2 + 10x - 24 = 0$$
$$(x + 12)(x - 2) = 0$$
$$x + 12 = 0 \quad \text{or} \quad x - 2 = 0$$
$$x = -12 \qquad\qquad x = 2$$

Reject the -12 because the border must have a positive width.
If the border is 2 feet, then the new rectangle will be 16 x 12 with an area of 192, which is twice 96.
Answer: The border should be 2 feet wide.

9) The area of the walk added to the area of the small rectangle equals the area of the large rectangle.

$10 \times 18 = 180$ is the area of the small rectangle.

$(18 + 2x)(10 + 2x)$ is the area of the large rectangle.

$$(18 + 2x)(10 + 2x) = 52 + 180$$
$$180 + 56x + 4x^2 = 232$$
$$4x^2 + 56x - 52 = 0$$
$$x^2 + 14x - 13 = 0$$
$$x = \frac{-(14) \pm \sqrt{(14)^2 - 4(1)(-13)}}{2(1)}$$
$$= \frac{-14 \pm \sqrt{248}}{2}$$
$$= \frac{-14 \pm 2\sqrt{62}}{2}$$

9) continued

$$x = \frac{-14 \pm 2\sqrt{62}}{2} \qquad \text{or} \qquad x = \frac{-14 - 2\sqrt{62}}{2}$$

$= -7 + \sqrt{62}$

Reject this because the answer should be positive.

This seems like an impractical width for a pool walk and it is tedious to check. But this is the answer: The walk is $-7 + \sqrt{62} \approx 0.9$ meters wide.

13) It takes some patience to sort this problem out. Begin by recording the information algebraically.

Let r represent the stillwater rate.

$r + 4$ represents the downstream rate because the current adds to the boat's speed.

$r - 4$ represents the upstream rate because the current subtracts from the boat's speed.

Now if time is represented by t, there will be too many variables. Notice that the distance of 36 miles has not been used yet and recall $t = \frac{d}{r}$.

$\frac{36}{r + 4}$ represents the downstream travel time.

$\frac{36}{r - 4}$ represents the upstream travel time.

The time for the downstream trip was $\frac{3}{4}$ hr. less than the upstream trip:

$$\frac{36}{r + 4} = \frac{36}{r - 4} - \frac{3}{4}$$

Multiply each term by the LCD: $4(r + 4)(r - 4)$

$$36 \cdot 4(r - 4) = 36 \cdot 4(r + 4) - 3(r + 4)(r - 4)$$
$$144(r - 4) = 144(r + 4) - 3(r^2 - 16)$$
$$144r - 576 = 144r + 576 - 3r^2 + 48$$
$$3r^2 = 1200$$
$$r^2 = 400$$
$$r = 20$$

Answers: The stillwater rate is 20 mph.

Downstream time: $\frac{36}{r + 4} = \frac{36}{24} = 1\frac{1}{2}$ hours

Upstream time: $\frac{36}{r - 4} = \frac{36}{16} = 2\frac{1}{4}$ hours.

17) $f(x) = 16x^2 - 64x + 100$

$= 16(x^2 - 4x + \quad) + 100$

$= 16(x^2 - 4x + 4) + 100 - 64$

$= 16(x - 2)^2 + 36$

$a = 16 > 0$, so the parabola opens up, vertex $(2, 36)$.

Minimum: 36 when $x = 2$.

21) $f(x) = -x\left(\frac{2}{3} + x\right)$

$= -\frac{2}{3}x - x^2$

$= -(x^2 + \frac{2}{3}x + \quad)$

$= -(x^2 + \frac{2}{3}x + \frac{1}{9}) + \frac{1}{9}$

$= -(x + \frac{1}{3})^2 + \frac{1}{9}$

$a = -1 < 0$, so the parabola opens down, vertex $(-\frac{1}{3}, \frac{1}{9})$.

Maximum: $\frac{1}{9}$ when $x = -\frac{1}{3}$

25) Let x represent one of the numbers.

$12 - x$ represents the other number.

$x(12 - x)$ represents their product.

Find the maximum for $f(x) = x(12 - x)$

$f(x) = 12x - x^2$

$= -(x^2 - 12x + \quad)$

$= -(x^2 - 12x + 36) + 36$

$= -(x - 6)^2 + 36$

$a = -1 < 0$, so the parabola opens down with a maximum of 36 when $x = 6$.

If $x = 6$, then $12 - x = 6$.

Answer: The numbers are 6 and 6.

29) Make a sketch. If the two ends of the rectangle are both of measure x, then the remaining side will be $40 - 2x$. $f(x) = x(40 - 2x)$ represents the area. Find the maximum of this function.

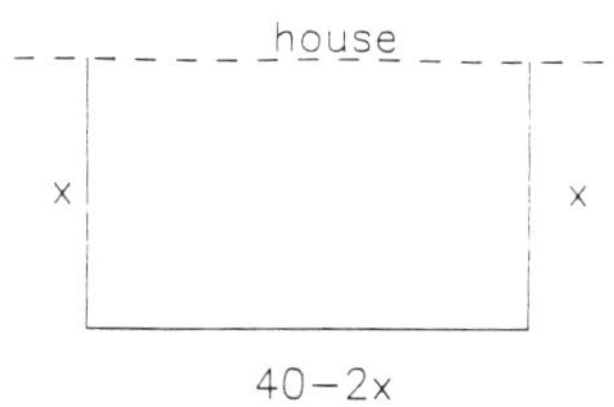

29) continued

$f(x) = -2x^2 + 40x$

$= -2(x^2 + 20x + \quad)$

$= -2(x^2 + 20x + 100) + 200$

$= -2(x + 10)^2 + 200$

$a = -2 < 0$, so the parabola opens down, with a maximum of 400 when $x = 10$.

If $x = 10$, $40 - 2x = 20$.

Answer: The rectangle should be 10 feet by 20 feet.

33) $h = 128t - 16t^2$ and h is to be 192. Solve for t.

$192 = 128t - 16t^2$

$0 = -16t^2 + 128t - 192$

$0 = t^2 - 8 + 12$

$0 = (t - 6)(t - 2)$

$t - 6 = 0$ or $t - 2 = 0$

$t = 6$ $\qquad$ $t = 2$

Answer: The object will be at 192 feet at 2 seconds (going up) and again at 6 seconds (falling down).

37) The area of the rectangle is given by

$$A = xy$$

To express this area as a function of one variable we may begin by writing y in terms of x. Since the bottom right triangle with height y and base 2 - x is similar to the large right triangle with height 4 and base 2, we obtain this proportion

$$\frac{y}{2 - x} = \frac{4}{2}$$

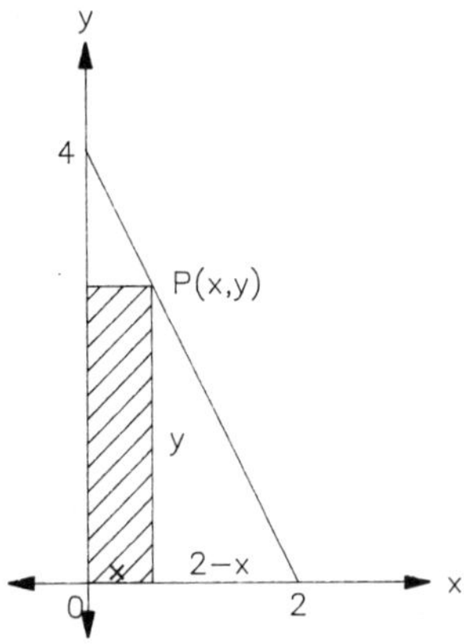

Solve this equation for y to obtain $y = 4 - 2x$ (note that $0 < x < 2$) and substitute into $A = xy$ to get $A(x) = x(4 - 2x)$. Now complete the square.

37) continued

$$A(x) = x(4 - 2x)$$
$$= -2x^2 + 4x$$
$$= -2(x^2 - 2x)$$
$$= -2(x^2 - 2x + 1) + 2$$
$$= -2(x - 1)^2 + 2$$

The maximum area is obtained when $x = 1$ and $y = 4 - 2x = 4 - 2(1) = 2$, and the coordinates for the rectangle of maximum area are $P(1, 2)$.

41) $y = 2x^2 - 6x + 9$

$\frac{-b}{2a} = \frac{-(-6)}{2 \cdot 2} = \frac{6}{4} = \frac{3}{2}$

$\frac{4ac - b^2}{4a} = \frac{4(2)(9) - (-6)^2}{4(2)} = \frac{72 - 36}{8} = \frac{36}{8} = \frac{9}{2}$

Vertex: $\left(\frac{3}{2}, \frac{9}{2}\right)$

$a = 2 > 0$, so $\frac{9}{2}$ is a minimum.

45) $y + \frac{2}{3}x^2 = 9$

$y = -\frac{2}{3}x^2 + 9$

$\frac{-b}{2a} = \frac{0}{2\left(-\frac{2}{3}\right)} = 0$

$\frac{4ac - b^2}{4a} = \frac{4\left(-\frac{2}{3}\right)(9) - 0}{4\left(-\frac{2}{3}\right)} = \frac{-24}{-\frac{8}{3}} = (24) \cdot \frac{3}{8} = 9$

Vertex: $(0, 9)$

$a = -\frac{2}{3} < 0$, so 9 is a maximum.

Section 4.5 Conic Sections: The Circle

1) a) $x^2 + y^2 = 25$ $C(0, 0); r = 5$

b) $x^2 + y^2 = 16$ $C(0, 0); r = 4$

c) $x^2 + y^2 = 4$ $C(0, 0); r = 2$

d) $x^2 + y^2 = 1$ $C(0, 0); r = 1$

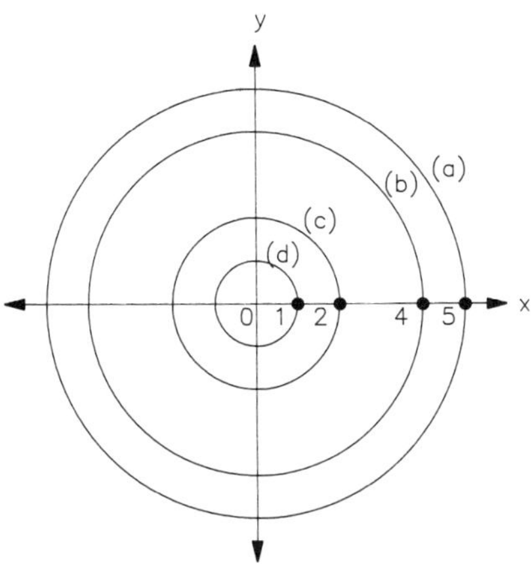

5) $x^2 - 2x + y^2 - 6y = -9$

$(x^2 - 2x + \quad) + (y^2 - 6y + \quad) = -9$

$(x^2 - 2x + 1) + (y^2 - 6y + 9) = -9 + 1 + 9$

$(x - 1)^2 + (y - 3)^2 = 1$

C(1, 3); $r = 1$

9) $x^2 - 8x + y^2 = -14$

$(x^2 - 8x + \quad) + y^2 = -14$

$(x^2 - 8x + 16) + y^2 = -14 + 16$

$(x - 4)^2 + y^2 = 2$

C(4, 0); $r = \sqrt{2}$

13) $16x^2 + 24x + 16y^2 - 32y = 119$

$x^2 + \frac{3}{2}x + y^2 - 2y = \frac{119}{16}$

$(x^2 + \frac{3}{2}x + \quad) + (y^2 - 2y + \quad) = \frac{119}{16}$

$(x^2 + \frac{3}{2}x + \frac{9}{16}) + (y^2 - 2y + 1) = \frac{119}{16} + \frac{9}{16} + \frac{16}{16}$

$(x + \frac{3}{4})^2 + (y - 1)^2 = \frac{144}{16} = 9$

$C\left(-\frac{3}{4}, 1\right)$; $r = 3$

17) C(-3, 3); $r = \sqrt{7}$

$[x - (-3)]^2 + (y - 3)^2 = (\sqrt{7})^2$

$(x + 3)^2 + (y - 3)^2 = 7$

21) $x^2 + y^2 = 4$ C(0, 0); $r = 2$

The lines with undefined slope will be vertical, perpendicular to the x-axis, so tangent at (r, 0) = (2, 0) and (-r, 0) = (-2, 0). These lines are $x = 2$ and $x = -2$.

25) $x^2 + 4x + y^2 - 6y = 60; \quad (6, 0)$

$$(x^2 + 4x + \quad) + (y^2 - 6y + \quad) = 60$$

$$(x^2 + 4x + 4) + (y^2 - 6y + 9) = 60 + 4 + 9$$

$$(x + 2)^2 + (y - 3)^2 = 73$$

$C(-2, 3); \quad r = \sqrt{73}$

Find the slope of the radius from (-2, 3) to (6, 0):

$$\frac{0 - 3}{6 + 2} = -\frac{3}{8}$$

The slope of the tangent line: $\frac{8}{3}$

Using this slope, the point (6, 0) and the point-slope form of the equation, the line is

$$y - 0 = \frac{8}{3}(x - 6)$$

$$y = \frac{8}{3}x - 16$$

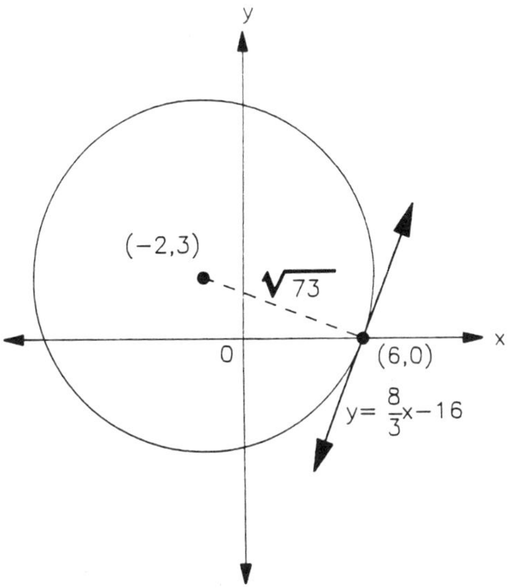

29) $P(-2, 4); \quad Q(3, -8)$

$$\left(\frac{-2 + 3}{2}, \frac{4 + (-8)}{2} \right) = \left(\frac{1}{2}, -2 \right)$$

33) $x^2 + y^2 = 80$ $C(0, 0)$; $r = 4\sqrt{5}$

Substitute $x = 4$ and solve for y.

$$(4)^2 + y^2 = 80$$
$$16 + y^2 = 80$$
$$y^2 = 64$$
$$y = \pm 8$$

Since the point must be in the first quadrant, it is (4, 8).

Find the slope of the radius from (0, 0) to (4, 8).

$$m = \frac{8}{4} = 2$$

The slope of the tangent line: $-\frac{1}{2}$

Use the slope, the point (4, 8) and the point-slope form of the equation. The line is

$$(y - 8) = -\frac{1}{2}(x - 4)$$
$$y - 8 = -\frac{1}{2}x + 2$$
$$y = -\frac{1}{2}x + 10$$

37) The triangle formed when the board is in its original position is a right triangle with a 13-foot hypotenuse and one 5-foot side. By the Pythagorean theorem, the other side measures

$$\sqrt{(13)^2 - (5)^2} = 12 \text{ feet}$$

So originally the top of the board touches the wall at 12 feet up from the ground. As the board slides down, this height can be expressed as $(12 - x)$. In the new position, the board is $(y + 5)$ feet from the wall. Therefore, a new right triangle is formed which has one side $(12 - x)$, another side $(y + 5)$, and the hypotenuse 13.

a) Applying the Pythagorean theorem again, obtain y as a function of x:

$$(y + 5)^2 + (12 - x)^2 = (13)^2$$
$$(y + 5)^2 = 169 - (12 - x)^2$$
$$y + 5 = \sqrt{169 - (144 - 24x + x^2)}$$
$$y = -5 + \sqrt{25 + 24x - x^2}$$

b) Let $x = 7$ and evaluate.

$$y = -5 + \sqrt{25 + 24(7) - (7)^2} = -5 + \sqrt{144}$$
$$= -5 + 12 = 7$$

Section 4.6 Conic Sections: The Ellipse and the Parabola

1) $\frac{x^2}{25} + \frac{y^2}{16} = 1$

Because 25 is the larger denominator, it is a^2. The major axis is parallel to the x axis because a^2 is with x^2.

$a^2 = 25$ $\quad b^2 = 16$ $\quad c^2 = a^2 - b^2$

$a = 5$ $\quad b = 4$ $\quad c^2 = 25 - 16 = 9$

$c = 3$

Center: (0, 0)

Vertices: (-5, 0), (5, 0)

Foci: (-3, 0), (3, 0)

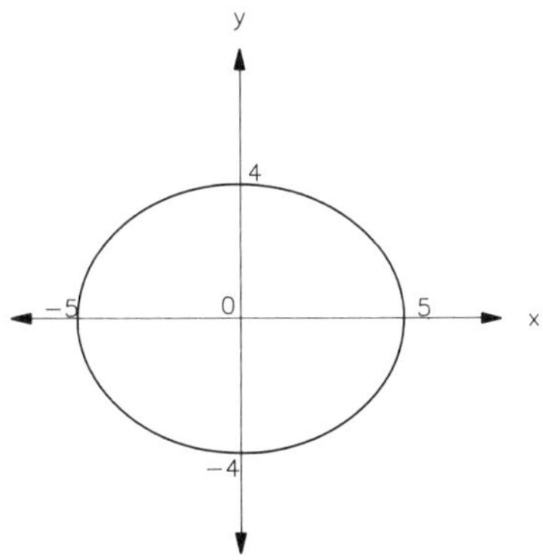

5) $\frac{x^2}{4} + \frac{y^2}{36} = 1$

Because 36 is the larger denominator, it is a^2. The major axis is parallel to the y-axis because a^2 is with y^2.

$a^2 = 36$ $\quad b^2 = 4$ $\quad c^2 = a^2 - b^2$

$a = 16$ $\quad b = 2$ $\quad = 36 - 4 = 32$

$c = 4\sqrt{2}$

Center: (0, 0)

Vertices: (0, 6), (0, -6)

Foci: $(0, 4\sqrt{2})$, $(0, -4\sqrt{2})$

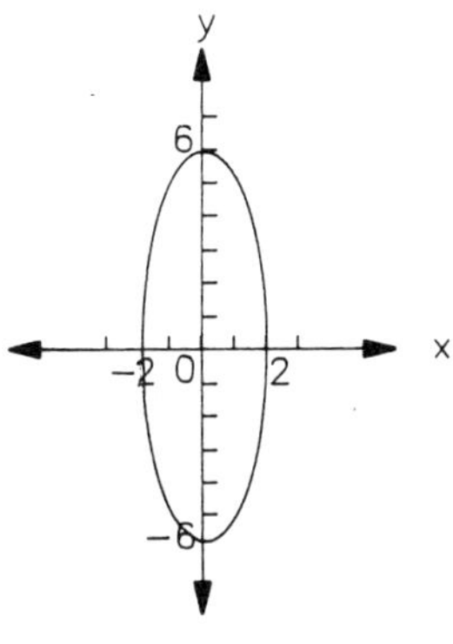

9) $4x^2 + 9y^2 = 36$

$$\frac{x^2}{9} + \frac{y^2}{4} = 1$$

Because 9 is the larger denominator, it is a^2. The major axis is parallel to the x-axis because a^2 is with x^2.

$a^2 = 9$ $\quad$ $b^2 = 4$ $\quad$ $c^2 = a^2 - b^2$

$a = 3$ $\quad$ $b = 2$ $\quad$ $= 9 - 4 = 5$

$c = \sqrt{5}$

Center: (0, 0)

Vertices: (-3, 0), (3, 0)

Foci: $(-\sqrt{5}, 0)$, $(\sqrt{5}, 0)$

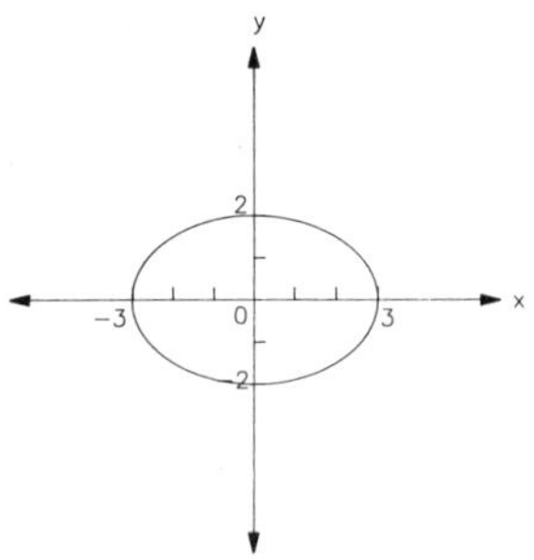

13) $$\frac{(x - 1)^2}{9} + \frac{(y + 2)^2}{4} = 1$$

Because 9 is the larger denominator, it is a^2. The major axis is parallel to the x-axis because a^2 is with x^2.

$a^2 = 9$ $\quad$ $b^2 = 4$ $\quad$ $c^2 = a^2 - b^2$

$a = 3$ $\quad$ $b = 2$ $\quad$ $= 9 - 4 = 5$

$c = \sqrt{5}$

Center: (h, k) = (1, -2)

Vertices: (h - a), k) = (1 - 3, -2) = (-2, -2)

(h + a, k) = (1 + 3, -2) = (4, -2)

Foci: $(h - c, k) = (1 - \sqrt{5}, -2)$

$(h + c, k) = (1 + \sqrt{5}, -2)$

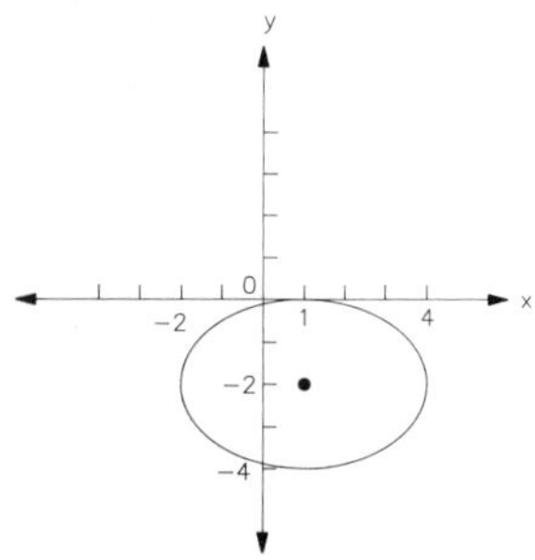

17) $c(0, 0)$; $2a = 10$; $2b = 6$

Horizontal axis is parallel to x-axis, so a^2 goes with x^2.

$$a = 5 \qquad b = 3 \qquad \frac{x^2}{25} + \frac{y^2}{9} = 1$$
$$a^2 = 25 \qquad b^2 = 9$$

21) Vertices $(0, \pm 5)$; foci $(0, \pm 3)$

Since the x-coordinate of all vertices and foci is 0, the y-axis is the major axis, so a^2 goes with y^2. The vertices are at $(0, \pm a)$ so $a = 5$, and $a^2 = 25$. The foci are at $(0, \pm c)$, so $c = 3$ and $c^2 = 9$.

$$b^2 = a^2 - c^2$$
$$b^2 = 25 - 9 = 16$$

$$\frac{x^2}{16} + \frac{y^2}{25} = 1$$

25) $x^2 = \frac{1}{4}y$

$4p = \frac{1}{4}$

$p = \frac{1}{16}$

x is squared, so the directrix is parallel to the x-axis.

Focus: $\left(0, \frac{1}{16}\right)$

Directrix: $y = -\frac{1}{16}$

29) $y^2 = 2x$

$4p = 2$

$p = \frac{1}{2}$

y is squared, so the directrix is parallel to the y-axis.

Focus: $\left(\frac{1}{2}, 0\right)$

Directrix: $x = -\frac{1}{2}$

33) $-9x = 6y^2$

$-\frac{9}{6}x = y^2$

$-\frac{3}{2}x = y^2$

$4p = -\frac{3}{2}$

$p = -\frac{3}{8}$

y is squared, so the directrix is parallel to the y-axis.

Focus: $\left(-\frac{3}{8}, 0\right)$

Directrix: $x = \frac{3}{8}$

37) Directrix $y = -\frac{2}{3}$; vertex (0, 0).

Because the vertex is at (0, 0), the directrix is $y = -p$, so $p = \frac{2}{3}$ and $4p = \frac{8}{3}$. The focus is at $(0, p) = (0, \frac{2}{3})$. The directrix is parallel to the x-axis, so x is squared.

$x^2 = \frac{8}{3}y$

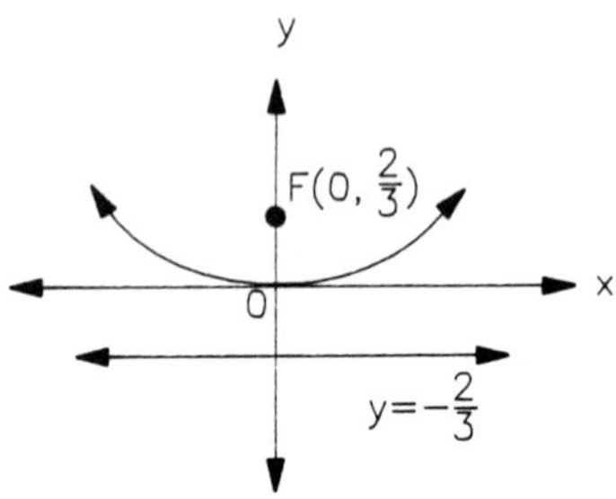

41) Vertex (0, 0)

If the axis is horizontal, the directrix is vertical, parallel to the y-axis, so y is squared, and the equation is of the form $y^2 = 4px$. The point (12, -6) satisfies the equation. Substitute, and solve for p.

$$y^2 = 4px$$
$$(-6)^2 = 4p(12)$$
$$36 = 48p$$
$$\frac{3}{4} = p$$
$$y^2 = 4\left(\frac{3}{4}\right)x = 3x$$

Focus at $(p, 0) = (\frac{3}{4}, 0)$.

Directrix $x = -p$, $x = -\frac{3}{4}$.

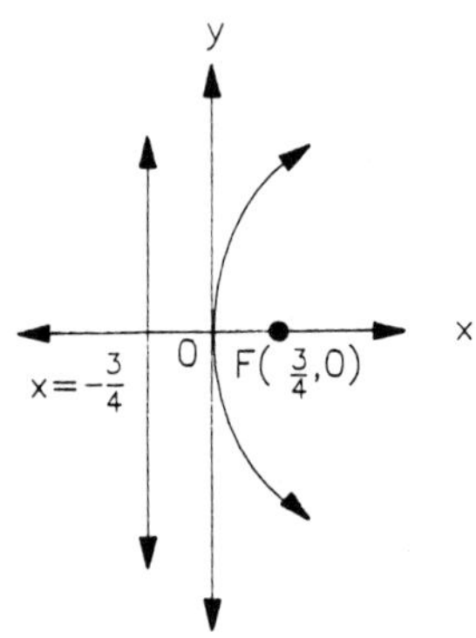

45) $(x - 2)^2 = 4(y + 5)$

$h = 2,\quad k = -5,\quad 4p = 4$ so $p = 1$

Vertex: $(h, k) = (2, -5)$

Axis: $x = h = 2$

Focus: $(h, k + p) = (2, -5 + 1) = (2, -4)$

Directrix: $y = k - p,\quad y = -5 - 1,\quad y = -6$

49) Vertex: $(-3, -2)$ Directrix: $x = -\frac{9}{2}$

Because the directrix is parallel to the y-axis, this equation will be in the form

$$(x - h) = 4p(y - k)^2$$

$$h = -3 \qquad k = -2$$

The distance from the directrix to the vertex will give p. From $x = -3$ to $x = -\frac{9}{2}$ is $\frac{3}{2} = p$. The equation is $4\left(\frac{3}{2}\right)(x + 3) = (y+2)^2$

$$6(x + 3) = (y + 2)^2$$

The focus is at $(h + p, k) = (-3 + \frac{3}{2}, -2) = (-\frac{3}{2}, -2)$.

The axis is $y = k$, $y = -2$.

53) $25x^2 + y^2 - 12y = -11$

$25x^2 + y^2 - 12y + 36 = 25$

$25x^2 + (y - 6)^2 = 25$

$$\frac{x^2}{1} + \frac{(y - 6)^2}{25} = 1$$

This is an ellipse with horizontal major axis since 25 is the larger denominator.

$a^2 = 25 \qquad b^2 = 1 \qquad c^2 = a^2 - b^2$

$a = 5 \qquad b = 1 \qquad\quad = 25 - 1$

$c = 2\sqrt{6}$

Center: $(0, 6)$

Vertices: $(0, 1), (0, 11)$

Foci: $(0, 6 - 2\sqrt{6}), (0, 6 + 2\sqrt{6})$

Section 4.7 Conic Sections: The Hyperbola

1) $\frac{x^2}{25} - \frac{y^2}{9} = 1$

$a^2 = 25$ $\quad$ $b^2 = 9$ $\quad$ $c^2 = a^2 + b^2$

$a = 5$ $\quad$ $b = 3$ $\quad$ $= 25 + 9 = 34$

$c = \sqrt{34}$

Since the y^2 term is preceded by a minus sign the transverse axis is horizontal.

Center: (0, 0) Vertices: (5, 0), (-5, 0)

Foci: $(\sqrt{34}, 0)$, $(-\sqrt{34}, 0)$ Asymptotes: $y = \pm \frac{3}{5}x$

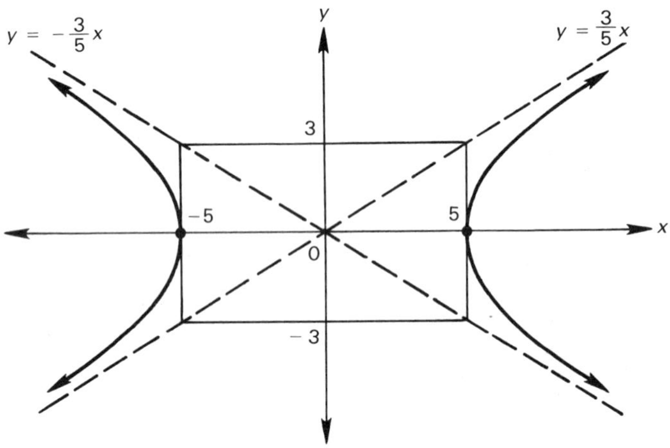

5) $\frac{y^2}{36} - \frac{x^2}{9} = 1$

$a^2 = 36$ $\quad$ $b^2 = 9$ $\quad$ $c^2 = a^2 + b^2$

$a = 6$ $\quad$ $b = 3$ $\quad$ $= 36 + 9$

$= 45$

$c = 3\sqrt{5}$

Since the x^2 term is preceded by a minus sign the transverse axis is vertical.

Center: (0, 0) Vertices: (0, 6), (0, -6)

Foci: $(0, 3\sqrt{5})$, $(0, -3\sqrt{5})$ Asymptotes: $y = \pm 2x$

5) continued

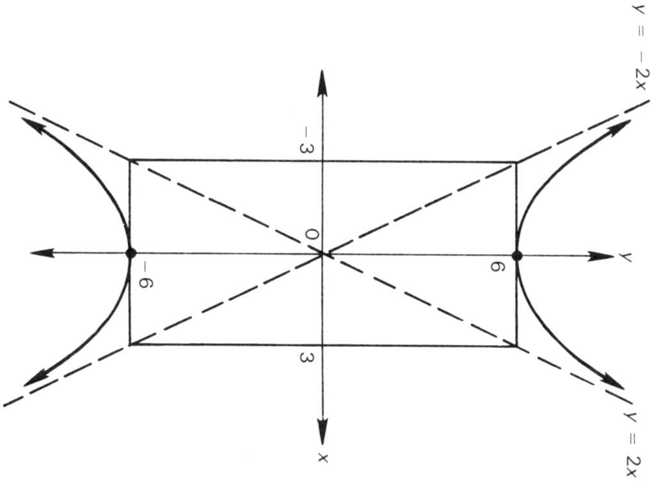

9) $4x^2 - 9y^2 = 36$

$\frac{4}{36}x^2 - \frac{9}{36}y^2 = \frac{36}{36}$

$\frac{x^2}{9} - \frac{y^2}{4} = 1$

$a^2 = 9$	$b^2 = 4$	$c^2 = a^2 + b^2$
$a = 3$	$b = 2$	$= 9 + 4$
		$= 13$
		$c = \sqrt{13}$

The minus sign precedes the y^2 term so the transverse axis is horizontal.

Center: (0, 0) Vertices: (3, 0), (-3, 0) Foci: $(\sqrt{13}, 0)$, $(-\sqrt{13}, 0)$

Asymptotes: $y = \pm \frac{2}{3}x$

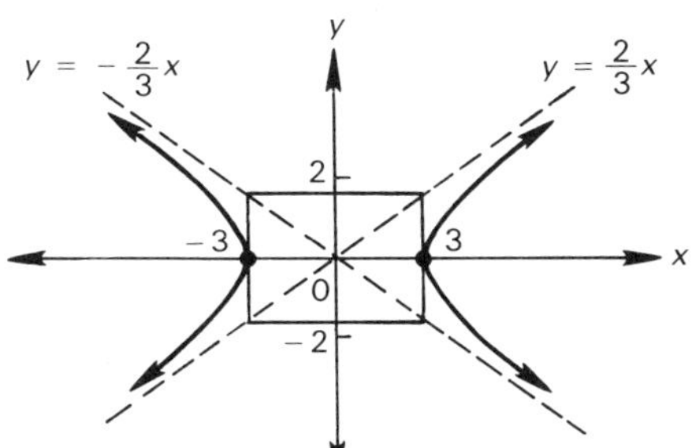

13) $\frac{(x - 2)^2}{9} - \frac{(y + 1)^2}{4} = 1$

$a^2 = 9$ $\quad$ $b^2 = 4$ $\quad$ $c^2 = a^2 + b^2$

$a = 3$ $\quad$ $b = 2$ $\quad$ $= 9 + 4 = 13$

$c = \sqrt{13}$

The minus sign precedes the fraction with the variable y, the transverse axis is horizontal.

Center: (2, -1)

Vertices: $(h + a, k) = (2 + 3, -1) = (5, -1)$

$(h - a, k) = (2 - 3, -1) = (-1, -1)$

Foci: $(h + c, k) = (2 + \sqrt{13}, -1)$ Asymptotes: $y + 1 = \pm \frac{2}{3}(x - 2)$

$(h - c, k) = (2 - \sqrt{13}, -1)$

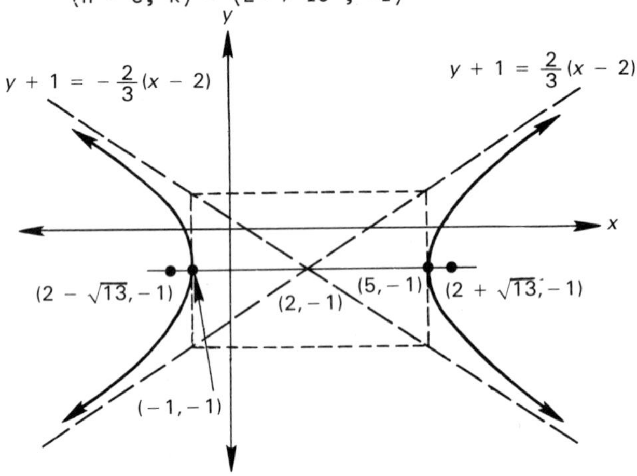

17) Center: (0, 0) Foci: (±6, 0) Vertices: (±4, 0)

The foci and vertices have the same y-coordinate, 0, so the major axis is the x-axis.

The vertices are at $(\pm a, 0) = (\pm 4, 0)$, so $a = 4$, $a^2 = 16$.

The foci are at $(\pm c, 0) = (\pm 6, 0)$, so $c = 6$, $c^2 = 36$.

$$c^2 = a^2 + b^2$$
$$36 = 16 + b^2$$
$$20 = b^2$$

$$\frac{x^2}{16} - \frac{y^2}{20} = 1$$

21) Center: (0, 0) Asymptotes: $y = \pm \frac{1}{2}x$ Vertices: (±4, 0)

Because the vertices are on the x-axis with the center, the transverse axis is horizontal.

The vertices are at (±a, 0) = (±4, 0), so a = 4, $a^2 = 16$.

The asymptotes have slope $\pm \frac{b}{a} = \pm \frac{1}{2}$

$$\frac{b}{a} = \frac{b}{4} = \frac{1}{2}$$

$$2b = 4$$

$$b = 2$$

$$b^2 = 4$$

$$\frac{x^2}{16} - \frac{y^2}{4} = 1$$

25) Both variables are squared, one with a positive coefficient and one with a negative coefficient, making this a hyperbola. The axis is horizontal since the minus sign precedes the y^2 term.

$a^2 = 36$ $\quad$ $b^2 = 25$ $\quad$ $c^2 = a^2 + b^2$

$a = 6$ $\quad$ $b = 5$ $\quad$ $= 36 + 25$

$= 61$

$c = \sqrt{61}$

Center: (0, 0) Vertices: (±6, 0) Foci: $(\pm\sqrt{61}, 0)$

Asymptotes: $y = \pm \frac{b}{a}x$, $y = \pm \frac{5}{6}x$

29) $25x^2 + 4y^2 = 100$

$$\frac{x^2}{4} + \frac{y^2}{25} = 1$$

Both variables are squared and both have positive coefficients, making this an <u>ellipse</u>. The larger denominator is with y^2, so the major axis is vertical.

$a^2 = 25$ $\quad$ $b^2 = 4$ $\quad$ $c^2 = a^2 - b^2$

$a = 5$ $\quad$ $b = 2$ $\quad$ $c^2 = 25 - 4 = 21$

$c = \sqrt{21}$

Center: (0, 0)

Vertices: (0, 5), (0, -5)

Foci: $(0, \sqrt{21})$, $(0, -\sqrt{21})$

33) Circle.

Center: (2, -1)

Radius: 2

37) $16y^2 - 9x^2 = 144$

$$\frac{y^2}{9} - \frac{x^2}{16} = 1$$

Both variables are squared, one with a positive coefficient and one with a negative coefficient, making this a hyperbola. The transverse axis is vertical since the minus sign precedes the x^2 term.

$a^2 = 9$ $\quad b^2 = 16$ $\quad$ Vertices: $(0, \pm 3)$

$a = 3$ $\quad b = 4$

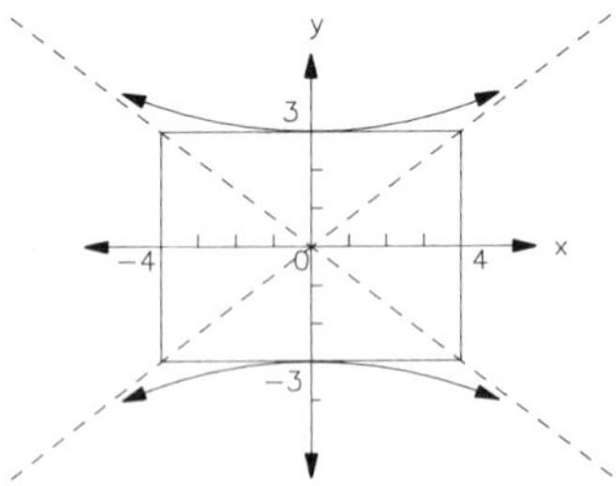

41) $\dfrac{(x - 1)^2}{64} + \dfrac{(y - 2)^2}{36} = 1$

Both variables are squared and both have positive coefficients, making this an <u>ellipse</u> with center at (1, 2). The larger denominator is with x^2, so the major axis is horizontal.

$a^2 = 64$ $\quad b^2 = 36$ $\quad$ Vertices: (9, 2), (-7, 2)

$a = 8$ $\quad b = 6$

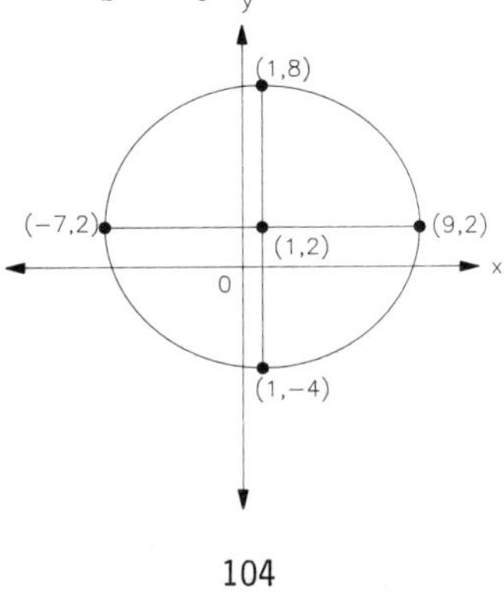

45) $(x + 3)^2 + (y - 2)^2 = 9$

Circle with center at (-3, 2) and radius 3.

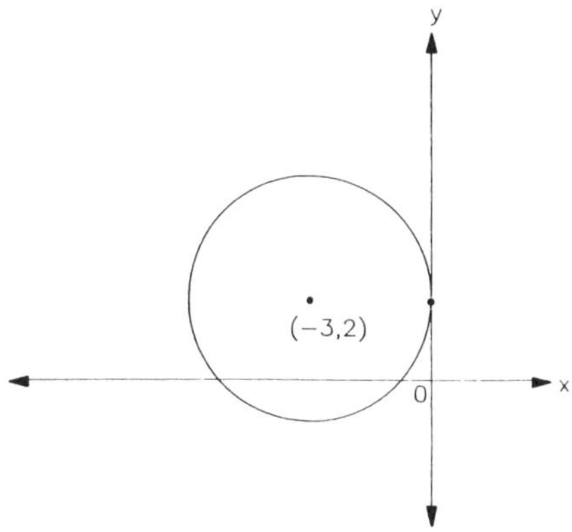

49) $x^2 + 4y^2 + 2x - 3 = 0$

$(x^2 + 2x + \quad) + 4y^2 = 3$

$(x^2 + 2x + 1) + 4y^2 = 4$

$(x + 1)^2 + 4y^2 = 4$

$$\frac{(x + 1)^2}{4} + \frac{y^2}{1} = 1$$

Ellipse; center (-1, 0); vertices (-3, 0) and (1, 0);

Foci $(-1 - \sqrt{3}, 0)$ and $(-1 + \sqrt{3}, 0)$

53) $x^2 + 6x + y^2 + 10y = 2$

$(x^2 + 6x + \quad) + (y^2 + 10y + \quad) = 2$

$(x^2 + 6x + 9) + (y^2 + 10y + 25) = 2 + 9 + 25$

$(x + 3)^2 + (y + 5)^2 = 36$

Circle; center (-3, -5); r = 6.

Section 4.8 Solving Nonlinear Systems

Note: One method of solving each exercise is shown here. Most of the problems can be solved by alternate methods. Try to find methods that are efficient.

1) $y = -x^2 - 4x + 1$

$y = 2x + 10$

Set the two values of y equal to each other.

$$2x + 10 = -x^2 - 4x + 1$$
$$x^2 + 6x + 9 = 0$$
$$(x + 3)^2 = 0$$
$$x = -3$$

Substitute this x-value into $y = 2x + 10$.

$$y = 2(-3) + 10 = -6 + 10 = 4$$

The only point common to both graphs is (-3, 4).

Graph $y = -x^2 - 4x + 1$
$$= -(x^2 + 4x + \quad) + 1$$
$$= -(x^2 + 4x + 4) + 1 + 4$$
$$= -(x + 2)^2 + 5$$

Graph $y = 2x + 10$

$m = 2 \qquad b = 10$

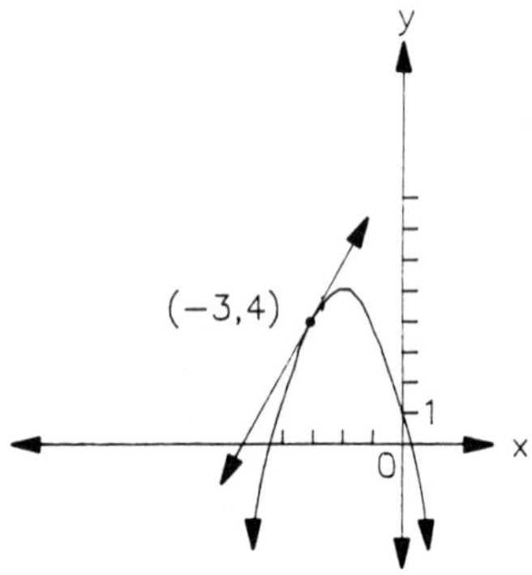

5) $y = x^2 + 6x + 6$

$y = -x^2 - 6x + 6$

The quickest method is to add the two equations.

$$\begin{array}{l} y = x^2 + 6x + 6 \\ \underline{y = -x^2 - 6x + 6} \\ 2y = \qquad\qquad 12 \\ \ \ y = 6 \end{array}$$

Substitute this y-value into the first equaiton.

$$(6) = x^2 + 6x + 6$$
$$0 = x^2 + 6x$$
$$0 = x(x + 6)$$
$$x = 0 \qquad \text{or} \qquad x + 6 = 0$$
$$x = -6$$

5) continued

The graphs intersect at two points: (0, 6) and (-6, 6).

Graph $y = x^2 + 6x + 6$

$= (x^2 + 6x + \quad) + 6$

$= (x^2 + 6x + 9) + 6 - 9$

$= (x + 3)^2 - 3$

Graph $y = -x^2 - 6x + 6$

$= -(x^2 + 6x + \quad) + 6$

$= -(x^2 + 6x + 9) + 6 + 9$

$= -(x + 3)^2 + 15$

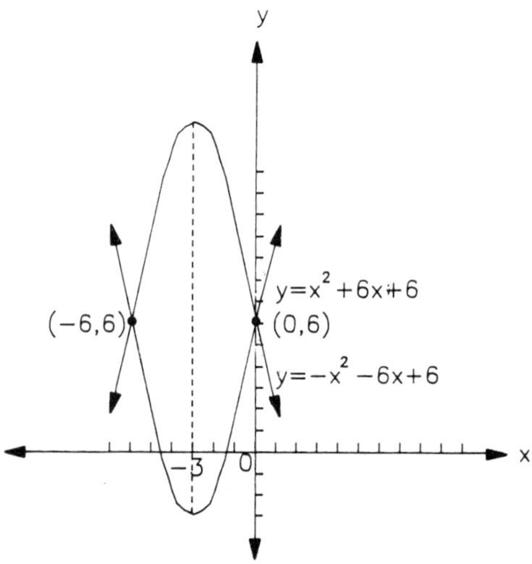

9) $y = x^2$

$y = -x^2 + 8x - 16$

Set the two values for y equal to each other.

$x^2 = -x^2 + 8x - 16$

$2x^2 - 8x + 16 = 0$

$x^2 - 4x + 8 = 0$

Use the discriminant to check for real solutions.

$b^2 - 4ac = 16 - 4 \cdot 1 \cdot 8 = -16$

no solutions

13) $y - x = 0$

$(x - 2)^2 + (y + 5)^2 = 25$

Solve the first equation for y and substitute into the second equation.

$y = x$

$(x - 2)^2 + [(x) + 5]^2 = 25$

$x^2 - 4x + 4 + x^2 + 10x + 25 = 25$

$2x^2 + 6x + 4 = 0$

$x^2 + 3x + 2 = 0$

$(x + 1)(x + 2) = 0$

$(x + 1) = 0 \quad \text{or} \quad (x + 2) = 0$

$x = -1 \qquad\qquad x = -2$

Substitute these values into $y = x$.

$y = -1 \qquad\qquad y = -2$

Solutions: $(-1, -1)$, $(-2, -2)$

17) $4x^2 + y^2 = 4$

$2x - y = 2$

Solve the second equation for y and substitute into the first equation.

$2x - y = 2$

$2x - 2 = y$

$4x^2 + (2x - 2)^2 = 4$

$4x^2 + 4x^2 - 8x + 4 = 4$

$8x^2 - 8x = 0$

$8x(x - 1) = 0$

$8x = 0 \quad \text{or} \quad x - 1 = 0$

$x = 0 \qquad\qquad x = 1$

Substitute these values into the second equation.

$2(0) - y = 2 \qquad\qquad 2(1) - y = 2$

$-2 = y \qquad\qquad 0 = y$

Solutions: $(0, -2)$, $(1, 0)$

21) $2x^2 + y^2 = 11$

$x^2 - 2y^2 = -2$

Multiply the second equation by (-2) and add the two equations.

$$\begin{array}{r} 2x^2 + y^2 = 11 \\ -2x^2 + 4y^2 = 4 \\ \hline 5y^2 = 15 \\ y^2 = 15 \\ y = \sqrt{3} \quad \text{or} \quad -\sqrt{3} \end{array}$$

Substitute these values into the second equation.

$$x^2 - 2(\sqrt{3})^2 = -2 \qquad x^2 - 2(-\sqrt{3})^2 = -2$$
$$x^2 - 2 \cdot 3 = -2 \qquad x^2 - 2 \cdot 3 = -2$$
$$x^2 - 6 = -2 \qquad x^2 - 6 = -2$$
$$x^2 = 4 \qquad x^2 = 4$$
$$x = \pm 2 \qquad x = \pm 2$$

Note that there are four solutions:

$(2, \sqrt{3})$, $(-2, \sqrt{3})$, $(2, -\sqrt{3})$, $(-2, -\sqrt{3})$

25) $(x - 1)^2 + y^2 = 1$

$x^2 + (y - 1)^2 = 1$

Since each left side is equal to 1, one way to tackle this is to set the two expressions equal to each other.

$$(x - 1)^2 + y^2 = x^2 + (y - 1)^2$$
$$x^2 - 2x + 1 + y^2 = x^2 + y^2 - 2y + 1$$
$$-2x = -2y$$
$$x = y$$

Now substitute x for y in the first equaiton.

$$(x - 1)^2 + x^2 = 1$$
$$x^2 - 2x + 1 + x^2 = 1$$
$$2x^2 - 2x = 0$$
$$2x(x - 1) = 0$$
$$2x = 0 \quad \text{or} \quad x - 1 = 0$$
$$x = 0 \qquad x = 1$$

Since $y = x$, then the solutions are (0, 0) and (1, 1).

29) $y \geq -x^2 - 4x + 1$
$y \leq 2x + 10$

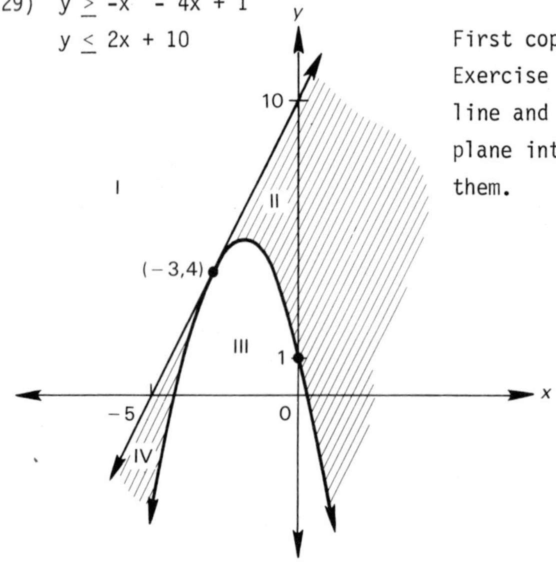

First copy the graph from Exercise 1 The line and the parabola divide the plane into four regions. Number them.

From each region choose one point (the easiest one you can think of!) and substitute it into each inequality to see whether true statements result.

I. (0, 12)

$12 \geq -(0)^2 - 4(0) + 1$ $\quad$ $12 \leq 2(0) + 10$

$12 \geq + 1$ $\quad$ $12 \leq 10$

TRUE $\quad$ FALSE

Region I is not part of the solution.

II. (0, 9)

$9 \geq -(0)^2 - 4(0) + 1$ $\quad$ $9 \leq 2(0) + 10$

$9 \geq 1$ $\quad$ $9 \leq 10$

TRUE $\quad$ TRUE

Region II is part of the solution.

III. (0, 0)

$0 \geq -(0)^2 - 4(0) + 1$ $\quad$ $0 \leq 2(0) + 10$

$0 \geq 1$ $\quad$ $0 \leq 10$

FALSE $\quad$ TRUE

Region III is not part of the solution.

29) continued

IV. (0, 4)

$4 \geq -(0)^2 - 4(0) + 1$ $\qquad$ $4 \leq 2(0) + 10$

$4 \geq 1$ $\qquad$ $4 \leq 10$

TRUE $\qquad$ TRUE

Region IV is part of the solution.

Because the symbols $\geq$ and $\leq$ are used, the boundaries are part of the solution along with regions II and IV. Shade the solution.

Note: Another way to solve this problem is to observe that the graph of $y \geq -x^2 - 4x + 1$ consists of all points on or above the parabola $y = -x^2 - 4x + 1$, and the graph of $y \leq 2x + 10$ consists of all points on or below the line $y = 2x + 10$. The set of points in the shaded regions and on the curves is the graph of the system since these points satisfy both inequalities.

33) $y \geq x^2 + 6x + 6$

$y \leq -x^2 - 6x + 6$

First copy the graph from Exercise 5.

The two parabolas divide the plane into four regions:

I. Above both parabolas

II. Below parabola 2 and above parabola 1

III. Above parabola 2 and below parabola 1

IV. Below both parabolas

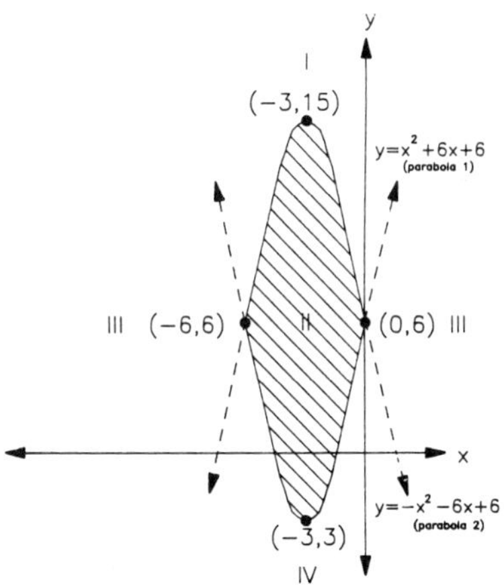

Since $y \geq x^2 + 6x + 6$ calls for all points on or above parabola 1, and $y \leq -x^2 - 6x + 6$ calls for all points on or below parabola 2, the graph of the system consists of region II and the points on both arcs of the parabolas between the points (-6, 6) and (0, 6).

Sample Test Questions for Chapter 4

Part I

1) $y = (x - 2)^2 + 3$
   Parabola
   Vertex (2, 3)
   a > 0, opens up

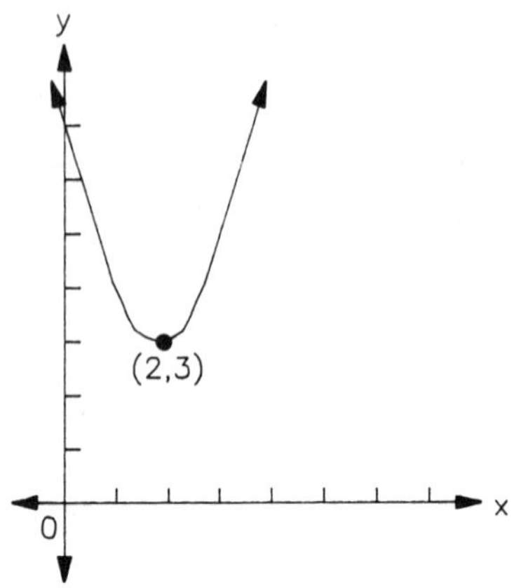

2) $y = f(x) = x^2 - 9$
   Parabola
   Vertex (0, -9)
   a > 0, opens up

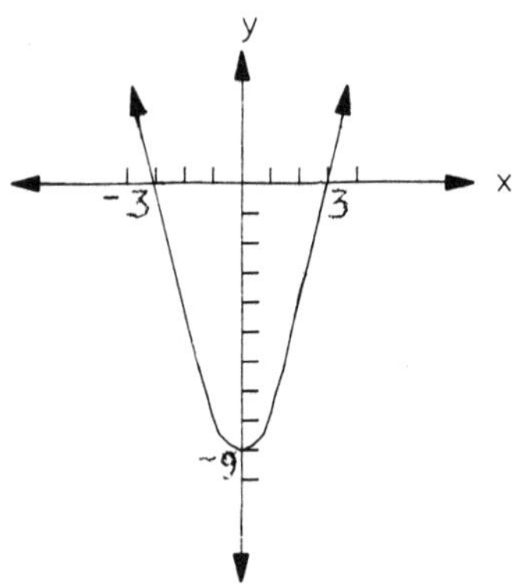

3) $y = -5x^2 + 20x - 1$

a) $y = -5(x^2 - 4x + \quad) - 1$

$= -5(x^2 - 4x + 4) - 1 + 20$

$= -5(x - 2)^2 + 19$

b) (2, 19)

c) $x = 2$

d) $a < 0$ The parabola opens down, $y = 19$ is the maximum.

Domain: all reals. Range: $y \leq 19$

4) $3x^2 - 8x - 3 = 0$

$$x = \frac{-(-8) \pm \sqrt{(-8)^2 - 4(3)(-3)}}{2(3)}$$

$$= \frac{8 \pm \sqrt{100}}{6}$$

$$= \frac{8 \pm 10}{6}$$

$x = \frac{18}{6} = 3$ or $x = \frac{-2}{6} = -\frac{1}{3}$

5) $y = -x^2 + 4x + 7$

x-intercepts: $0 = -x^2 + 4x + 7$

$$x = \frac{-(4) \pm \sqrt{(4)^2 - 4(-1)(7)}}{2(-1)}$$

$$= \frac{-4 \pm \sqrt{44}}{-2}$$

$$= \frac{-4 \pm 2\sqrt{11}}{-2}$$

$$= 2 \pm \sqrt{11}$$

6) $y = x^2 + 3x + 1$

$b^2 - 4ac = (3)^2 - 4(1)(1)$

$= 5$

two irrational numbers

7) $y = 6x^2 + 5x - 6$

$b^2 - 4ac = (5)^2 - 4(6)(-6)$

$= 169$

two rational numbers

8) $f(x) = -\frac{1}{2}x^2 - 6x + 2$

$= -\frac{1}{2}(x^2 - 12x + \quad) + 2$

$= -\frac{1}{2}(x^2 - 12x + 36) + 2 + 18$

$= -\frac{1}{2}(x - 6)^2 + 20$

Vertex: (-6, 20) $a < 0$, so 20 is a maximum for y and occurs at $x = -6$.

9) Since PQRC is a rectangle, $PQ = CR = x$.

Adding all the parts of the partial perimeter:

$$x + QR + x + 4 = 28$$
$$QR = 24 - 2x$$

The area of the rectangle:

$$A(x) = x(24 - 2x)$$
$$= 24x - 2x^2$$

Since $a < 0$, the vertex of this parabola will be the desired maximum.

$$y = -2x^2 + 24x$$
$$= -2(x^2 - 12x + \quad)$$
$$= -2(x^2 - 12x + 36) + 72$$
$$= -2(x - 6)^2 + 72$$

Vertex: (6, 72)

The area will be at a maximum when $x = 6$ feet.

10) $h = 64t - 16t^2$

$a < 0$, so the vertex will be the desired maximum.

$$h = -16(t^2 - 4t + \quad)$$
$$= -16(t^2 - 4t + 4) + 64$$
$$= -16(t - 2)^2 + 64$$

Vertex: (2, 64)

The maximum height reached is 64 feet.

11) a) $(x - 3)^2 + (y + 4)^2 = 25$

$C(3, -4) \qquad r = 5$

b) The slope of the radius to (6, 0):

$$\frac{0 + 4}{6 - 3} = \frac{4}{3}$$

Slope of tangent line: $-\frac{3}{4}$

$$y - 0 = -\frac{3}{4}(x - 6)$$
$$y = -\frac{3}{4}x + \frac{9}{2}$$

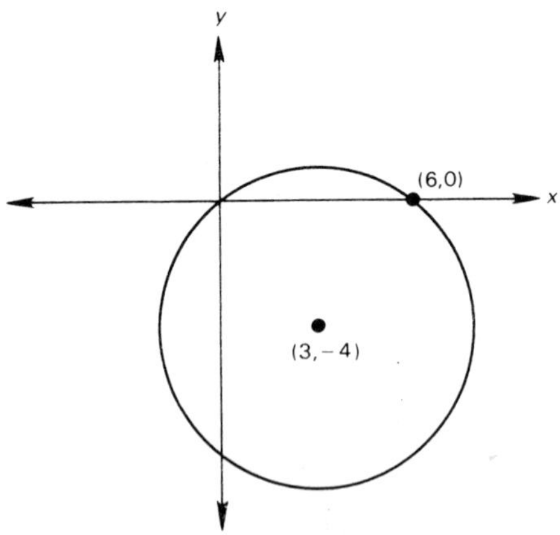

12) $4x^2 + 4x + 4y^2 - 56y = -97$

$4(x^2 + x + \quad) + 4(y^2 - 14y + \quad) = -97$

$4\left(x^2 + x + \frac{1}{4}\right) + 4(y^2 - 14y + 49) = -97 + 1 + 196$

$4\left(x + \frac{1}{2}\right)^2 + 4(y - 7)^2 = 100$

$\left(x + \frac{1}{2}\right)^2 + (y - 7)^2 = 25$

$C\left(-\frac{1}{2}, 7\right), \quad r = 5$

13) A(-2, 5), B(3, -4)

$d = \sqrt{(3 + 2)^2 + (-4 - 5)^2}$

$= \sqrt{25 + 81}$

$= \sqrt{106}$

14) P(1, -5), Q(-3, 3)

The center of the circle will be the midpoint of the line segment.

$= \left(\frac{1 - 3}{2}, \frac{-5 + 3}{2}\right)$

$= \left(\frac{-2}{2}, \frac{-2}{2}\right)$

$= (-1, -1)$

The radius is equal to the distance from the center to either point P or Q. Using P:

$d = \sqrt{(-1 - 1)^2 + (-1 + 5)^2}$

$= \sqrt{4 + 16}$

$= \sqrt{20}$

The equation uses C(-1, -1) and $r = \sqrt{20}$:

$$(x + 1)^2 + (y + 1)^2 = 20$$

15) $4x^2 + 16x - y^2 + 6y = -3$

$4(x^2 + 4x + \quad) - (y^2 - 6y + \quad) = -3$

$4(x^2 + 4x + 4) - (y^2 - 6y + 9) = -3 + 16 - 9$

$4(x + 2)^2 - (y - 3)^2 = 4$

$\frac{(x + 2)^2}{1} - \frac{(y - 3)^2}{4} = 1$

Hyperbola

Center: (-2, 3)

15) continued

Since the minus sign in the standard form precedes the fraction with the variable y, the transverse axis is horizontal.

Vertices: $(-2 + 1, 3) = (-1, 3)$ (note that $a = 1$)

$(-2 - 1, 3) = (-3, 3)$

asymptotes: $y - 3 = \pm 2(x + 2)$ (note that $\frac{b}{a} = \frac{2}{1} = 2$)

Foci: $(-2 + \sqrt{5}, 3)$ (Note that $c = \sqrt{1 + 4} = \sqrt{5}$)

$(-2 - \sqrt{5}, 3)$

16) $9x^2 - 18x + 4y^2 + 16y = 11$

$9(x^2 - 2x + \quad) + 4(y^2 + 4y + \quad) = 11$

$9(x^2 - 2x + 1) + 4(y^2 + 4y + 4) = 11 + 9 + 16$

$9(x - 1)^2 + 4(y + 2)^2 = 36$

$$\frac{(x - 1)^2}{4} + \frac{(y + 2)^2}{9} = 1$$

Ellipse

Center: $(1, -2)$

The large denominator is with y^2 so the major axis is vertical.

$a = 3$ $b = 2$, $c = \sqrt{9 - 4} = \sqrt{5}$

Vertices: $(1, -2 + 3) = (1, 1)$ Foci: $(1, -2 - \sqrt{5})$

$(1, -2 - 3) = (1, -5)$ $(1, -2 + \sqrt{5})$

17) $C(0, 0)$, major axis $= 2a = 8$, minor axis $= 2b = 6$

$a = 4$ $b = 3$

$a^2 = 16$ $b^2 = 9$

Because the major axis is horizontal, a goes with x^2.

$$\frac{x^2}{16} + \frac{y^2}{9} = 1$$

18) $C(0, 0)$ Foci: $(\pm 8, 0)$ Vertices $(\pm 6, 0)$

The foci and vertices have the same y-coordinate, so the transverse axis is horizontal and x^2 will be positive.

Foci are at $(\pm c, 0) = (\pm 8, 0)$, so $c = 8$, $c^2 = 64$

Vertices are at $(\pm a, 0) = (\pm 6, 0)$, so $a = 6$, $a^2 = 36$.

$b^2 = c^2 - a^2$

$b^2 = 64 - 36$

$b^2 = 28$

$$\frac{x^2}{36} - \frac{y^2}{28} = 1$$

19) Since the parabola $y = -\frac{1}{8}x^2$ can be written as $x^2 = -8y$, which is of the form $x^2 = 4py$, the focus is at (0, p) and the directrix is $y = -p$. In this case $4p = -8$ so that $p = -2$. Thus (0, -2) is the focus and the directrix has equation $y = 2$.

20) directrix $x = \frac{2}{3}$, vertex at (0, 0)

The directrix is a vertical line, so y will be the squared variable. The distance from the vertex to the directrix is $-p = \frac{2}{3}$, so $p = -\frac{2}{3}$.

$$y^2 = 4\left(-\frac{2}{3}\right)x$$

$$y^2 = -\frac{8}{3}x$$

Focus at (p, 0) = $\left(-\frac{2}{3}, 0\right)$

21) $y = (x - 2)^2 - 2$ Parabola: vertex (2, -2), opens up

$(x - 2)^2 + (y - 2)^2 = 4$ Circle: C(2, 2), r = 2

Because $(x - 2)^2$ appears in both equations, solve the first for $(x - 2)^2$.

$$y = (x - 2)^2 - 2$$

$$y + 2 = (x - 2)^2$$

Substitute into the second equation

$$(y + 2) + (y - 2)^2 = 4$$

$$y + 2 + y^2 - 4y + 4 = 4$$

$$y^2 - 3y + 2 = 0$$

$$(y - 1)(y - 2) = 0$$

$y - 1 = 0$ or $y - 2 = 0$

$y = 1$ or $y = 2$

Substitute these values into the first equation.

$1 = (x - 2)^2 - 2$	$2 = (x - 2)^2 - 2$
$3 = (x - 2)^2$	$4 = (x - 2)^2$
$\pm 3\sqrt{3} = x - 2$	$\pm 2 = x - 2$
$\sqrt{3} = x - 2$ or $-\sqrt{3} = x - 2$	$2 = x - 2$ or $-2 = x - 2$
$2 + \sqrt{3} = x$ $2 - \sqrt{3} = x$	$4 = x$ or $0 = x$

21) continued

Four solutions: $(2 + \sqrt{3}, 1)$, $(2 - \sqrt{3}, 1)$, $(4, 2)$, $(0, 2)$

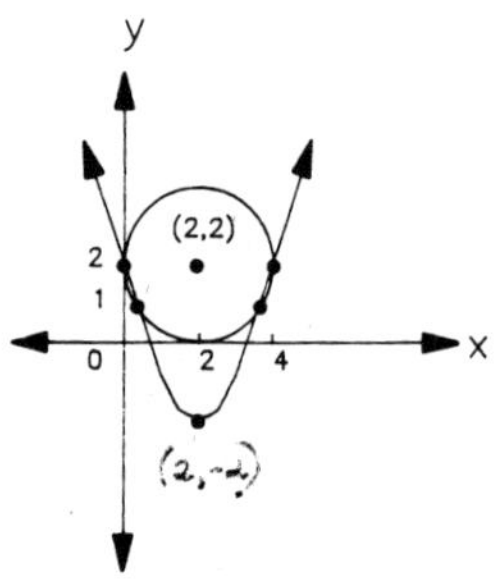

22) $2x^2 - 3y^2 = 15$

$3x^2 + 2y^2 = 29$

Multiply the first equation by 3 and the second by -2 and add.

$$\begin{aligned} 6x^2 - 9y^2 &= 45 \\ -6x^2 - 4y^2 &= -58 \\ \hline -13y^2 &= -13 \\ y^2 &= 1 \\ y &= \pm 1 \end{aligned}$$

Substitute these values into the second equation

$$\begin{aligned} 3x^2 + 2(1)^2 &= 29 \\ 3x^2 + 2 &= 29 \\ 3x^2 &= 27 \\ x^2 &= 9 \\ x &= \pm 3 \end{aligned} \qquad \begin{aligned} 3x^2 + 2(-1)^2 &= 29 \\ 3x^2 + 2 &= 29 \\ 3x^2 &= 27 \\ x^2 &= 9 \\ x &= \pm 3 \end{aligned}$$

Four solutions: $(3, 1)$, $(3, -1)$, $(-3, 1)$, $(-3, -1)$.

23) $y^2 - 2x^2 = 16$

$y - x = 2$

Solve the second equation for y.

$y = x + 2$

Substitute into the first equation.

$(x + 2)^2 - 2x^2 = 16$

$x^2 + 4x + 4 - 2x^2 = 16$

$-x^2 + 4x + 4 = 16$

$x^2 - 4x + 12 = 0$

Use the discriminant to test for real roots:

$b^2 - 4ac = (-4)^2 - 4(1)(12)$

$= -32$

No solutions.

Part II

1) (c)

2) (e)

The vertex is (-1, -3), and $a = -2 < 0$, so the parabola opens down and $y = -3$ is a maximum.

Range: $y \leq -3$

3) (a)

$$\begin{aligned} y &= x^2 - 4x + 1 \\ &= (x^2 - 4x + \quad) + 1 \\ &= (x^2 - 4x + 4) + 1 - 4 \\ &= (x - 2)^2 - 3 \end{aligned}$$

4) (b)

5) (a)

$$\begin{aligned} f(x) &= x^2 - 8x + 10 \\ &= (x^2 - 8x + \quad) + 10 \\ &= (x^2 - 8x + 16) + 10 - 16 \\ &= (x - 4)^2 - 6 \end{aligned}$$

Vertex: (4, -6)

6) (d)

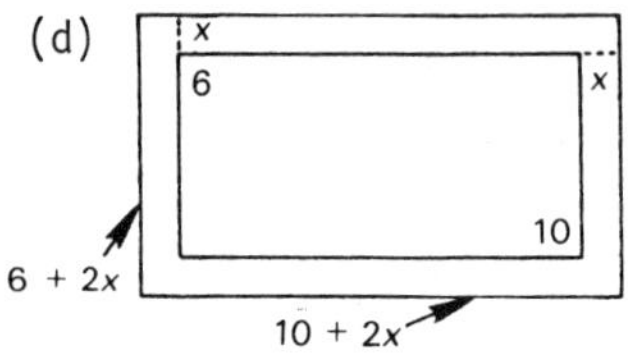

Original dimensions: 6 by 10

Original area: 6 x 10 = 60

New dimensions: $(6 + 2x)$ by $(10 + 2x)$

New area: $(6 + 2x)(10 + 2x)$

The original area plus 80 equal the new area.

$$\begin{aligned} 60 + 80 &= 4x^2 + 32x + 60 \\ 0 &= 4x^2 + 32x - 80 \\ 0 &= x^2 + 8x - 20 \end{aligned}$$

7) (c)

$$y = -4x^2 + 20x - 25$$

$$\begin{aligned} b^2 - 4ac &= (20)^2 - 4(-4)(-25) \\ &= 0 \end{aligned}$$

8) (a)

$3x^2 + 6x + 2 = 0$

$$x = \frac{-(-6) \pm \sqrt{(6)^2 - 4(3)(2)}}{2(3)}$$

$$= \frac{-6 \pm \sqrt{12}}{6}$$

$$= \frac{-6 \pm 2\sqrt{3}}{6}$$

$$= \frac{-3 \pm \sqrt{3}}{3}$$

9) (d)

$x^2 + y^2 + 3y = -\frac{1}{4}$

$x^2 + (y^2 + 3y + \quad) = -\frac{1}{4}$

$x^2 + \ y^2 + 3y + \frac{9}{4} \ = -\frac{1}{4} + \frac{9}{4}$

$x^2 + (y + \frac{3}{2})^2 = \frac{8}{4} = 2$

C: $(0, -\frac{3}{2})$ $\qquad r = \sqrt{2}$

10) (c)

foci at $(0, \pm 3)$ major axis = $2a = 10$, $a = 5$

The foci are on the y axis, so the major axis is vertical and a^2 goes with y^2. The foci aré 3 units up and 3 units down from $y = 0$, so the center is $(0, 0)$, and $c = 3$.

$$a^2 - c^2 = b^2$$
$$25 - 9 = b^2$$
$$16 = b^2$$

$$\frac{x^2}{16} + \frac{y^2}{25} = 1$$

11) (b)

$\frac{x^2}{10} - \frac{y^2}{15} = 1$ hyperbola

The minus sign precedes the y^2 term, so the transverse axis is horizontal.

foci at $(\pm c, 0)$

$$c^2 = a^2 + b^2$$
$$c^2 = 10 + 15$$
$$c^2 = 25$$
$$c = 5$$

12) (a)

$-2y^2 = x$

$y^2 = -\frac{1}{2}x$

y is the squared variable, so the axis is horizontal.

$4p = -\frac{1}{2}$

$p = -\frac{1}{8}$

Focus at $(p, 0) = \left(-\frac{1}{8}, 0\right)$

Directrix $x = -p$, $x = \frac{1}{8}$

13) (c)

Center: (-3, 2), Foci: (-4, 2), (-2, 2)

minor axis = 2b = 4, b = 2, $b^2 = 4$

The center and the foci have the same y coordinate, making the major axis horizontal, so a^2 goes with x^2.

To find c, note that the center is at

(-3, 2) = (h, k), so h = -3 and k = 2.

The foci are at (h + c, k) = (-2, 2) and (h - c, k) = (-4, 2).

$h + c = -2$

$-3 + c = -2$

$c = 1$

$c^2 = 1$

To get a:

$a^2 = b^2 + c^2$

$a^2 = 2^2 + 1^2 = 5$

$$\frac{(x + 3)^2}{5} + \frac{(y - 2)^2}{4} = 1$$

14) (d)

Center: (-2, 1) Vertices: (1, 1), (-5, 1)

Foci: $\sqrt{13}$ units from center, $c = \sqrt{13}$, $c^2 = 13$

The center and vertices have the same y-coordinates so the transverse axis is horizontal.

The center is at (-2, 1) = (h, k), so h = -2, k = 1.

14) continued

The vertices are at $(h + a, k) = (1, 1)$ and $(h - a, k) = (-5, 1)$.

$$h + a = 1$$
$$-2 + a = 1$$
$$a = 3$$
$$a^2 = 9$$

To find b:

$$b^2 = c^2 - a^2$$
$$b^2 = 13 - 9$$
$$b^2 = 4$$

$$\frac{(x + 2)^2}{9} - \frac{(y - 1)^2}{4} = 1$$

15) (d)

$x^2 + 8y = 80$

$x^2 + (y - 5)^2 = 25$

Solve the first equation for x^2.

$$x^2 = 80 - 8y$$

Substitute into the second equation.

$$(80 - 8y) + (y - 5)^2 = 25$$
$$80 - 8y + y^2 - 10y + 25 = 25$$
$$y^2 - 18y + 80 = 0$$
$$(y - 8)(y - 10) = 0$$

$y - 8 = 0$ or $y - 10 = 0$

$y = 8$ $\qquad$ $y = 10$

Substitute these values into the first equation.

$x^2 + 8(8) = 80$ $\qquad$ $x^2 + 8(10) = 80$

$x^2 + 64 = 80$ $\qquad$ $x^2 + 80 = 80$

$x^2 = 16$ $\qquad$ $x^2 = 0$

$x = \pm 4$ $\qquad$ $x = 0$

Three solutions: $(4, 8)$, $(-4, 8)$, $(0, 10)$.

CHAPTER 5
POLYNOMIAL AND RATIONAL FUNCTIONS

Section 5.1 Hints for Graphing

1)

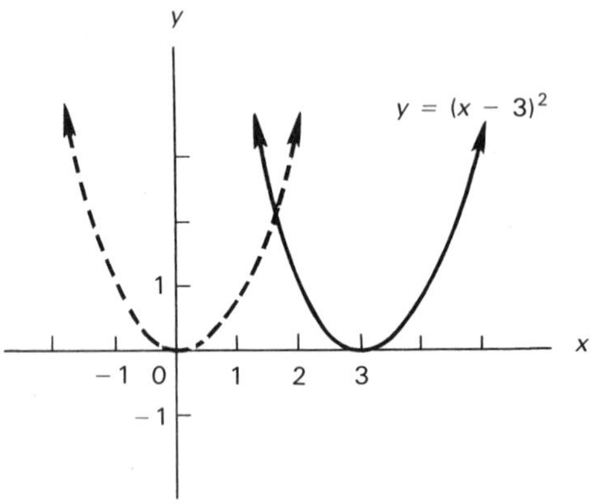

5) $f(x) = x^3$

$g(x) = \frac{1}{2}x^3$

$h(x) = \frac{1}{4}x^3$

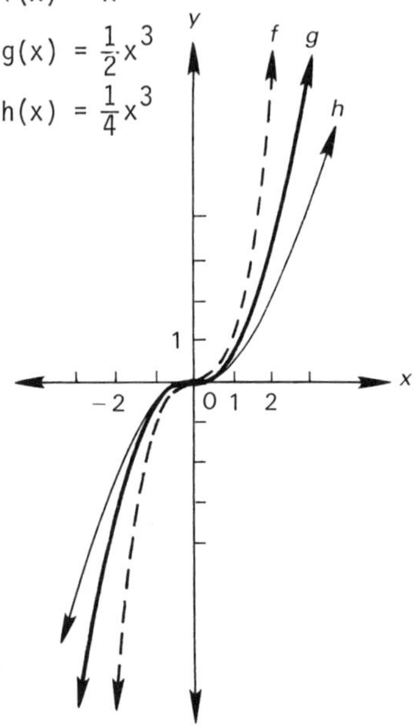

9) $y = |(x + 1)^3|$

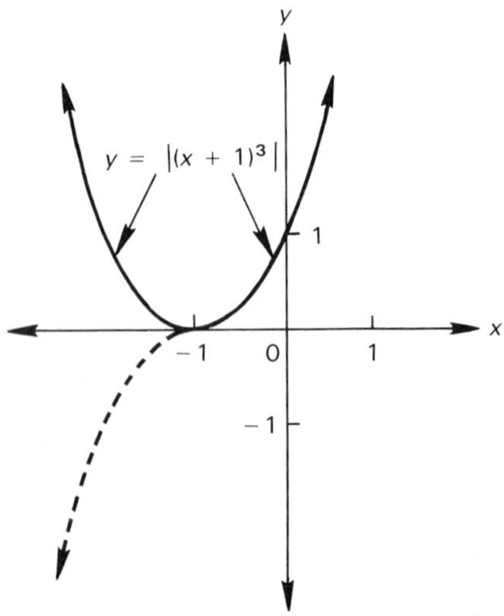

13) $y = f(x) = (x - 1)^3 + 2$

$y = x^3$ is shifted right one unit, up two units.

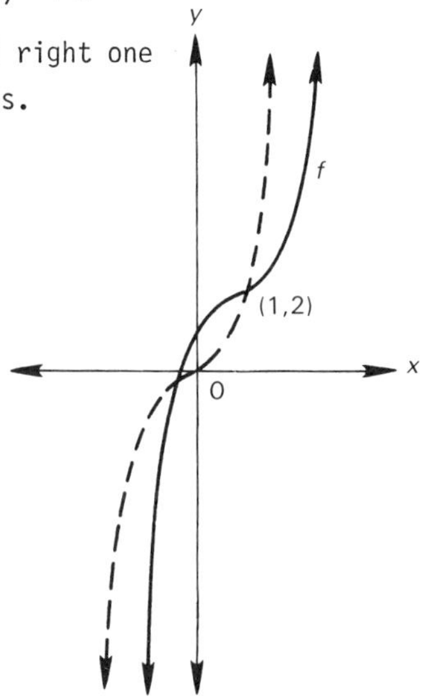

17) $y = f(x) = -(x + 1)^3 - 1$

$y = x^3$ is reflected through the x-axis then shifted one unit left and one unit down.

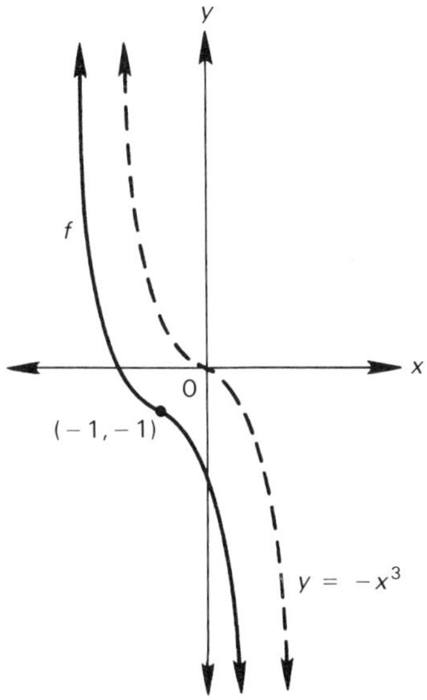

21) $y = |x|$ has been shifted $\frac{3}{4}$ unit left.

$y = \left|x + \frac{3}{4}\right|$

25) $y = x^3 + 3x^2 + 3x + 1$

$y = (x + 1)^3$

Shift $y = x^3$ one unit left.

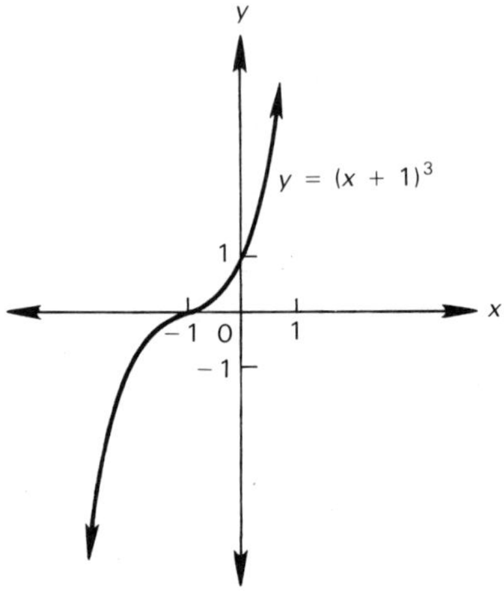

29) $f(x) = x^3; \quad f(3) = 3^3 = 27$

$$\frac{f(x) - f(3)}{x - 3} = \frac{x^3 - 27}{x - 3} = \frac{\cancel{(x - 3)}(x^2 + 3x + 9)}{\cancel{x - 3}} = x^2 + 3x + 9$$

Section 5.2 Graphing Some Special Rational Functions

1) $y = \frac{1}{x}, \quad y = \frac{2}{x}$

Another way to write $y = \frac{2}{x}$ is $y = 2\left(\frac{1}{x}\right)$. This form shows that the y-values will be twice the y-values for $y = \frac{1}{x}$.

x	$\frac{1}{x}$	$2\left(\frac{1}{x}\right)$
-2	$-\frac{1}{2}$	-1
-1	-1	-2
$-\frac{1}{2}$	-2	-4
$\frac{1}{2}$	2	4
1	1	2
2	$\frac{1}{2}$	1

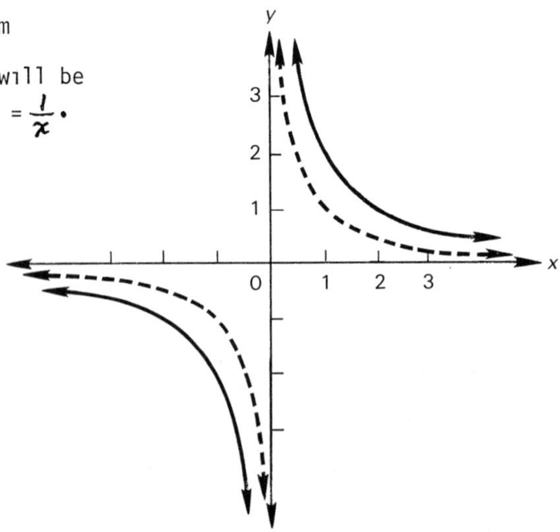

5) $y = \frac{1}{x}$, $y = \frac{1}{2x}$

Rewrite $y = \frac{1}{2x}$ as $y = \frac{1}{2}\left(\frac{1}{x}\right)$.
These y-values will be 1/2 the y-values for $y = \frac{1}{x}$.

x	$\frac{1}{x}$	$\frac{1}{2}\left(\frac{1}{x}\right)$
-2	$-\frac{1}{2}$	$-\frac{1}{4}$
-1	-1	$-\frac{1}{2}$
$-\frac{1}{2}$	-2	-1
$\frac{1}{2}$	2	1
1	1	$\frac{1}{2}$
2	$\frac{1}{2}$	$\frac{1}{4}$

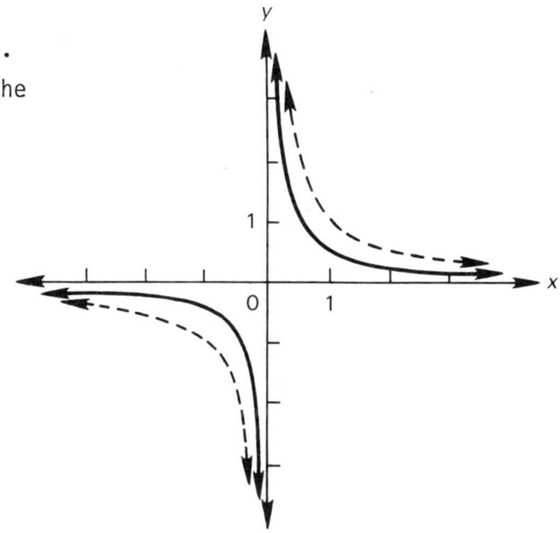

9) $y = -\frac{1}{x} + 2$

Reflect $\frac{1}{x}$ through the x-axis and shift it up 2 units.
Asymptotes: $x = 0$, $y = 2$

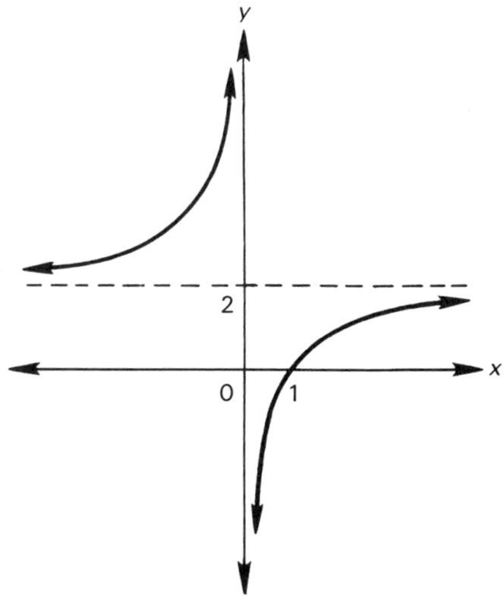

13) $y = -\frac{1}{x - 2} + 1$

Reflect $\frac{1}{x}$ through the x-axis then shift it right 2 units and up 1 unit.

Asymptotes: $x = 2, \quad y = 1$

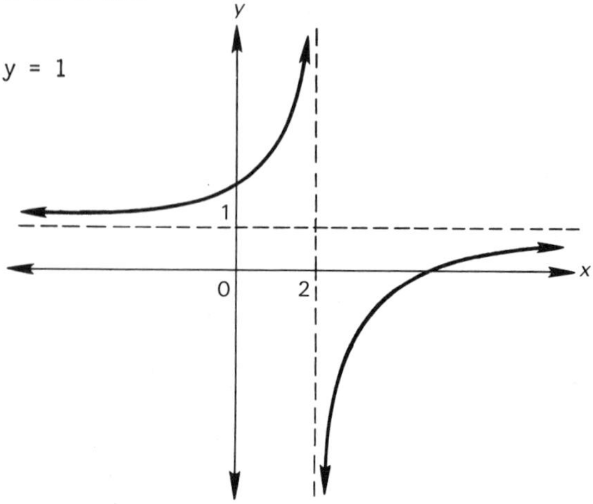

17) $y = \frac{1}{|x - 2|}$

Shift $\frac{1}{x}$ 2 units to the right.

Take the part that is below the x-axis and reflect it through the axis.

Asymptotes: $x = 2, \quad y = 0$

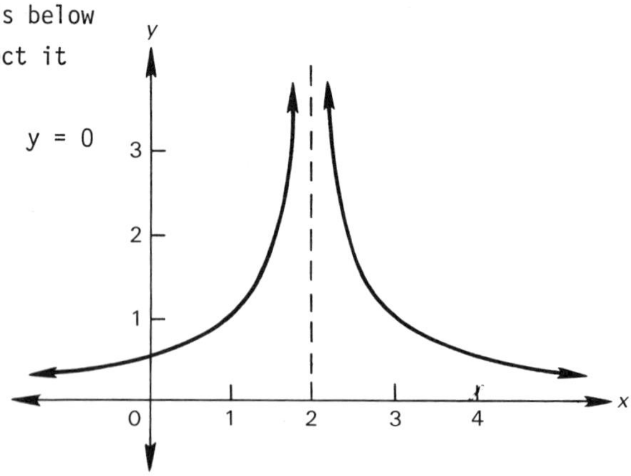

21) $xy - y = 1$

$y(x - 1) = 1$

$$y = \frac{1}{x - 1}$$

Shift $y = \frac{1}{x}$ right 1 unit.

Asymptotes: $x = 1$, $y = 0$

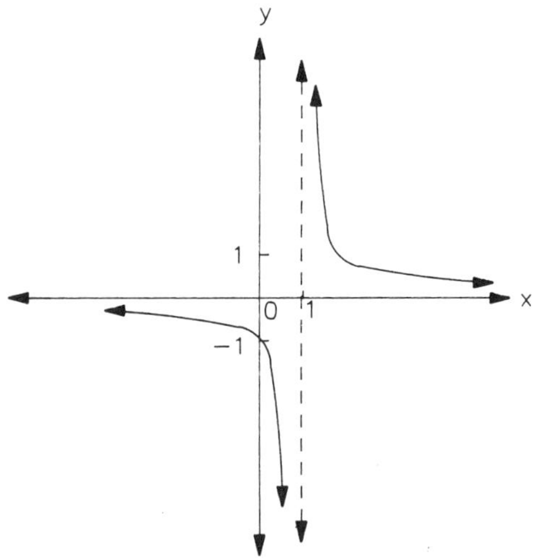

25) $y = \dfrac{x^2 - x - 6}{x - 3}$

$y = \dfrac{(x - 3)(x + 2)}{x - 3}$

$y = x + 2$, $x \neq 3$

Hole at $x = 3$, $y = 3 + 2 = 5$

No asymptotes.

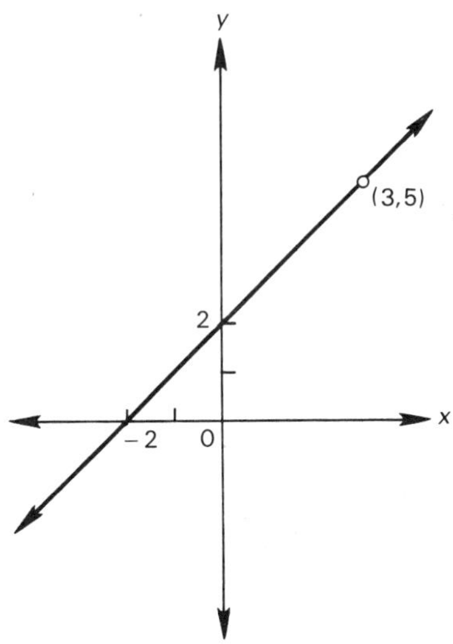

29) $y = \dfrac{x^3 - 8}{x - 2}$

$y = \dfrac{\cancel{(x - 2)}(x^2 + 2x + 4)}{\cancel{x - 2}}$

$y = x^2 + 2x + 4, \; x \neq 2$

$y = (x + 1)^2 + 3$

Hole at x = 2,

$y = (2 + 1)^2 + 3 = 12$

No asymptotes.

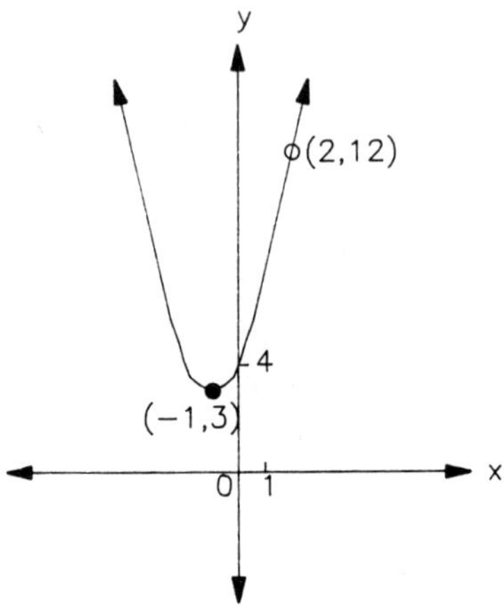

33) $f(x) = \dfrac{1}{x}$

$f(3) = \dfrac{1}{3}$

$$\frac{f(x) - f(3)}{x - 3} = \frac{\frac{1}{x} - \frac{1}{3}}{x - 3}$$

$$= \frac{\frac{3 - x}{3x}}{x - 3}$$

$$= \frac{-\cancel{(x - 3)}}{3x} \cdot \frac{1}{\cancel{x - 3}}$$

$$= -\frac{1}{3x}$$

Section 5.3 Polynomial and Rational Functions

1) $f(x) = (x - 1)(x - 2)(x - 3)$

$x - 1 = 0$ $\quad x - 2 = 0$ $\quad x - 3 = 0$

$x = 1$ $\quad x = 2$ $\quad x = 3$

Interval	$(-\infty, 1)$	$(1, 2)$	$(2, 3)$	$(3, \infty)$
Sign of $(x - 1)$	-	+	+	+
Sign of $(x - 2)$	-	-	+	+
Sign of $(x - 3)$	-	-	-	+
Sign of $f(x)$	-	+	-	+

$f(x) < 0$ on $(-\infty, 1)$ and on $(2, 3)$

$f(x) > 0$ on $(1, 2)$ and on $(3, \infty)$

5) $f(x) = (x^2 + 2)(x - 4)(x + 1) > 0$

$x^2 + 2 = 0$ $\quad x - 4 = 0$ $\quad x + 1 = 0$

never $\quad x = 4$ $\quad x = -1$

$x^2 + 2 > 0$ always

Interval	$(-\infty, -1)$	$(-1, 4)$	$(4, \infty)$
Sign of $(x^2 + 2)$	+	+	+
Sign of $(x - 4)$	-	-	+
Sign of $(x + 1)$	-	+	+
Sign of $f(x)$	+	-	+

$f(x) > 0$ on $(-\infty, -1)$ and on $(4, \infty)$

9) $f(x) = (x + 3)(x + 1)(x - 2)$

x-intercepts: $x + 3 = 0$ $\quad x + 1 = 0$ $\quad x - 2 = 0$

$x = -3$ $\quad x = -1$ $\quad x = 2$

9) continued

Interval	$(-\infty, -3)$	$(-3, -1)$	$(-1, 2)$	$(2, \infty)$
Sign of $(x + 3)$	-	+	+	+
Sign of $(x + 1)$	-	-	+	+
Sign of $(x - 2)$	-	-	-	+
Sign of $f(x)$	-	+	-	+

x	y
-4	-18
-2	4
0	-6
3	24

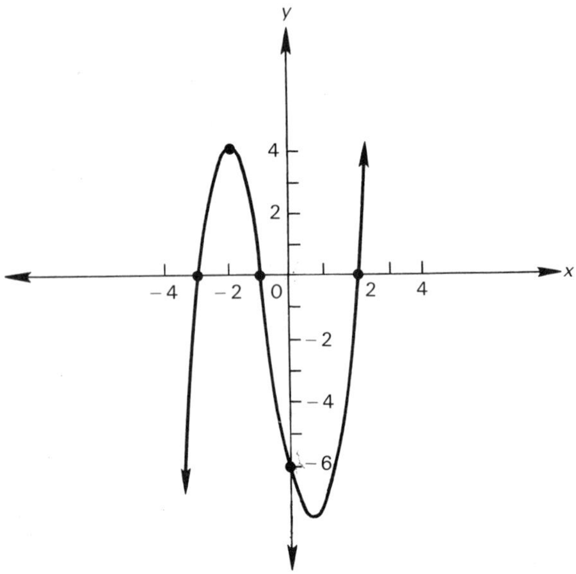

13) $f(x) = x^3 - 4x = x(x^2 - 4) = x(x + 2)(x - 2)$

x-intercepts: $x = 0$ $\quad x + 2 = 0$ $\quad x - 2 = 0$

$x = -2$ $\quad x = 2$

Interval	$(-\infty, -2)$	$(-2, 0)$	$(0, 2)$	$(2, \infty)$
Sign of $x + 2$	-	+	+	+
Sign of x	-	-	+	+
Sign of $(x - 2)$	-	-	-	+
Sign of $f(x)$	-	+	-	+

13) continued

Note $f(-x) = (-x)^3 - 4(-x) = -x^3 + 4x = -f(x)$.

Therefore, f(x) symmetric through (0, 0).

x	y
-3	-15
-1	3
1	-3
3	15

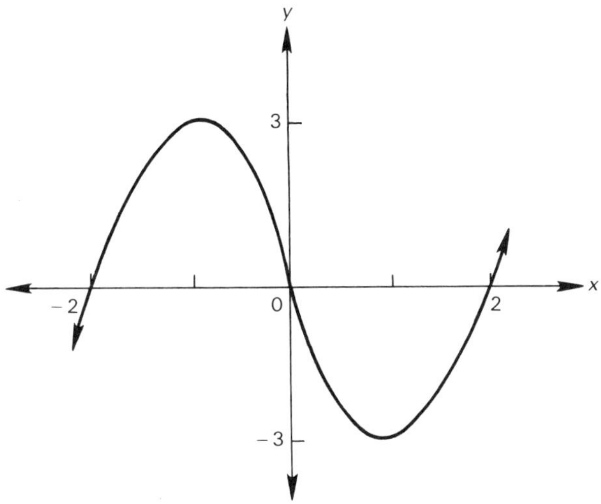

17) $f(x) = -x^3 - x^2 + 6x = -x(x^2 + x - 6) = -x(x + 3)(x - 2)$

x-intercepts $-x = 0$ $x + 3 = 0$ $x - 2 = 0$

$x = 0$ $x = -3$ $x = 2$

Interval	$(-\infty, -3)$	$(-3, 0)$	$(0, 2)$	$(2, \infty)$
Sign of $-x$	+	+	-	-
Sign of $(x + 3)$	-	+	+	+
Sign of $(x - 2)$	-	-	-	+
Sign of $f(x)$	+	-	+	-

17) continued

x	y
-4	24
-3	0
-1	-6
1	4
3	-18

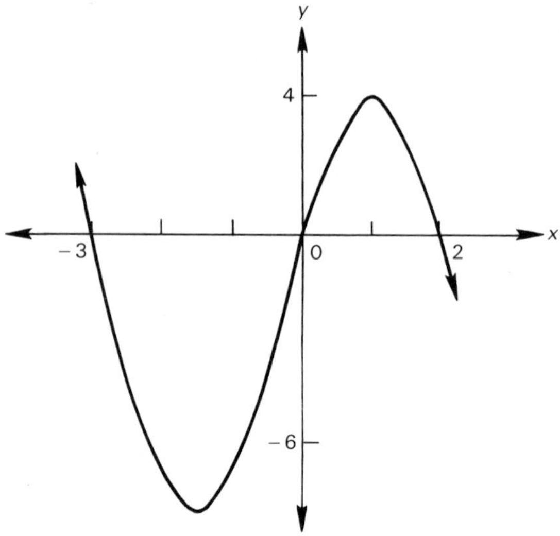

21) $f(x) = x^4 - 6x^3 + 8x^2 = x^2(x^2 - 6x + 8) = x^2(x - 4)(x - 2)$

x-intercepts $x^2 = 0$ $\quad x - 4 = 0$ $\quad x - 2 = 0$

$x = 0$ $\quad x = 4$ $\quad x = 2$

Interval	$(-\infty, 0)$	$(0, 2)$	$(2, 4)$	$(4, \infty)$
Sign of x^2	+	+	+	+
Sign of $(x - 4)$	-	-	-	+
Sign of $(x - 2)$	-	-	+	+
Sign of $f(x)$	+	+	-	+

Because there is no change of sign at $x = 0$, the graph does not cross the x-axis at this point but is tangent to it.

x	y
-1	15
1	3
3	-9
5	75

21) continued

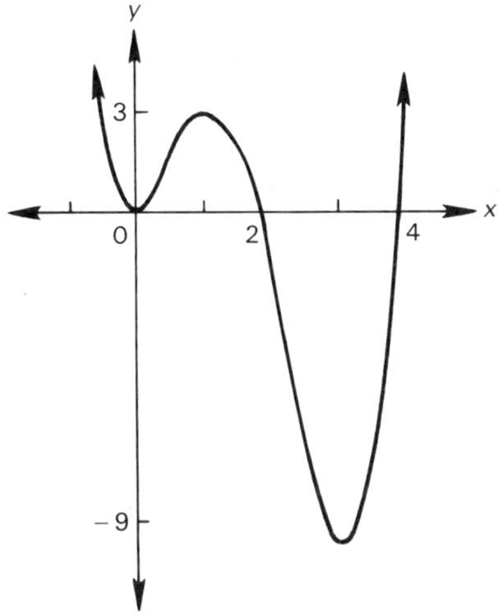

25) $f(x) = \frac{1}{4 - x^2} = \frac{1}{(2 - x)(2 + x)}$

Vertical asymptotes: $(2 - x)(2 + x) = 0$

$2 - x = 0 \qquad 2 + x = 0$

$2 = x \qquad -2 = x$

Horizontal asymptote: numerator degree 0 < denominator degree 2

so y = 0 is an asymptote.

Interval	$(-\infty, -2)$	$(-2, 2)$	$(2, \infty)$
Sign of (2 - x)	+	+	-
Sign of (2 + x)	-	+	+
Sign of f(x)	-	+	-

Note that $f(-x) = \frac{1}{4 - (-x)^2} = \frac{1}{4 - x^2} = f(x)$

so the graph is symmetric around the y-axis.

x	y
-3	$-\frac{1}{5}$
-2.1	-2.44
-1.9	2.56
-1	$\frac{1}{3}$
0	$\frac{1}{4}$
1	$\frac{1}{3}$

25) continued

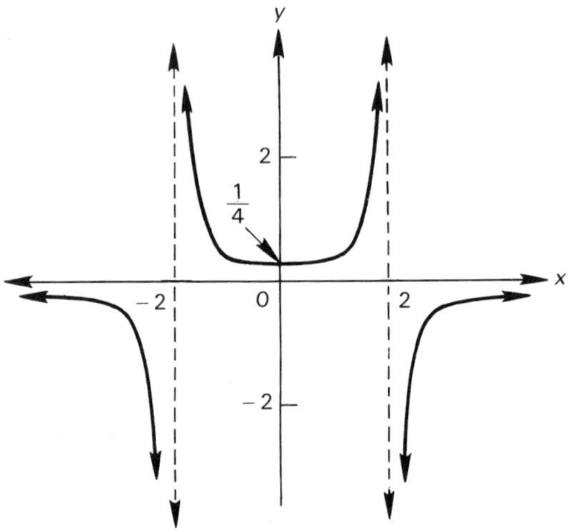

29) $f(x) = \dfrac{3}{x^2 + 1}$

Vertical asymptotes: none, $x^2 + 1 > 0$ always.

Also $f(x) > 0$ always.

Horizontal asymptote: numerator degree 0 < denominator degree 2

$y = 0$ is an asymptote

$f(-x) = \dfrac{3}{(-x)^2 + 1} = \dfrac{3}{x^2 + 1} = f(x)$, so the graph is symmetric through the y-axis.

y-intercept: $\dfrac{3}{0^2 + 1} = 3$

$y = 3$ is also the maximum value, because the denominator is smallest when $x = 0$.

x	y
1	$\frac{3}{2}$
2	$\frac{3}{5}$
3	$\frac{3}{10}$

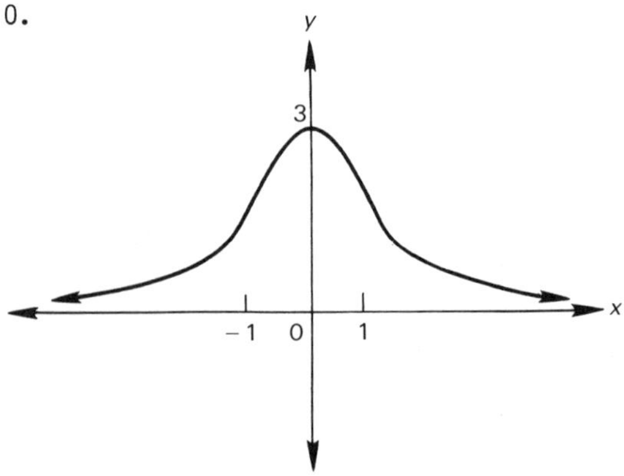

33) $f(x) = \dfrac{x - 1}{x + 3}$

Vertical asymptote: $x + 3 = 0$

$$x = -3$$

Horizontal asymptote: numerator degree 1 = denominator degree 1

$$y = \frac{a_n}{b_n} = \frac{1}{1} = 1$$

y-intercept: $\dfrac{0 - 1}{0 + 3} = -\dfrac{1}{3}$

x-intercept: $0 = \dfrac{x - 1}{x + 3}$

$$x - 1 = 0$$

$$x = 1$$

Interval	$(-\infty, -3)$	$(-3, -1)$	$(1, \infty)$
Sign of $(x - 1)$	-	-	+
Sign of $(x + 3)$	-	+	+
Sign of $f(x)$	+	-	+

x	y
-5	3
-4	4
-3.1	41
-2.9	-39
-2	-3
-1	-1
0	$-\frac{1}{3}$
1	0
2	$\frac{1}{5}$

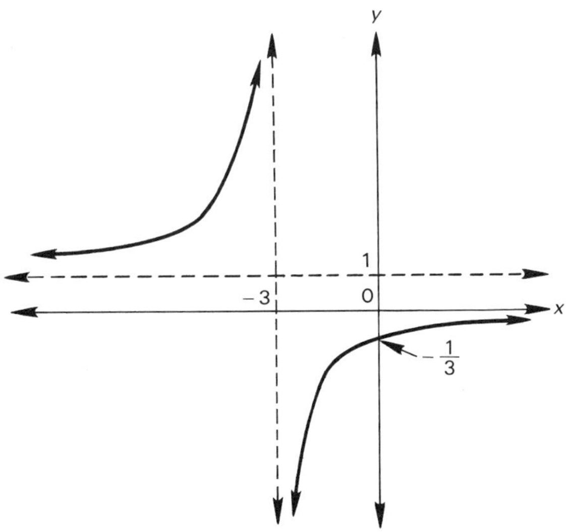

37) $f(x) = \dfrac{9}{3x^2 + 6} = \dfrac{3}{x^2 + 2}$

Vertical asymptotes: none, $x^2 + 2 >$ always.

Also $f(x) > 0$ always.

Horizontal asymptote: numerator degree $0 <$ denominator degree 2

$y = 0$ is an asymptote

$f(-x) = \dfrac{3}{(-x)^2 + 2} = \dfrac{3}{x^2 + 2} = f(x)$, so the graph is symmetric through the y-axis.

y-intercept: $\dfrac{3}{0^2 + 2} = \dfrac{3}{2}$

$y = \frac{3}{2}$ is also the maximum value, because the denominator is smallest when $x = 0$.

x	y
1	1
2	$\frac{1}{2}$
3	$\frac{3}{11}$

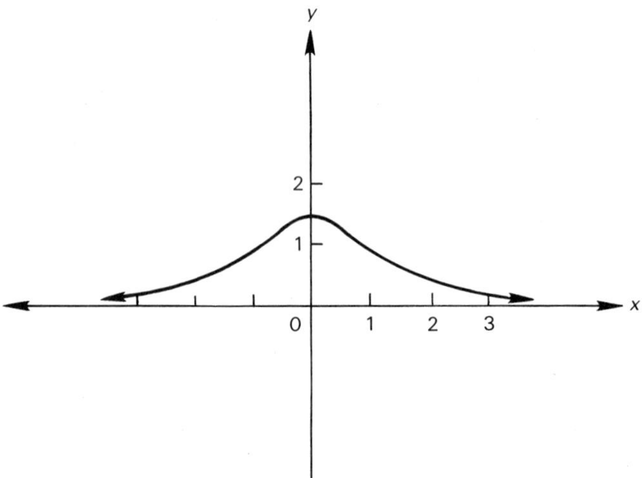

41) $g(x) = 2(2x - 1)(x - 2) + 2(x - 2)^2$

$= 2(x - 2)(2x - 1 + x - 2)$

$= 2(x - 2)(3x - 3)$

$= 2(x - 2)\,3(x - 1) = 6(x - 2)(x - 1)$

$x - 2 = 0 \qquad x - 1 = 0$

$x = 2 \qquad x = 1$

41) continued

Interval	$(-\infty, 1)$	$(1, 2)$	$(2, \infty)$
Sign of $(x - 2)$	-	-	+
Sign of $(x - 1)$	-	+	+
Sign of $f(x)$	+	-	+

$g(x) > 0$ on $(-\infty, 1)$ and on $(2, \infty)$

$g(x) < 0$ on $(1, 2)$

45) $$g(x) = 2\left(\frac{x - 5}{x + 2}\right)\frac{(x + 2) - (x - 5)}{(x + 2)^2}$$

$$= 2\left(\frac{x - 5}{x + 2}\right)\left(\frac{7}{(x + 2)^2}\right)$$

$$= \frac{2(7)(x - 5)}{(x + 2)^3}$$

$$= \frac{14(x - 5)}{(x + 2)^3}$$

Vertical asymptote at $x = -2$.

y-intercept at $x = 5$

Interval	$(-\infty, -2)$	$(-2, 5)$	$(5, \infty)$
Sign of $(x - 5)$	-	-	+
Sign of $(x + 2)^3$	-	+	+
Sign of $f(x)$	+	-	+

$g(x) > 0$ on $(-\infty, -2)$ and on $(5, \infty)$

$g(x) < 0$ on $(-2, 5)$

Section 5.4 Equations and Inequalities with Fractions

1) $\frac{x}{2} - \frac{x}{5} = 6$

$$10(\tfrac{x}{2}) - 10(\tfrac{x}{5}) = 10(6)$$
$$5x - 2x = 60$$
$$3x = 60$$
$$\tfrac{1}{3}(3x) = \tfrac{1}{3}(60)$$
$$x = 20$$

Check:

$$\frac{20}{2} - \frac{20}{5} = 6$$
$$10 - 4 = 6$$
$$6 = 6$$

5) $\frac{5}{x} - \frac{3}{4x} = 1$

$$4x\left(\frac{5}{x}\right) - 4x\left(\frac{3}{4x}\right) = 4x(1)$$
$$20 - 3 = 4x$$
$$17 = 4x$$
$$\frac{17}{4} = x$$

Check:

$$\frac{5}{\frac{17}{4}} - \frac{3}{4\left(\frac{17}{4}\right)} = 1$$
$$\frac{20}{17} - \frac{3}{17} = 1$$
$$\frac{17}{17} = 1$$
$$1 = 1$$

9) $\frac{x + 4}{2x - 10} = \frac{8}{7}$

$$7(x + 4) = 8(2x - 10)$$
$$7x + 28 = 16x - 80$$
$$108 = 9x$$
$$12 = x$$

Check:

$$\frac{12 + 4}{2(12) - 10} = \frac{8}{7}$$
$$\frac{16}{14} = \frac{8}{7}$$
$$\frac{8}{7} = \frac{8}{7}$$

13) $\frac{x + 1}{x + 10} = \frac{1}{2x}$

$$2x(x + 1) = 1(x + 10)$$
$$2x^2 + 2x = x + 10$$
$$2x^2 + x - 10 = 0$$
$$(x - 2)(2x + 5) = 0$$
$$x - 2 = 0 \quad \text{or} \quad 2x + 5 = 0$$
$$x = 2 \qquad\qquad x = -\frac{5}{2}$$

Check: $x = 2$

$$\frac{2 + 1}{2 + 10} = \frac{1}{2(2)}$$
$$\frac{3}{12} = \frac{1}{4}$$
$$\frac{1}{4} = \frac{1}{4}$$

Check $x = -\frac{5}{2}$

$$\frac{-\frac{5}{2} + 1}{-\frac{5}{2} + 10} = \frac{1}{2\left(-\frac{5}{2}\right)}$$
$$\frac{-\frac{3}{2}}{\frac{15}{2}} = -\frac{1}{5}$$
$$-\frac{3}{15} = -\frac{1}{5}$$
$$-\frac{1}{5} = -\frac{1}{5}$$

17) $\dfrac{5}{x^2 - 9} = \dfrac{3}{x + 3} - \dfrac{2}{x - 3}$

$$\frac{5}{(x - 3)(x + 3)} = \frac{3}{x + 3} - \frac{2}{x - 3}$$

$$(x - 3)(x + 3)\frac{5}{(x - 3)(x + 3)} = (x - 3)(x + 3)\frac{3}{x + 3} - (x - 3)(x + 3)\frac{2}{x - 3}$$

$$5 = 3(x - 3) - 2(x + 3)$$

$$5 = 3x - 9 - 2x - 6$$

$$5 = x - 15$$

$$20 = x$$

Check:

$$\frac{5}{(20)^2 - 9} = \frac{3}{20 + 3} - \frac{2}{20 - 3}$$

$$\frac{5}{391} = \frac{3}{23} - \frac{2}{17}$$

$$\frac{5}{391} = \frac{51}{391} - \frac{46}{391}$$

$$\frac{5}{391} = \frac{5}{391}$$

21) $\dfrac{2x - 5}{x + 1} - \dfrac{3}{x^2 + x} = 0$

$$\frac{2x - 5}{x + 1} - \frac{3}{x(x + 1)} = 0$$

$$x(x + 1)\frac{2x - 5}{x + 1} - x(x + 1)\frac{3}{x(x + 1)} = 0$$

$$2x^2 - 5x - 3 = 0$$

$$(2x + 1)(x - 3) = 0$$

$2x + 1 = 0$ or $x - 3 = 0$

$x = -\frac{1}{2}$ $\qquad$ $x = 3$

Check: $x = -\frac{1}{2}$

$$\frac{2\left(-\frac{1}{2}\right) - 5}{-\frac{1}{2} + 1} - \frac{3}{\left(-\frac{1}{2}\right)^2 - \frac{1}{2}} = 0$$

$$\frac{-6}{\frac{1}{2}} - \frac{3}{-\frac{1}{4}} = 0$$

$$-12 + 12 = 0 \qquad 0 = 0$$

Check: $x = 3$

$$\frac{2(3) - 5}{3 + 1} - \frac{3}{3^2 + 3} = 0$$

$$\frac{1}{4} - \frac{3}{12} = 0$$

$$0 = 0$$

25) $\dfrac{3}{2x^2 - 3x - 2} - \dfrac{x + 2}{2x + 1} = \dfrac{2x}{10 - 5x}$

$$\frac{3}{(2x + 1)(x - 2)} - \frac{x + 2}{2x + 1} = \frac{2x}{-5(x - 2)}$$

$$-5(2x + 1)(x - 2)\frac{3}{(2x + 1)(x - 2)} - (-5)(2x + 1)(x - 2)\frac{x + 2}{2x + 1}$$

$$= (-5)(2x + 1)(x - 2)\frac{2x}{-5(x - 2)}$$

$$-15 - (-5)(x - 2)(x + 2) = 2x(2x + 1)$$

$$-15 + 5(x^2 - 4) = 4x^2 + 2x$$

$$-15 + 5x^2 - 20 = 4x^2 + 2x$$

$$x^2 - 2x - 35 = 0$$

$$(x - 7)(x + 5) = 0$$

$$x - 7 = 0 \quad \text{or} \quad x + 5 = 0$$

$$x = 7 \qquad\qquad x = -5$$

Check: $x = 7$

$$\frac{3}{2(7)^2 - 3(7) - 2} - \frac{7 + 2}{2(7) + 1} = \frac{2(7)}{10 - 5(7)}$$

$$\frac{3}{75} - \frac{9}{15} = \frac{14}{-25}$$

$$\frac{3}{75} - \frac{45}{75} = -\frac{42}{75}$$

$$-\frac{42}{75} = -\frac{42}{75}$$

Check: $x = -5$

$$\frac{3}{2(-5)^2 - 3(-5) - 2} - \frac{(-5) + 2}{2(-5) + 1} = \frac{2(-5)}{10 - 5(-5)}$$

$$\frac{3}{63} - \frac{-3}{-9} = \frac{-10}{35}$$

$$\frac{3}{63} - \frac{21}{63} = \frac{-18}{63}$$

$$\frac{-18}{63} = \frac{-18}{63}$$

29) $S = \pi(r_1 + r_2)s$

$S = \pi r_1 s + \pi r_2 s$

$S - \pi r_2 s = \pi r_1 s$

$\dfrac{S - \pi r_2 s}{\pi s} = r_1$

$\dfrac{S}{\pi s} - r_2 = r_1$

33) $\dfrac{1}{f} = \dfrac{1}{m} + \dfrac{1}{p}$

$fmp \; \dfrac{1}{f} = fmp \; \dfrac{1}{m} + fmp \; \dfrac{1}{p}$

$mp = fp + fm$

$mp - fm = fp$

$m(p - f) = fp$

$m = \dfrac{fp}{p - f}$

37) $\dfrac{x + 3}{4} - \dfrac{x}{2} > 1$

$4\left(\dfrac{x + 3}{4}\right) - 4\left(\dfrac{x}{2}\right) > 4(1)$

$x + 3 - 2x > 4$

$-x + 3 > 4$

$-x > 1$

$x < -1$

41) $\dfrac{2x + 3}{x} < 1$

$\dfrac{2x + 3}{x} - 1 < 0$

$\dfrac{2x + 3 - x}{x} < 0$

$\dfrac{x + 3}{x} < 0$

$x = 0$ $\quad$ $x + 3 = 0$

$x = -3$

Interval	$(-\infty, -3)$	$(-3, 0)$	$(0, \infty)$
Sign of $x + 3$	-	+	+
Sign of x	-	-	+
Sign of $\dfrac{x + 3}{x}$	+	-	+

Since $\dfrac{2x + 3}{x} < 1$ is equivalent to

$\dfrac{x + 3}{x} < 0$, $\dfrac{2x + 3}{x} < 1$ when $-3 < x < 0$.

45) Let x represent the number

$\dfrac{11 - x}{23 - x} = \dfrac{2}{5}$

$5(11 - x) = 2(23 - x)$

$55 - 5x = 46 - 2x$

$9 = 3x$

$3 = x$

Answer: The number is **3**.

Check:

$\dfrac{11 - 3}{23 - 3} = \dfrac{8}{20} = \dfrac{2}{5}$

49) Let x represent the larger fraction.

$\frac{x}{2}$ represents the smaller fraction.

$$x + \frac{x}{2} = \frac{2}{3}$$

$$6(x) + 6\left(\frac{x}{2}\right) = 6\left(\frac{2}{3}\right)$$

$$6x + 3x = 4$$

$$9x = 4$$

$$x = \frac{4}{9}$$

$$\frac{x}{2} = \frac{2}{9}$$

Check:

$$\frac{4}{9} + \frac{2}{9} = \frac{6}{9} = \frac{2}{3}$$

Answer: The fractions are $\frac{4}{9}$ and $\frac{2}{9}$.

53) Let x represent the number.

$$\frac{561}{x} = 29 + \frac{10}{x}$$

$$\frac{561}{x} - \frac{10}{x} = 29$$

$$\frac{551}{x} = 29$$

$$551 = 29x$$

$$19 = x$$

Answer: The number is 19.

Check:

$$\frac{561}{19} = 29 + \frac{10}{19}$$

$$\frac{561}{19} = \frac{551 + 10}{19}$$

$$\frac{561}{19} = \frac{561}{19}$$

57) Proof:

$$\frac{a}{b} = \frac{c}{d}$$

$$\frac{a}{b} + 1 = \frac{c}{d} + 1$$

$$\frac{a}{b} + \frac{b}{b} = \frac{c}{d} + \frac{d}{d}$$

$$\frac{a + b}{b} = \frac{c + d}{d}$$

61) Let x represent the point-value of the grade earned in the fifth course.

First determine how many credits in the fifth course.

Total credits in the other courses: $4 + 3 + 3 + 2 = 12$

$15 - 12 = 3$, so the fifth course is a 3-credit course.

To get an expression for the total of points, multiply the credits of each course by the point-value of the expected grade.

$A(4) + A(3) + B(3) + D(2) + ?(3) =$

$4 \cdot 4 + 4 \cdot 3 + 3 \cdot 3 + 1 \cdot 2 + x \cdot 3 = 3x + 39$

Divide this expression by the total credits, 15, and set equal to the desired average expressed as a fraction. $3.4 = \frac{17}{5}$

$$\frac{3x + 39}{15} = \frac{17}{5}$$

$$5(3x + 39) = 15(17)$$

$$15x + 195 = 255$$

$$15x = 60$$

$$x = 4$$

If the point-value of the grade is 4, the student needs an A in the fifth course.

Section 5.5 Variation

1) $p = 4s, \quad k = 4$

5) $z = kxy^3$

9) $w = \dfrac{kx^2}{yz}$

13) $u = kvw$

$$2 = k(15)\left(\frac{2}{3}\right)$$

$$2 = 10k$$

$$\frac{1}{5} = k$$

17) $s = k\ell w^2$

$\frac{10}{3} = k(12)\left(\frac{5}{6}\right)^2$

$\frac{10}{3} = k(12)\left(\frac{25}{36}\right)$

$\frac{10}{3} = \frac{50}{6}k$

$\frac{2}{5} = k$

$s = \frac{2}{5}(15)\left(\frac{9}{4}\right)^2$

$s = 6\left(\frac{81}{16}\right)$

$s = \frac{243}{8}$

21) $F = kx$

$\frac{F}{x} = k$, so $\frac{F_1}{x_1} = \frac{F_2}{x_2}$ can be used to solve this problem.

Let F_2 represent the unknown force

$\frac{2.4}{1.8} = \frac{F_2}{3}$

$7.2 = 1.8F_2$

$4 = F_2$

4 pounds are required.

25) s is traditionally used for distance in this formula.

$s = kt^2$

$\frac{s}{t^2} = k$, so $\frac{s_1}{t_1^2} = \frac{s_2}{t_2^2}$ can be used to solve this problem.

Let s_2 represent the unknown distance.

$\frac{256}{4^2} = \frac{s_2}{7^2}$

$\frac{256}{16} = \frac{s_2}{49}$

$12544 = 16s_2$

$784 = s_2$

The object will fall 784 feet in 7 seconds.

29) $R = \frac{k\ell}{d^2}$ $\quad k \approx 0.00819$ from exercise 28.

$R = \frac{0.00819(100)}{(0.4)^2} = \frac{0.819}{0.16} = 5.12$ ohms

$R = \frac{0.00819(100)}{(0.04)^2} = \frac{0.819}{0.0016} = 512$ ohms

Section 5.6 Division of Polynomials and Synthetic Division

1) $x + 3 \overline{\smash{)}\, x^3 - 2x^2 - 13x + 6}$, quotient $x^2 - 5x + 2$

$$\begin{array}{l} x^3 + 3x^2 \\ \hline -5x^2 - 13x \\ -5x^2 - 15x \\ \hline 2x + 6 \\ 2x + 6 \\ \hline 0 \end{array}$$

Check:

$(x + 3)(x^2 - 5x + 2)$

$= x^3 - 2x^2 - 13x + 6$

5) $x - 2 \overline{\smash{)}\, x^3 + 5x^2 - 7x + 8}$, quotient $x^2 + 7x + 7 \quad r = 22$

$$\begin{array}{l} x^3 - 2x^2 \\ \hline 7x^2 - 7x \\ 7x^2 - 14x \\ \hline 7x + 8 \\ 7x - 14 \\ \hline 22 \end{array}$$

Check:

$(x - 2)(x^2 + 7x + 7) + 22$

$= x^3 + 7x^2 + 7x - 2x^2 - 14x - 14 + 22$

$= x^3 + 5x^2 - 7x + 8$

9) $x^2 - 3x + 5 \overline{\smash{)}\, x^3 - x^2 - x + 10}$, quotient $x + 2$

$$\begin{array}{l} x^3 - 3x^2 + 5x \\ \hline 2x^2 - 6x + 10 \\ 2x^2 - 6x + 10 \\ \hline 0 \end{array}$$

Check:

$(x + 2)(x^2 - 3x + 5)$

$= x^3 - 3x^2 + 5x + 2x^2 - 6x + 10$

$= x^3 - x^2 - x + 10$

13) $(2x^3 + x^2 - 3x + 7) \div (x + 1)$

$c = -1$

$$\begin{array}{r|rrrr} -1 & 2 & +1 & -3 & +7 \\ & & -2 & 1 & +2 \\ \hline & 2 & -1 & -2 & +9 \end{array}$$

$Q = 2x^2 - x - 2, \quad r = 9$

Check:

$(x + 1)(2x^2 - x - 2) + 9$

$= 2x^3 - x^2 - 2x + 2x^2 - x - 2 + 9$

$= 2x^3 + x^2 - 3x + 7$

17) $(x^4 - 3x^3 + 7x^2 - 2x + 1) \div (x + 2)$

$c = -2$

$$\begin{array}{r|rrrrr} -2 & 1 & -3 & +7 & -2 & +1 \\ & & -2 & +10 & -34 & +72 \\ \hline & 1 & -5 & +17 & -36 & 73 \end{array}$$

$Q = x^3 - 5x^2 + 17x - 36, \quad r = 73$

Check:

$(x + 2)(x^3 - 5x^2 + 17x - 36)$

$= x^4 - 5x^3 + 17x^2 - 36x + 2x^3 - 10x^2$

$\quad + 34x - 72 + 73$

$= x^4 - 3x^3 + 7x^2 - 2x + 1$

21) $(x^3 - 27) \div (x - 3)$

$c = 3$

$$\begin{array}{r|l}
+3 & 1 + 0 + 0 - 27 \\
 & \quad\ + 3 + 9 + 27 \\
\hline
 & 1 + 3 + 9 \;|\; \underline{0}
\end{array}$$

$Q = x^2 + 3x + 9, \quad r = 0$

This quotient can also be found by factoring:

$$\frac{x^3 - 27}{x - 3} = \frac{(x^2 + 3x + 9)\cancel{(x - 3)}}{\cancel{x - 3}} = x^2 + 3x + 9$$

25) $(x^4 - 16) \div (x - 2)$

$c = 2$

$$\begin{array}{r|l}
+2 & 1 + 0 + 0 + 0 - 16 \\
 & \quad\ +2 + 4 + 8 + 16 \\
\hline
 & 1 + 2 + 4 + 8 \;|\; \underline{0}
\end{array}$$

$Q = x^3 + 2x^2 + 4x + 8, \quad r = 0$

This quotient can also be found by factoring:

$$\frac{x^4 - 16}{x - 2} = \frac{(x^2 + 4)(x + 2)\cancel{(x - 2)}}{\cancel{x - 2}}$$

$$= x^3 + 2x^2 + 4x + 8$$

29) $x^4 - \frac{1}{2}x^3 + \frac{1}{3}x^2 - \frac{1}{4}x + \frac{1}{5} \div (x - 1)$

$c = 1$

$$\begin{array}{r|l}
1 & 1 - \frac{1}{2} + \frac{1}{3} - \frac{1}{4} + \frac{1}{5} \\
 & \quad\ +1 + \frac{1}{2} + \frac{5}{6} + \frac{7}{12} \\
\hline
 & 1 + \frac{1}{2} + \frac{5}{6} + \frac{7}{12} \;|\; \underline{\frac{47}{60}}
\end{array}$$

$Q = x^3 + \frac{1}{2}x^2 + \frac{5}{6}x + \frac{7}{12}, \quad r = \frac{47}{60}$

33) $(6x^3 + 7x^2 + x + 8) \div (2x + 3)$

$$2x + 3 = 2\left(x + \frac{3}{2}\right) \qquad c = -\frac{3}{2}$$

$$\frac{6x^3 + 7x^2 + x + 8}{2x + 3} = \frac{1}{2}\left(\frac{6x^3 + 7x^2 + x + 8}{x + \frac{3}{2}}\right) =$$

$$\begin{array}{r|rrrr} -\frac{3}{2} & 6 & +7 & +1 & +8 \\ & & -9 & +3 & -6 \\ \hline & 6 & -2 & +4 & \underline{2} \end{array}$$

$$\frac{1}{2}(6x^2 - 2x + 4) + \frac{1}{2}\left(\frac{2}{x + \frac{3}{2}}\right)$$

$$= 3x^2 - x + 2 + \frac{2}{2x + 3}$$

$$Q = 3x^2 - x + 2, \qquad r = 2$$

Section 5.7 The Remainder, Factor, and Rational Root Theorem

1) $f(x) = x^3 - x^2 + 3x - 2;$ find $f(2)$

$$\begin{array}{r|rrrr} 2 & 1 & -1 & +3 & -2 \\ & & +2 & +2 & +10 \\ \hline & 1 & +1 & +5 & \underline{+8} \end{array} = \text{remainder} = f(2)$$

5) $f(x) = x^5 - x^3 + 2x^2 + x - 3;$ find $f(1)$

$$\begin{array}{r|rrrrr} 1 & 1 & -1 & +2 & +1 & -3 \\ & & +1 & +0 & +2 & +3 \\ \hline & 1 & +0 & +2 & +3 & \underline{0} \end{array} = \text{remainder} = f(1)$$

9) $f(x) = 2x^3 + 3x^2 - 5x + 1 \div (x - 3)$

remainder $= f(3) = 2(3)^3 + 3(3)^2 - 5(3) + 1$

$= 54 + 27 - 15 + 1$

$= 67$

13) $p(x) = x^3 + 6x^2 + 11x + 6$

$(x + 1)$ is a factor if $p(-1) = 0$

$$\begin{array}{r|rrrr} -1 & 1 & +6 & +11 & +6 \\ & & -1 & -5 & -6 \\ \hline & 1 & +5 & +6 & \underline{0} \end{array} = r$$

$$x^3 + 6x^2 + 11x + 6 = (x + 1)(x^2 + 5x + 6) = (x + 1)(x + 2)(x + 3)$$

17) $p(x) = -x^3 + 7x + 6$

$(x + 2)$ is a factor if $p(-2) = 0$.

$$\begin{array}{r|rrrr} -2 & -1 & +0 & +7 & +6 \\ & & +2 & -4 & -6 \\ \hline & -1 & +2 & +3 & \underline{0} \end{array} = r$$

$$\begin{aligned} -x^3 + 7x + 6 = (x + 2)(-x^2 + 2x + 3) &= (x + 2)(x + 1)(-x + 3) \\ &= -(x + 2)(x + 1)(x - 3) \end{aligned}$$

21) $p(x) = x^4 + 4x^3 + 3x^2 - 4x - 4$

$x + 2$ is a factor if $p(-2) = 0$.

$$\begin{array}{r|rrrrr} -2 & 1 & +4 & +3 & -4 & -4 \\ & & -2 & -4 & +2 & +4 \\ \hline & 1 & +2 & -1 & -2 & \underline{0} \end{array}$$

$$\begin{aligned} x^4 + 4x^3 + 3x^2 - 4x - 4 &= (x + 2)(x^3 + 2x^2 - x - 2) \\ &= (x + 2)[x^2(x + 2) - (x + 2)] \\ &= (x + 2)(x + 2)(x^2 - 1) \\ &= (x + 2)^2(x + 1)(x - 1) \end{aligned}$$

25) $x^3 + x^2 - 21x - 45 = 0$

$n = 3$, so there are 3 roots. First find any rational roots. The leading coefficient is 1, so all denominators are ± 1. Possible rational roots (factors of 45): ± 1, ± 3, ± 5, ± 9, ± 15, ± 45.

Try $c = 1$

$$\begin{array}{r|rrrr} 1 & 1 & +1 & -21 & -45 \\ & & +1 & +2 & +19 \\ \hline & 1 & +2 & -19 & \underline{-26} \end{array}$$

Try $c = -1$

$$\begin{array}{r|rrrr} -1 & 1 & +1 & -21 & -45 \\ & & -1 & +0 & +21 \\ \hline & 1 & +0 & -21 & \underline{-14} \end{array}$$

25) continued

Try $c = 3$

$$\begin{array}{r|rrrr} 3 & 1 & +1 & -21 & -45 \\ & & +3 & +12 & -27 \\ \hline & 1 & +4 & -9 & \underline{-72} \end{array}$$

Try $c = -3$

$$\begin{array}{r|rrrr} -3 & 1 & +1 & -21 & -45 \\ & & -3 & +6 & +45 \\ \hline & 1 & -2 & -15 & \underline{0} \end{array}$$

$(x + 3)$ is a factor.

$$0 = x^3 + x^2 - 21x - 45 = (x + 3)(x^2 - 2x - 15)$$
$$= (x + 3)(x - 5)(x + 3) = (x + 3)^2(x - 5)$$

$x + 3 = 0$ $\quad$ $x - 5 = 0$

$x = -3$ $\quad$ or $\quad$ $x = 5$

(double root)

29) $x^4 + 3x^2 + 3x^2 + x = 0$

$n = 4$, so there are 4 roots. First find any rational roots.

Begin by factoring out x:

$x(x^3 + 3x^2 + 3x + 1) = 0$

One root is $x = 0$.

Find the roots of $x^3 + 3x^2 + 3x + 1 = 0$.

The leading coefficient is 1 and the constant is 1.

The only possibilities are $c = \pm 1$.

Try $c = 1$

$$\begin{array}{r|rrrr} 1 & 1 & +3 & +3 & +1 \\ & & +1 & +4 & +7 \\ \hline & 1 & +4 & +7 & \underline{+8} \end{array}$$

Try $c = -1$

$$\begin{array}{r|rrrr} -1 & 1 & +3 & +3 & +1 \\ & & -1 & -2 & -1 \\ \hline & 1 & +2 & +1 & \underline{0} \end{array}$$

$(x + 1)$ is a Factor

$$0 = x^4 + 3x^3 + 3x^2 + x = x(x + 1)(x^2 + 2x + 1)$$
$$= x(x + 1)(x + 1)^2 = x(x + 1)^3$$

$x = 0$ $\quad$ or $\quad$ $x + 1 = 0$

$x = -1$

roots: -1, 0

(The work for this problem would be shortened if $x^3 + 3x^2 + 3x + 1$ is recognized as an expansion of $(a + b)^3$.)

33) $x^4 + 2x^3 - 7x^2 - 18x - 18 = 0$

n = 4, so there are 4 roots. First find any rational roots.

The leading coefficient is 1, so all denominators are ±1.

Possible reational roots (factors of 18): ±1, ±2, ±3, ±6, ±9, ±18.

```
Try c = 1                              Try c = -1
 1 | 1 + 2 - 7 - 18 - 18               -1 | 1 + 2 - 7 - 18 - 18
   |     +1 + 3 -  4 + 22                 |     -1 - 1 -  8 + 26
   ------------------------               ------------------------
     1 + 3 - 4 + 22  | 6                    1 + 1 + 8 - 26  | 8

Try c = 2                              Try c = -2
 2 | 1 + 2 - 7 - 18 - 18               -2 | 1 + 2 - 7 - 18 - 18
   |     +2 + 8 -  2 + 40                 |     -2 + 0 + 14 +  8
   ------------------------               ------------------------
     1 + 4 - 1 - 20 | 22                    1 + 0 - 7 -  4 | -10

Try c = 3
 3 | 1 + 2 - 7 - 18 - 18
   |     +3 +15 + 24 + 18
   ------------------------
     1 + 5 + 8 + 6   | 0
```

(x - 3) is a factor

$0 = x^4 - 2x^3 - 7x^2 - 18x - 18 = (x - 3)(x^3 + 5x^2 + 8x + 6)$

To factor $x^3 + 5x^2 + 8x + 6$, test the remaining possible rational roots, after observing that 3 is not a root again.

```
Try c = -3
 -3 | 1 + 5 + 8 + 6
    |    -3 - 6 - 6
    ------------------
      1 + 2 + 2 | 0
```

(x + 3) is a factor

$0 = x^4 - 2x^3 - 7x^2 - 18x - 18 = (x - 3)(x + 3)(x^2 + 2x + 2)$

$x - 3 = 0$ or $x + 3 = 0$ or $x^2 + 2x + 2 = 0$

$x = 3$ $\quad$ $x = -3$ $\quad$ Solve by formula:

$$x = \frac{(-2) \pm\sqrt{(2)^2 - 4(1)(2)}}{2(1)}$$

Two real roots: 3, -3

$$x = \frac{-2 \pm\sqrt{-4}}{2}$$

$$x = -1 \pm i$$

37) $2x^3 - 5x - 3 = 0$

$n = 3$, so there are 3 roots. First find any rational roots.

Possible numerators (factors of 3): $\pm 1, \pm 3$

Possible denominators (factors of 2): $\pm 1, \pm 2$

Possible rational roots: $\pm 1, \pm \frac{1}{2}, \pm 3, \pm \frac{3}{2}$

Try $c = 1$

$$\begin{array}{r|rrrr} 1 & 2 & +0 & -5 & -3 \\ & & +2 & +2 & -3 \\ \hline & 2 & +2 & -3 & \underline{-6} \end{array}$$

Try $c = -1$

$$\begin{array}{r|rrrr} -1 & 2 & +0 & -5 & -3 \\ & & -2 & +2 & +3 \\ \hline & 2 & -2 & -3 & \underline{0} \end{array}$$

$(x + 1)$ is a factor.

$2x^3 - 5x - 3 = (x + 1)(2x^2 - 2x - 3)$

$x + 1 = 0$ or $2x^2 - 2x - 3 = 0$

$x = -1$

Solve by formula

$$x = \frac{-(-2) \pm \sqrt{(-2)^2 - 4(2)(-3)}}{2(2)}$$

$$x = \frac{2 \pm 2\sqrt{7}}{4}$$

$$x = \frac{1 \pm \sqrt{7}}{2}$$

roots: $-1, \frac{1 + \sqrt{7}}{2}, \frac{1 - \sqrt{7}}{2}$

41) $-x^3 - 3x^2 + 24x + 80 = -1(x^3 + 3x^2 - 24 - 80)$

$n = 3$, so there are at most 3 factors.

The leading coefficient is 1, so possible denominators are ± 1.

Possible rational factors (factors of 80):

$\pm 1, \pm 2, \pm 4, \pm 5, \pm 8, \pm 10, \pm 15, \pm 20, \pm 40, \pm 80$

Try $c = 1$

$$\begin{array}{r|rrrr} 1 & 1 & +3 & -24 & -80 \\ & & -1 & +2 & -22 \\ \hline & -1 & +2 & -22 & \underline{-102} \end{array}$$

Try $c = -1$

$$\begin{array}{r|rrrr} -1 & 1 & +3 & -24 & -80 \\ & & -1 & -2 & +26 \\ \hline & 1 & +2 & -26 & \underline{-54} \end{array}$$

Try $c = 2$

$$\begin{array}{r|rrrr} 2 & 1 & +3 & -24 & -80 \\ & & +2 & +10 & -24 \\ \hline & 1 & +5 & -14 & \underline{-104} \end{array}$$

Try $c = -2$

$$\begin{array}{r|rrrr} -2 & 1 & +3 & -24 & -80 \\ & & -2 & -2 & +52 \\ \hline & 1 & +1 & -26 & \underline{28} \end{array}$$

41) continued

Try c = 4

```
4 | 1 + 3 - 24 - 80
  |     +4 + 28 + 16
  ------------------
    1 + 7 +  4 | -64
```

Try c = -4

```
-4 | 1 + 3 - 24 - 80
   |     -4 +  4 + 80
   ------------------
     1 - 1 - 20 | 0
```

(x + 4) is a factor.

$$-x^3 - 3x^2 + 24x + 8 = -(x + 4)(x^2 - x - 20)$$
$$= -(x + 4)(x + 4)(x - 5) = -(x + 4)^2(x - 5)$$

45) $p(x) = x^4 - 3x^3 + 5x^2 - x - 10 = 0$

n = 4, so there are 4 roots. First find any rational roots.

The leading coefficient is one, so all denominators are ±1.

Possible numerators (factors of 10): ±1, ±2, ±5, ±10.

Try c = 1

```
1 | 1 - 3 + 5 - 1 - 10
  |     +1 - 2 + 3 +  2
  ---------------------
    1 - 2 + 3 + 2 | -8
```

Try c = -1

```
-1 | 1 - 3 + 5 - 1 - 10
   |     -1 + 4 - 9 + 10
   ---------------------
     1 - 4 + 9 + 10 | 0
```

(x + 1) is a factor.

Divide the previous quotient, trying c = -1 again.

```
-1 | 1 - 4 + 9 - 10
   |     -1 + 5 - 14
   -----------------
     1 - 5 + 14 | -24
```

Try c = 2

```
2 | 1 - 4 + 9 - 10
  |     +2 - 4 + 10
  -----------------
    1 - 2 + 5 | 0
```

(x - 2) is a factor.

$$0 = x^4 - 3x^3 + 5x^2 - x - 10 = (x + 1)(x - 2)(x^2 - 2x + 5)$$

$x + 1 = 0$ or $x - 2 = 0$ or $0 = (x^2 - 2x + 5)$

$x = -1$ $\quad$ $x = 2$ $\quad$ Solve by formula:

$$x = \frac{-(-2) \pm \sqrt{(-2)^2 - 4(1)(5)}}{2(1)}$$

$$x = \frac{2 \pm \sqrt{-16}}{2}$$

$$x = \frac{2 \pm 4i}{2}$$

$$x = 1 \pm 2i$$

Two real roots: -1, 2

Two complex roots: 1 + 2i, 1 - 2i

49) $p(x) = x^5 - 9x^4 + 31x^3 - 49x^2 + 36x - 10 = 0$

n = 5, so there are 5 roots. First find any rational roots.

The leading coefficient is one, so all denominators are ±1.

Possible rational factors (factors of (0) = ±1, ±2, ±5, ±10.

Try c = 1

$$\begin{array}{r|rrrrrr} 1 & 1 & -9 & +31 & -49 & +36 & -10 \\ & & +1 & -8 & +23 & -26 & +10 \\ \hline & 1 & -8 & +23 & -26 & +10 & \boxed{0} \end{array}$$

(x - 1) is a factor.

Divide this quotient, trying c = 1 again.

$$\begin{array}{r|rrrrr} 1 & 1 & -8 & +23 & -26 & +10 \\ & & +1 & -7 & +16 & -10 \\ \hline & 1 & -7 & +16 & -10 & \boxed{0} \end{array}$$

(x - 1) is a factor twice.

Divide this quotient, trying c = 1 again.

$$\begin{array}{r|rrrr} 1 & 1 & -7 & +16 & -10 \\ & & +1 & -6 & +10 \\ \hline & 1 & -6 & +10 & \boxed{0} \end{array}$$

(x - 1) is a factor three times.

$$0 = x^5 - 9x^4 + 31x^3 - 49x^2 + 36x - 10 = (x - 1)^3(x^2 - 6x + 10)$$

$x - 1 = 0$ or $0 = x^2 - 6x + 10$

$x = 1$ Solve by formula

$$x = \frac{-(-6) \pm \sqrt{(-6)^2 - 4(1)(10)}}{2(1)}$$

$$x = \frac{6 \pm \sqrt{-4}}{2}$$

Roots: 1, 3 ± i $x = 3 \pm i$

53) If (x - 2) is a factor of $x^3 + bx^2 - 13x + 10$, then the remainder of the related synthetic division must be 0, and f(2) = 0.

53) continued

Here are two methods for finding b.

(1) Use $f(2) = 0$.

$$f(x) = x^3 + bx^2 - 13x + 10$$
$$f(2) = 8 + 4b - 26 + 10 = 0$$
$$4b - 8 = 0$$
$$4b = 8$$
$$b = 2$$

(2) Use synthetic division

$$\begin{array}{r|l} 2 & 1 + b - 13 + 10 \\ & \quad +2 + (2b + 4)(4b - 18) \\ \hline & 1 + (b+2) + (2b-9) + (4b-8) \end{array}$$

Since $r = 4b - 8 = 0$

$b = 2$

57) $y = 4x^3 - 7x^2 + 10$

$y = x^3 + 43x - 5$

Set the two right sides equal to each other.

$$4x^3 - 7x^2 + 10 = x^3 + 43x - 5$$
$$3x^3 - 7x^2 - 43x + 15 = 0$$

n = 3, so there are 3 roots. Search for rational roots first.

Possible numerators (factors of 15): ±1, ±3, ±5, ±15.

Possible denominators (factors of 3): ±1, ±3.

Possible rational roots: $\pm 1, \pm\frac{1}{3}, \pm 3, \pm 5, \pm\frac{5}{3}$

Try c = 1

$$\begin{array}{r|rrrr} 1 & 3 & -7 & -43 & +15 \\ & & +3 & -4 & -47 \\ \hline & 3 & -4 & -47 & \lvert -32 \end{array}$$

Try c = -1

$$\begin{array}{r|rrrr} -1 & 3 & -7 & -43 & +15 \\ & & -3 & +10 & +33 \\ \hline & 3 & -10 & -33 & \lvert 48 \end{array}$$

Try $c = \frac{1}{3}$

$$\begin{array}{r|rrrr} \frac{1}{3} & 3 & -7 & -43 & +15 \\ & & +1 & +2 & -\frac{41}{3} \\ \hline & 3 & -6 & -41 & \lvert +\frac{4}{3} \end{array}$$

Try $c = -\frac{1}{3}$

$$\begin{array}{r|rrrr} -\frac{1}{3} & 3 & -7 & -43 & +15 \\ & & -1 & +\frac{8}{3} & +\frac{121}{9} \\ \hline & 3 & -8 & -\frac{121}{3} & \lvert \frac{14}{9} \end{array}$$

Try c = 3

$$\begin{array}{r|rrrr} 3 & 3 & -7 & -43 & +15 \\ & & +9 & +6 & -51 \\ \hline & 3 & +2 & -17 & \lvert -36 \end{array}$$

Try c = -3

$$\begin{array}{r|rrrr} -3 & 3 & -7 & -43 & +15 \\ & & -9 & +48 & -15 \\ \hline & 3 & -16 & +5 & \lvert 0 \end{array}$$

(x + 3) is a factor.

57) continued

$$0 = 3x^3 - 7x^2 - 43x + 15 = (x - 3)(3x^2 - 16x + 5)$$
$$= (x + 3)(3x - 1)(x - 5)$$

$x + 3 = 0$ or $3x - 1 = 0$ or $x - 5 = 0$

$x = -3$ $\quad$ $x = \frac{1}{3}$ $\quad$ $x = 5$

When $x = -3$, $y = (-3)^3 + 43(-3) - 5 = -161$

When $x = \frac{1}{3}$, $y = \left(\frac{1}{3}\right)^3 + 43\left(\frac{1}{3}\right) - 5 = \frac{253}{27}$

When $x = 5$, $y = (5)^3 + 43(5) - 5 = 335$

Solutions: $(-3, -161)$, $\left(\frac{1}{3}, \frac{253}{27}\right)$, $(5, 335)$

Section 5.8 Decomposing Rational Funcitons

1) $\frac{2x}{(x + 1)(x - 1)} = \frac{A}{x + 1} + \frac{B}{x - 1}$

Multiply through by $(x + 1)(x - 1)$

$$2x = A(x - 1) + B(x + 1)$$

To eliminate A, let $x = 1$.

$$2 \cdot 1 = A(1 - 1) + B(1 + 1)$$
$$2 = 2B$$
$$1 = B$$

To eliminate B, let $x = -1$.

$$2(-1) = A(-1 - 1) + B(-1 + 1)$$
$$-2 = -2A$$
$$1 = A$$

$$\frac{2x}{(x + 1)(x - 1)} = \frac{1}{x + 1} + \frac{1}{x - 1}$$

5) $\frac{5x^2 + 9x - 56}{(x - 4)(x - 2)(x + 1)} = \frac{A}{x - 4} + \frac{B}{x - 2} + \frac{C}{x + 1}$

Multiply through by $(x - 4)(x - 2)(x + 1)$

$5x^2 + 9x - 56 = A(x - 2)(x + 1) + B(x - 4)(x + 1) + C(x - 4)(x - 2)$

To eliminate A and C, let $x = 2$.

$5(2)^2 + 9(2) - 56 = A(2 - 2)(2 + 1) + B(2 - 4)(2 + 1) + C(x - 4)(2 - 2)$

$$-18 = -6B$$
$$3 = B$$

5) continued

To eliminate B and C, let $x = 4$.

$5(4)^2 + 9(4) - 56 = A(4 - 2)(4 + 1) + B(4 - 4)(4 + 1)$
$+ C(4 - 4)(x - 2)$

$$60 = 10A$$
$$6 = A$$

To eliminate A and B, let $x = -1$.

$5(-1)^2 + 9(-1) - 56 = A(-1 -2)(-1 + 1) + B(-1 - 4)(-1 + 1)$
$+ C(-1 - 4)(-1 - 2)$

$$-60 = 15C$$
$$-4 = C$$

$$\frac{5x^2 + 9x - 56}{(x - 4)(x - 2)(x + 1)} = \frac{6}{x - 4} + \frac{3}{x - 2} - \frac{4}{x + 1}$$

9) $\dfrac{3x - 30}{15x^2 - 14x - 8} = \dfrac{3x - 30}{(5x + 2)(3x - 4)} = \dfrac{A}{5x + 2} + \dfrac{B}{3x - 4}$

Multiply through by $(5x + 2)(3x - 4)$

$$3x - 30 = A(3x - 4) + B(5x + 2)$$

To eliminate A, let $x = \frac{4}{3}$.

$$3\left(\frac{4}{3}\right) - 30 = A[3\left(\frac{4}{3}\right) - 4] + B[5\left(\frac{4}{3}\right) + 2]$$
$$-26 = \frac{26}{3}B$$
$$-3 = B$$

To eliminate B, let $x = -\frac{2}{5}$.

$$3\left(-\frac{2}{5}\right) - 30 = A[3\left(-\frac{2}{5}\right) - 4] + B[5\left(-\frac{2}{5}\right) + 2]$$
$$-\frac{156}{5} = -\frac{26}{5}A$$
$$6 = A$$

$$\frac{3x - 30}{15x^2 - 14x - 8} = \frac{6}{5x + 2} - \frac{3}{3x - 4}$$

13) $(x^3 - x + 2) \div (x^2 - 1)$

$$\begin{array}{r|l} x^2 - 1 & \overline{x^3 - x + 2} \\ & \underline{-(x^3 - x)} \\ & \qquad +2 \end{array}$$

$$\frac{x^3 - x + 2}{x^2 - 1} = x + \frac{2}{x^2 - 1}$$

Decompose: $\dfrac{2}{x^2 - 1} = \dfrac{2}{(x + 1)(x - 1)} = \dfrac{A}{x + 1} + \dfrac{B}{x - 1}$

Multiply through by $(x + 1)(x - 1)$.

$2 = A(x - 1) + B(x + 1)$

To eliminate A, let $x = 1$.

$2 = A(1 - 1) + B(1 + 1)$

$2 = 2B$

$1 = B$

To eliminate B, let $x = -1$.

$2 = A(-1 - 1) + B(-1 + 1)$

$2 = -2A$

$-1 = A$

$$\frac{x^3 - x + 2}{x^2 - 1} = x - \frac{1}{x + 1} + \frac{1}{x - 1}$$

17) $(10x^3 - 15x^2 - 35x) \div (x^2 - x - 6)$

$$\begin{array}{r|l} & 10x - 5 \\ x^2 - x - 6 & \overline{10x^3 - 15x^2 - 35x + 0} \\ & \underline{-(10x^3 - 10x^2 - 60x)} \\ & \qquad -5x^2 + 25x + 0 \\ & \qquad \underline{-(-5x^2 + 5x + 30} \\ & \qquad\qquad 20x - 30 \end{array}$$

$$\frac{10x^3 - 15x^2 - 35x}{x^2 - x - 6} = 10x - 5 + \frac{20x - 30}{x^2 - x - 6}$$

Decompose: $\dfrac{20x - 30}{x^2 - x - 6} = \dfrac{20x - 30}{(x - 3)(x + 2)} = \dfrac{A}{x - 3} + \dfrac{B}{x + 2}$

Multiply through by $(x - 3)(x + 2)$.

$20x - 30 = A(x + 2) + B(x - 3)$

17) continued

To eliminate A, let $x = -2$.

$$20(-2) - 30 = A(-2 + 2) + B(-2 - 3)$$
$$-70 = -5B$$
$$14 = B$$

To eliminate B, let $x = 3$.

$$20(3) - 30 = A(3 + 2) + B(3 - 3)$$
$$30 = 5A$$
$$6 = A$$

$$\frac{10x^3 - 15x^2 - 35x}{x^2 - x - 6} = 10x - 5 + \frac{6}{x - 3} + \frac{14}{x + 2}$$

21) $\dfrac{-4x}{(x^2 + 3)(x - 1)} = \dfrac{A}{x - 1} + \dfrac{Bx + C}{x^2 + 3}$

Multiply by $(x^2 + 3)(x - 1)$

$$-4x = A(x^2 + 3) + (Bx + C)(x - 1)$$

To eliminate B and C, let $x = 1$.

$$-4(1) = A(1 + 3) + (B \cdot 1 + C)(1 - 1)$$
$$-4 = 4A$$
$$-1 = A$$

To eliminate B, let $A = -1$ and $x = 0$.

$$-4(0) = -1(0 + 3) + B \cdot 0 + C)(0 - 1)$$
$$0 = -3 - C$$
$$C = -3$$

To solve for B, let $A = -1$, $C = -3$, and $x = -1$.

$$-4(-1) = -1(1 + 3) + [B(-1) - 3](-1 - 1)$$
$$4 = -4 + 2B + 6$$
$$2 = 2B$$
$$1 = B$$

$$\frac{-4x}{(x^2 + 3)(x - 1)} = \frac{-1}{x - 1} + \frac{x - 3}{x^2 + 3}$$

Sample Test Questions for Chapter 5

Part I

1) $f(x) = (x + 2)^3 - \frac{3}{2}$

Shift the graph of $y = x^2$ 2 units left and $\frac{3}{2}$ units down.

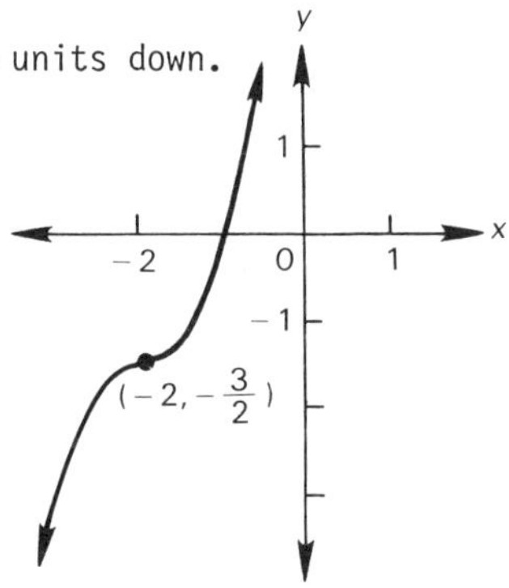

No asymptotes.

2) $f(x) = x^3$

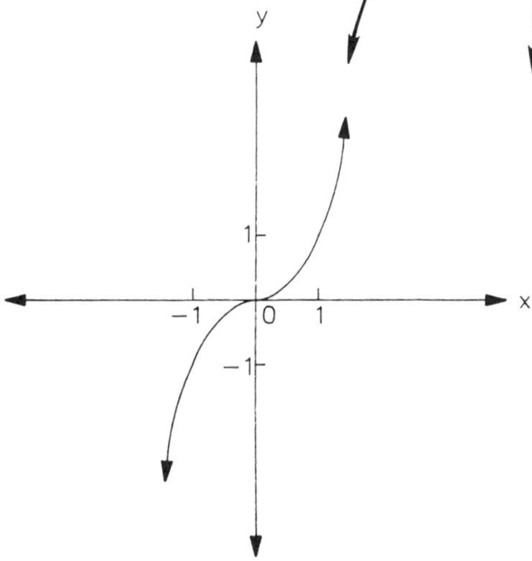

No asymptotes.

3) $y = f(x) = -\frac{1}{x - 2}$

Asymptotes: $x = 2$, $y = 0$.

Reflect $\frac{1}{x}$ through the x-axis and shift 2 units to the right.

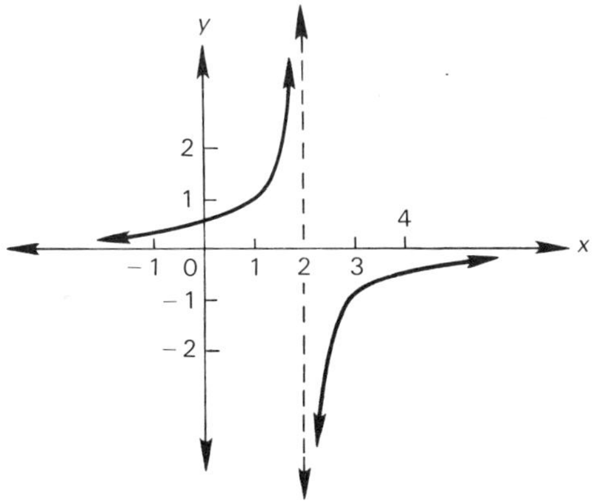

4) $y = f(x) = \dfrac{x - 1}{x^2 - x - 2} = \dfrac{x - 1}{(x - 2)(x + 1)}$

Asymptotes: $y = 0$, $x = 2$, $x = -1$.

Interval	$(-\infty, -1)$	$(-1, 1)$	$(1, 2)$	$(2, \infty)$
Sign of $x - 1$	-	-	+	+
Sign of $x - 2$	-	-	-	+
Sign of $x + 1$	-	+	+	+
Sign of $f(x)$	-	+	-	+

x	y
-2	$-\frac{3}{4}$
-1.1	-6.7
-0.9	6.6
0	$\frac{1}{2}$
1	0
1.9	-3.1
2.1	3.5
4	0.3

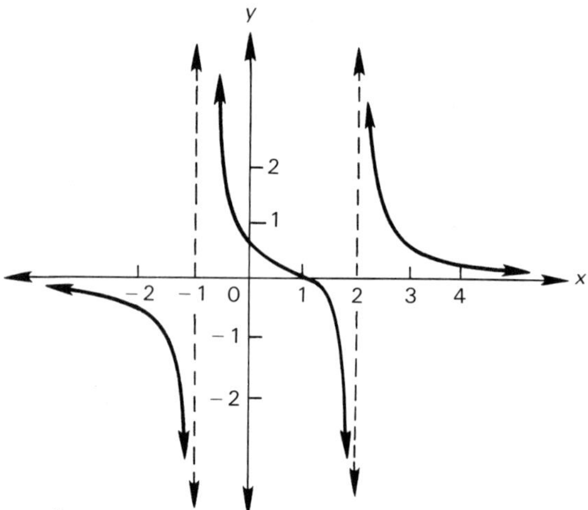

5) $y = x^3 - x^2 - 4x + 4 = x^2(x - 1) - 4(x - 1) = (x - 1)(x^2 - 4)$
$= (x - 1)(x - 2)(x + 2)$

No asymptotes. x-intercepts: $-2, 1, 2$

Interval	$(-\infty, -2)$	$(-2, 1)$	$(1, 2)$	$(2, \infty)$
Sign of $x - 1$	-	-	+	+
Sign of $x - 2$	-	-	-	+
Sign of $x + 2$	-	+	+	+
Sign of $f(x)$	-	+	-	+

5) continued

x	y
-3	-20
0	4
1.5	-0.125
3	10

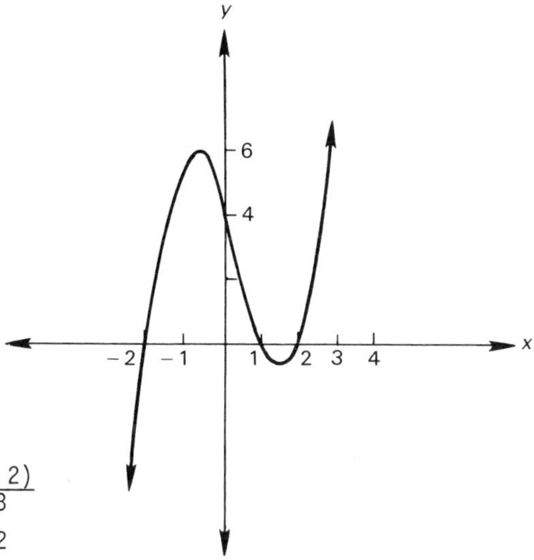

6) $f(x) = \dfrac{x^2 - 2x}{x + 3} = \dfrac{x(x - 2)}{x + 3}$

x-intercepts: -3, 0, 2

Interval	$(-\infty, -3)$	$(-3, 0)$	$(0, 2)$	$(2, \infty)$
Sign of x	-	-	+	+
Sign of (x - 2)	-	-	-	+
Sign of (x + 3)	-	+	+	+
Sign of f(x)	-	+	-	+

$f(x) < 0$ on $(-\infty, -3)$ and $(0, 2)$

$f(x) > 0$ on $(-3, 0)$ and $(2, \infty)$

7) $\dfrac{6}{x} = 2 + \dfrac{3}{x + 1}$

$$x(x + 1)\left(\frac{6}{x}\right) = x(x + 1)(2) + x(x + 1)\left(\frac{3}{x + 1}\right)$$

$$6x + 6 = 2x^2 + 2x + 3x$$

$$0 = 2x^2 - x - 6$$

$$= (2x + 3)(x - 2)$$

$2x + 3 = 0$ or $x - 2 = 0$

$x = -\dfrac{3}{2}$ $\qquad$ $x = 2$

8) $\frac{x}{2} - \frac{3x + 1}{3} > 2$

$6\left(\frac{x}{2}\right) - 6\left(\frac{3x + 1}{3}\right) > 6(2)$

$3x - 2(3x + 1) > 12$

$3x - 6x - 2 > 12$

$-3x > 14$

$x < -\frac{14}{3}$

9) Let x represent the larger piece. (10 - x) represents the smaller piece.

$\frac{10 - x}{x} = \frac{3}{4}$

$3x = 40 - 4x$

$7x = 40$

$x = 5\frac{5}{7}$

The larger piece is $5\frac{5}{7}$ feet.

10) Let x represent the time it would take Ellen to do the job alone.

$\frac{1}{x}$ is the part of the job she can do in 1 minute.

$\frac{1}{45}$ is the part of the job Dave does in 1 minute.

$\frac{1}{30}$ is the part both do together in 1 minute.

$\frac{1}{45} + \frac{1}{x} = \frac{1}{30}$

$90x\left(\frac{1}{45}\right) + 90x\left(\frac{1}{x}\right) = 90x\left(\frac{1}{30}\right)$

$2x + 90 = 3x$

$90 = x$

Ellen would need 90 minutes to do the job alone.

11. $z = \frac{kx}{y}$

$\frac{2}{3} = \frac{k(2)}{15}$

$\frac{15}{2} \cdot \frac{2}{3} = k$

$5 = k$

$z = \frac{5(4)}{10}$

$z = 2$

12) $V = kr^3$

$k = \frac{V}{r^3}$, so $\frac{V_1}{(r_1)^3} = \frac{V_2}{(r_2)^3}$

$\frac{36\pi}{3^3} = \frac{V_2}{6^3}$

$\frac{36\pi}{27} = \frac{V_2}{216}$

$\frac{4\pi}{3} = \frac{V_2}{216}$

$3V_2 = 864\pi$

$V_2 = 288\pi$ cubic inches

13) Long division of $5x^3 + 0x^2 - 3x + 6$ by $x - 2$:

$$\begin{array}{r l}
 & 5x^2 + 10x + 17 \\
x - 2 \,\big|\, & 5x^3 + 0x^2 - 3x + 6 \\
 & -(5x^3 - 10x^2) \\
 & \quad +10x^2 - 3x \\
 & \quad -(10x^2 - 20x) \\
 & \qquad 17x + 6 \\
 & \qquad -(17x - 34) \\
 & \qquad\qquad 40 = r
\end{array}$$

Check:

$$(x - 2)(5x^2 + 10x + 17) + 40 = 5x^3 + 10x^2 + 17x - 10x^2 - 20x - 34 + 40$$
$$= 5x^3 - 3x + 6$$

14)
$$\begin{array}{r|l}
-3 & 2 + 5 + 0 - 1 - 21 + 7 \\
 & \quad -6 + 3 - 9 + 30 - 27 \\
\hline
 & 2 - 1 + 3 - 10 + 9 \;|\; -20
\end{array}$$

Quotient: $2x^4 - x^3 + 3x^2 - 10x + 9$

Remainder: -20

15) $p(x) = 27x^4 - 36x^3 + 18x^2 - 4x + 1$

a)
$$\begin{array}{r|l}
\frac{1}{3} & 27 - 36 + 18 - 4 + 1 \\
 & \quad + 9 - 9 + 3 - \frac{1}{3} \\
\hline
 & 27 - 27 + 9 - 1 \;|\; \frac{2}{3}
\end{array}$$

Because $r = \frac{2}{3}$, $p\left(\frac{1}{3}\right) = \frac{2}{3}$.

b) $p\left(\frac{1}{3}\right) \neq 0$, therefore $x - \frac{1}{3}$ is not a factor of $p(x)$.

16)
$$\begin{array}{r|l}
2 & 1 - 4 + 7 - 12 + 12 \\
 & \quad +2 - 4 + 6 - 12 \\
\hline
 & 1 - 2 + 3 - 6 \;|\; 0
\end{array}$$

$r = 0$, therefore $(x - 2)$ is a factor of $p(x)$.

$$x^4 - 4x^3 + 7x^2 - 12x + 12 = (x - 2)(x^3 - 2x^2 + 3x - 6)$$
$$= (x - 2)[x^2(x - 2) + 3(x - 2)]$$
$$= (x - 2)(x - 2)(x^2 + 3) = (x - 2)^2(x^2 + 3)$$

17) $f(x) = x^4 + 5x^3 + 4x^2 - 3x + 9$

The leading coefficient is 1, so denominators are ± 1.

Possible rational factors (factors of 9): ± 1, ± 3, ± 9

Try $c = 1$

$$\begin{array}{r|rrrrr} 1 & 1 & +5 & +4 & -3 & +9 \\ & & +1 & +6 & +10 & +7 \\ \hline & 1 & +6 & +10 & +7 & \underline{16} \end{array}$$

Try $c = -1$

$$\begin{array}{r|rrrrr} -1 & 1 & +5 & +4 & -3 & +9 \\ & & -1 & -4 & +0 & +3 \\ \hline & 1 & +4 & +0 & -3 & \underline{12} \end{array}$$

Try $c = 3$

$$\begin{array}{r|rrrrr} 3 & 1 & +5 & +4 & -3 & +9 \\ & & +3 & +12 & +48 & +135 \\ \hline & 1 & +8 & +16 & +45 & \underline{144} \end{array}$$

Try $c = -3$

$$\begin{array}{r|rrrrr} -3 & 1 & +5 & +4 & -3 & +9 \\ & & -3 & -6 & +6 & -9 \\ \hline & 1 & +2 & -2 & +3 & \underline{0} \end{array}$$

$(x + 3)$ is a factor.

Continue by dividing the quotient.

Try $c = -3$ again.

$$\begin{array}{r|rrrr} -3 & 1 & +2 & -2 & +3 \\ & & -3 & +3 & -3 \\ \hline & 1 & -1 & +1 & \underline{0} \end{array}$$

$(x + 3)$ is a factor twice.

$x^4 - 5x^3 + 4x^2 - 3x + 9 = (x + 3)^2(x^2 - x + 1)$

18) $0 = p(x) = x^4 + 3x^3 - 3x^2 - 11x - 6$

$n = 4$, so there will be 4 roots.

First find any rational roots.

The leading coefficient is 1, so all denominators will be ± 1.

Possible rational factors (factors of 6): ± 1, ± 2, ± 3, ± 6.

Try $c = 1$

$$\begin{array}{r|rrrrr} 1 & 1 & +3 & -3 & -11 & -6 \\ & & +1 & +4 & +1 & -10 \\ \hline & 1 & +4 & +1 & -10 & \underline{-16} \end{array}$$

Try $c = -1$

$$\begin{array}{r|rrrrr} -1 & 1 & +3 & -3 & -11 & -6 \\ & & -1 & -2 & +5 & +6 \\ \hline & 1 & +2 & -5 & -6 & \underline{0} \end{array}$$

$(x + 1)$ is a factor.

Continue by dividing the quotient.

Try $c = -1$ again.

$$\begin{array}{r|rrrr} -1 & 1 & +2 & -5 & -6 \\ & & -1 & -1 & +6 \\ \hline & 1 & +1 & -6 & \underline{0} \end{array}$$

$(x + 1)$ is a factor twice.

18) continued

$$0 = x^4 + 3x^3 - 3x^2 - 11x - 6 = (x + 1)^2(x^2 + x - 6)$$
$$= (x + 1)^2(x + 3)(x - 2)$$

$x + 1 = 0$ or $x + 3 = 0$ or $x - 2 = 0$

$x = -1$ $\quad x = -3$ $\quad x = 2$

(double root)

19) $$\frac{x - 15}{x^2 - 25} = \frac{x - 15}{(x - 5)(x + 5)} = \frac{A}{x + 5} + \frac{B}{x - 5}$$

Multiply through by $(x + 5)(x - 5)$

$$x - 15 = A(x - 5) + B(x + 5)$$

To eliminate B, let $x = -5$.

$$(-5) - 15 = A(-5 - 5) + B(-5 + 5)$$
$$-20 = -10A$$
$$2 = A$$

To eliminate A, let $x = 5$.

$$(5) - 15 = A(5 - 5) + B(5 + 5)$$
$$-10 = 10B$$
$$-1 = B$$

$$\frac{x - 15}{x^2 - 25} = \frac{2}{x + 5} - \frac{1}{x - 5}$$

20) $$\frac{6x^2 - 2x + 2}{x^3 - 2x^2 - 5x + 6} = \frac{6x^2 - 2x + 2}{(x - 1)(x + 2)(x - 3)} = \frac{A}{x - 1} + \frac{B}{x + 2} + \frac{C}{x - 3}$$

Multiply through by $(x - 1)(x + 2)(x - 3)$

$$6x^2 - 2x + 2 = A(x + 2)(x - 3) + B(x - 1)(x - 3) + C(x - 1)(x + 2)$$

To eliminate B and C, let $x = 1$.

$$6(1)^2 - 2(1) + 2 = A(1 + 2)(1 - 3) + B(1 - 1)(1 - 3)$$
$$+ C(1 - 1)(1 + 2)$$
$$6 = -6A$$
$$-1 = A$$

To eliminate A and C, let $x = -2$.

$$6(-2)^2 - 2(-2) + 2 = A(-2 + 2)(-2 - 3) + B(-2 - 1)(-2 - 3)$$
$$+ C(-2 - 1)(-2 + 2)$$
$$30 = 15B$$
$$2 = B$$

20) continued

To eliminate A and B, let x = 3.

$$6(3)^2 - 2(3) + 2 = A(3 + 2)(3 - 3) + B(3 - 1)(3 - 3) + C(3 - 1)(3 + 2)$$

$$50 = 10C$$

$$5 = C$$

$$\frac{6x^2 - 2x + 2}{x^3 - 2x^2 - 5x + 6} = \frac{-1}{x - 1} + \frac{2}{x + 2} + \frac{5}{x - 3}$$

Sample Test Questions for Chapter 5

Part II

1) (d)

2) (b)

3) (a)

4) (c)

5) (b)

Interval	$(-\infty, -3)$	$(-3, -1)$	$(-1, 2)$	$(2, \infty)$
Sign of $(x + 3)$	-	+	+	+
Sign of $(x + 1)$	-	-	+	+
Sign of $(x - 2)$	-	-	-	+
Sign of $f(x)$	-	+	-	+

6) (e)

$$\frac{3x^2}{x^3 + x} = \frac{3x^2}{x(x^2 + 1)}$$

At no x-value will there be a zero denominator with a non-zero numerator.

7) (d)

8) (a)

$$c = \frac{a + 2b}{ab}$$

$$abc = 9 + 2b$$

$$abc - 2b = a$$

$$b(ac - 2) = a$$

$$b = \frac{a}{ac - 2}$$

9) (b)

$$\frac{9x + 14}{2x^2 - x - 6} - \frac{1}{x - 2} = \frac{2}{2x + 3}$$

$$\frac{9x + 14}{(2x+3)(x-2)} - \frac{1}{x - 2} = \frac{2}{2x + 3}$$

Multiply through by $(2x+3)(x-2)$

$$9x + 14 - 1(2x + 3) = 2(x - 2)$$

$$9x + 14 - 2x - 3 = 2x - 4$$

$$5x = -15$$

This is a linear equation. It will have one solution, $x = -3$ that checks in the given equation.

10) (a)

$s = kt^2$

$\frac{s}{t^2} = k$ so $\frac{s_2}{t_2^2} = \frac{s_1}{t_1^2}$

$\frac{s_2}{6^2} = \frac{64}{2^2}$

$\frac{s_2}{36} = \frac{64}{4}$

$\frac{s_2}{36} = 16$

$s_2 = 576$ feet

11) (d)

$a = \frac{kx}{y^2}$

$60 = \frac{k(3)}{\left(\frac{1}{2}\right)^2}$

$\frac{1}{4}(60) = 3k$

$15 = 3k$

$5 = k$

$z = \frac{5 \cdot 6}{4^2}$

$z = \frac{30}{16} = \frac{15}{8}$

12) (c)

$$\begin{array}{r|rrrr} -2 & 1 & +3 & -5 & +7 \\ & & -2 & -2 & +14 \\ \hline & 1 & +1 & -7 & \underline{21} \end{array}$$

13) (b)

Possible numerators (factors of 10): $\pm 1, \pm 2, \pm 5, \pm 10$

Possible denominators (factors of 2): $\pm 1, \pm 2$

Possible rational roots: $\pm 1, \pm \frac{1}{2}, \pm 2, \pm 5, \pm \frac{5}{2}, \pm 10$

14) (c)

$$\frac{6x^2 + x - 37}{(x - 3)(x + 2)(x - 1)} = \frac{A}{x - 3} + \frac{B}{x + 2} + \frac{C}{x - 1}$$

Multiply through by $(x - 3)(x + 2)(x - 1)$

$$6x^2 + x - 37 = A(x + 2)(x - 1) + B(x - 3)(x - 1) + C(x - 3)(x + 2)$$

To eliminate B and C, let $x = 3$.

$$6(3)^2 + 3 - 37 = A(3 + 2)(3 - 1) + B(3 - 3)(3 - 1) + C(3 - 3)(3 + 2)$$

$$20 = 10A$$

$$2 = A$$

15) (a)

If 2 is a double root, $(x - 2)^2$ is a factor.

$(x - 2)^2 = x^2 - 4x + 2$ Divide to find the other factors.

$$
\begin{array}{r|l}
 & x^2 - 2x + 2 \\
\hline
x^2 - 4x + 4 & x^4 - 6x^3 + 14x^2 - 16x + 8 \\
 & -\underline{(x^4 - 4x^3 + 4x^2)} \\
 & \quad 2x^3 + 10x^2 - 16x \\
 & \quad -\underline{(-2x^3 + 8x^2 + 8x)} \\
 & \qquad 2x^2 - 8x + 8 \\
 & \qquad -\underline{(2x^2 - 8x + 8)} \\
 & \qquad\qquad 0
\end{array}
$$

Solve $0 = x^2 - 2x + 2$ by formula

$$x = \frac{-(-2) \pm \sqrt{(-2)^2 - 4(1)(2)}}{2(1)} = \frac{2 \pm \sqrt{-4}}{2} = \frac{2 \pm 2i}{2} = 1 \pm i$$

CHAPTER 6

THE ALGEBRA OF FUNCTIONS

Section 6.1 Graphing Some Special Radical Functions

1) $f(x) = \sqrt{x}$

$g(x) = \sqrt{x - 1}$, shift f(x) 1 unit right.

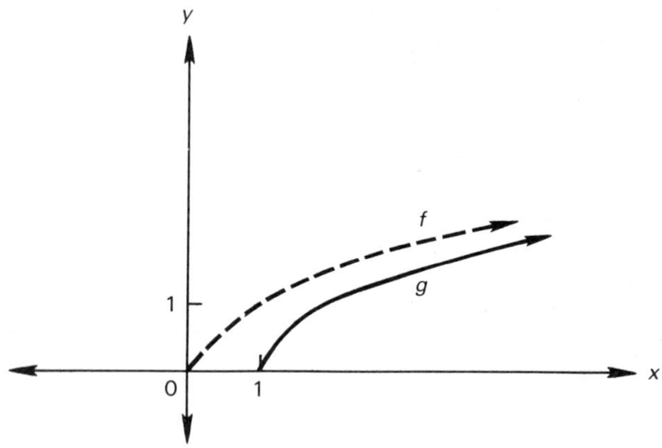

5) $f(x) = \sqrt[3]{x}$

$g(x) = \sqrt[3]{x + 2}$, shift f(x) 2 units left.

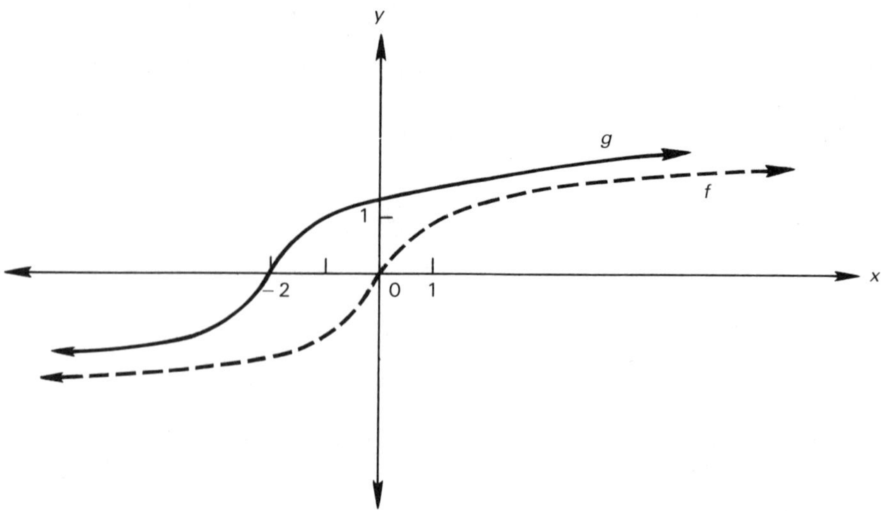

9) $f(x) = \sqrt{x + 2}$

$x + 2 \geq 0$

$x \geq -2$

No asymptotes.

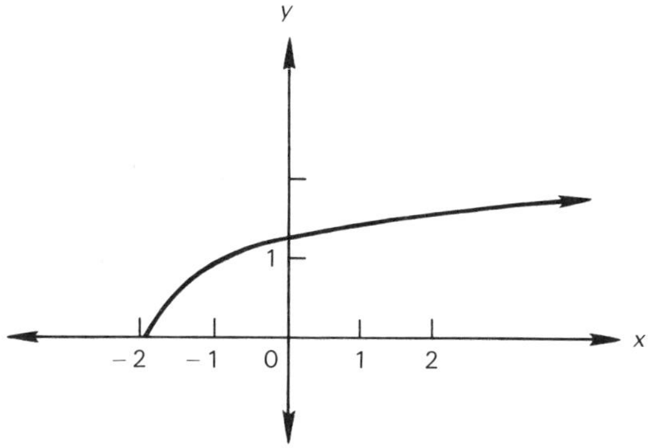

13) $f(x) = \sqrt{-x}$

$-x \geq 0$

$x \leq 0$

No asymptotes.

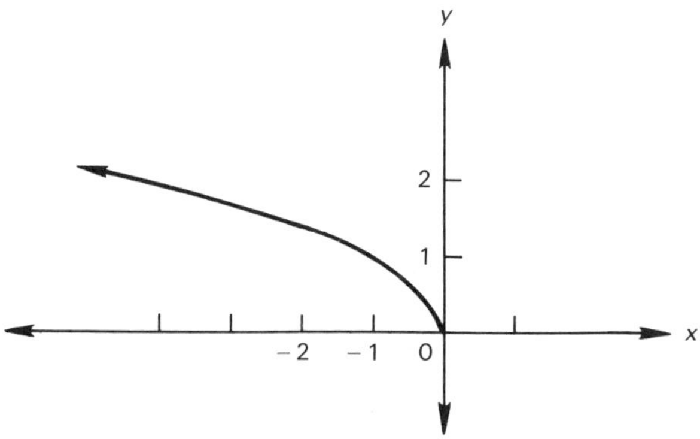

17) $f(x) = -x^{\frac{1}{3}}$

Domain: all reals.

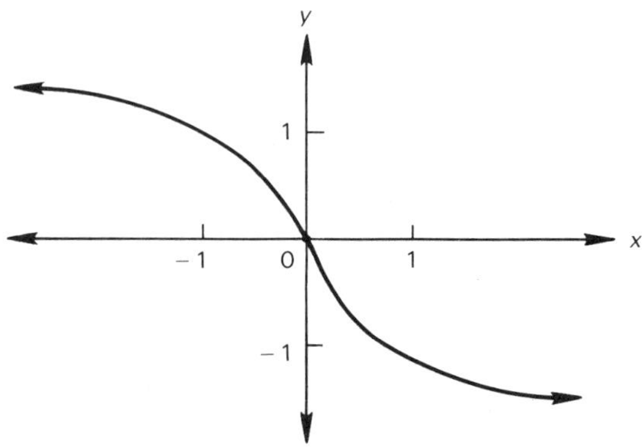

21) $f(x) = \frac{1}{\sqrt[3]{x}}$

a) $f(-x) = \frac{1}{\sqrt[3]{-x}} = \frac{1}{-\sqrt[3]{x}} = -\frac{1}{\sqrt[3]{x}} = -f(x)$

b) All $x \neq 0$

c)

x	$\frac{1}{27}$	$\frac{1}{8}$	1	8
y	3	2	1	$\frac{1}{2}$

d) $x = 0$; $y = 0$

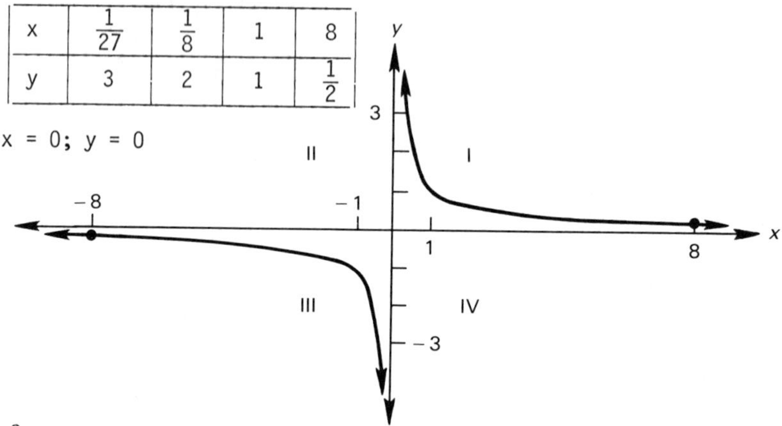

25) $y = x^3$

Interchange variables: $x = y^3$

Solve for y: $\sqrt[3]{x} = y$

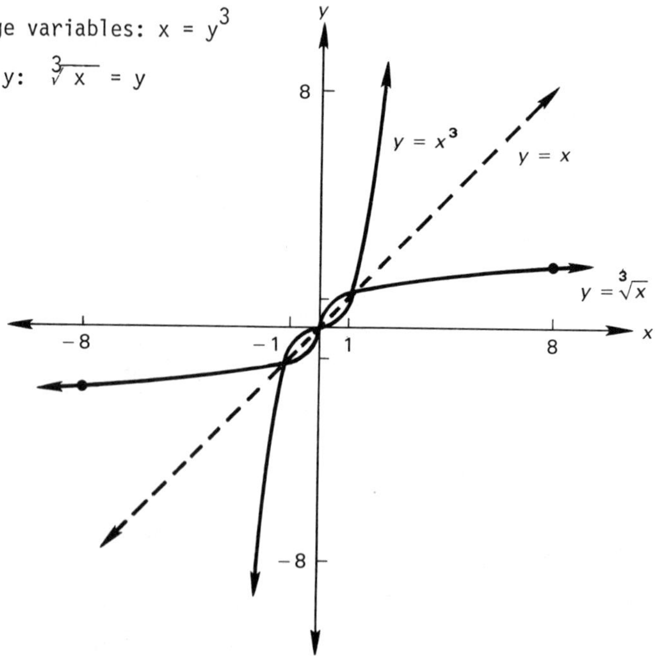

29) $f(x) = \sqrt{x}$ $f(4 + h) = \sqrt{4 + h}$ $f(4) = \sqrt{4} = 2$

$$\frac{f(4 + h) - f(4)}{h} = \frac{\sqrt{4 + h} - 2}{h} = \frac{\sqrt{4 + h} - 2}{h} \cdot \frac{\sqrt{4 + h} + 2}{\sqrt{4 + h} + 2}$$

$$= \frac{4 + h - 4}{h(\sqrt{4 + h} + 2)} = \frac{\not h}{\not h(\sqrt{4 + h} + 2)} = \frac{1}{\sqrt{4 + h} + 2}$$

33) Consider any point on the curvy $y = \frac{1}{\sqrt{x}}$. The point has coordinates (x, y). Substitute $\frac{1}{\sqrt{x}}$ for y to get coordinates $\left(x, \frac{1}{\sqrt{x}}\right)$. The distance from the origin (0, 0) may be found with the distance formula.

$$d(x) = \sqrt{(x - 0)^2 + \left(\frac{1}{\sqrt{x}} - 0\right)^2} = \sqrt{x^2 + \frac{1}{x}}$$

Section 6.2 Radical Equations

1) $f(x) = \sqrt{x - 1} - 4$

Find the x-intercepts.

$$0 = \sqrt{x - 1} - 4$$
$$4 = \sqrt{x - 1}$$
$$4^2 = (\sqrt{x - 1})^2$$
$$16 = x - 1$$
$$17 = x$$

Find the domain: $x - 1 \geq 0$

$$x \geq 1$$

5) $f(x) = \sqrt{x^2 - 5x - 6}$

Find the x-intercepts:

$$0 = \sqrt{x^2 - 5x - 6}$$
$$0^2 = \left(\sqrt{x^2 - 5x - 6}\right)^2$$
$$0 = x^2 - 5x - 6$$
$$0 = (x - 6)(x + 1)$$

$x - 6 = 0$ or $x + 1 = 0$

$x = 6$ $x = -1$

5) continued

Find the domain: $x^2 - 5x - 6 \geq 0$

$(x - 6)(x + 1) \geq 0$

Find the intervals.

$x - 6 = 0$ $\quad$ $x + 1 = 0$

$x = 6$ $\quad$ $x = -1$

Interval	$(-\infty, -1)$	$(-1, 6)$	$(6, \infty)$
Sign of $x - 6$	-	-	+
Sign of $x + 1$	-	+	+
Sign of $(x - 6)(x + 1)$	+	-	+

$x^2 - 5x - 6 \geq 0$ when $x \leq -1$ or $x \geq 6$

9) $(3x + 1)^{1/2} = (2x + 6)^{1/2}$

$$\left((3x + 1)^{\frac{1}{2}}\right)^2 = [(2x + 6)^{1/2}]^2$$

$3x + 1 = 2x + 6$

$x = 5$

Check:

$(3 \cdot 5 + 1)^{\frac{1}{2}} = (2 \cdot 5 + 6)^{\frac{1}{2}}$

$16^{\frac{1}{2}} = 16^{\frac{1}{2}}$

$4 = 4$

true

$x = 5$

13) $\sqrt{x^2 + \frac{1}{2}} = \frac{1}{\sqrt{3}}$

$$\left(\sqrt{x^2 + \frac{1}{2}}\right)^2 = \left(\frac{1}{\sqrt{3}}\right)^2$$

$x^2 + \frac{1}{2} = \frac{1}{3}$

$x^2 = -\frac{1}{6}$

No solutions. Any real number squared is nonnegative.

17) $\dfrac{1}{\sqrt{2x - 1}} = \dfrac{3}{\sqrt{5 - 3x}}$

$$\left(\frac{1}{\sqrt{2x - 1}}\right)^2 = \left(\frac{3}{\sqrt{5 - 3x}}\right)^2$$

$$\frac{1}{2x - 1} = \frac{9}{5 - 3x}$$

$$1\,(5 - 3x) = 9(2x - 1)$$

$$5 - 3x = 18x - 9$$

$$14 = 21x$$

$$\frac{14}{21} = x$$

$$\frac{2}{3} = x$$

Check:

$$\frac{1}{\sqrt{2 \cdot \frac{2}{3} - 1}} = \frac{3}{\sqrt{5 - 3 \cdot \frac{2}{3}}}$$

$$\frac{1}{\sqrt{\frac{4}{3} - 1}} = \frac{3}{\sqrt{5 - 2}}$$

$$\frac{1}{\sqrt{\frac{1}{3}}} = \frac{3}{\sqrt{3}}$$

$$\frac{1}{\frac{1}{\sqrt{3}}} = \frac{3\sqrt{3}}{3}$$

$$\sqrt{3} = \sqrt{3}$$

True

$x = \frac{2}{3}$

21) $\sqrt{x} + \sqrt{x - 5} = 5$

$$\sqrt{x - 5} = 5 - \sqrt{x}$$

$$(\sqrt{x - 5})^2 = (5 - \sqrt{x})^2$$

$$x - 5 = 25 - 10\sqrt{x} + x$$

$$10\sqrt{x} = 30$$

$$\sqrt{x} = 3$$

$$x = 9$$

Check:

$$\sqrt{9} + \sqrt{9 - 5} = 5$$

$$\sqrt{9} + \sqrt{4} = 5$$

$$3 + 2 = 5$$

true

$x = 9$

25) $\sqrt{4x + 1} + \sqrt{x + 7} = 6$

$$\sqrt{4x + 1} - 6 = -\sqrt{x + 7}$$

$$(\sqrt{4x + 1} - 6)^2 = (-\sqrt{x + 7})^2$$

$$4x + 1 - 12\sqrt{4x + 1} + 36 = x + 7$$

$$3x + 30 = 12\sqrt{4x + 1}$$

$$x + 10 = 4\sqrt{4x + 1}$$

$$(x + 10)^2 = (4\sqrt{4x + 1})^2$$

$$x^2 + 20x + 100 = 16(4x + 1)$$

$$x^2 + 20x + 100 = 64x + 16$$

$$x^2 - 44x + 84 = 0$$

Check $x = 42$:

$$\sqrt{4 \cdot 42 + 1} + \sqrt{42 + 7} = 6$$

$$\sqrt{169} + \sqrt{49} = 6$$

$$13 + 7 = 6$$

False

Check $x = 2$:

$$\sqrt{4 \cdot 2 + 1} + \sqrt{2 + 7} = 6$$

$$\sqrt{9} + \sqrt{9} = 6$$

$$3 + 3 = 6$$

true

$x = 2$

25) continued

$(x - 42)(x - 2) = 0$

$x - 42 = 0$ or $x - 2 = 0$

$x = 42$ or $x = 2$

29) $x = 8 - 2\sqrt{x}$

$2\sqrt{x} = 8 - x$

$4x = 64 - 16x + x^2$

$0 = x^2 - 20x + 64$

$0 = (x - 4)(x - 16)$

$x - 4 = 0$ or $x - 16 = 0$

$x = 4$ or $x = 16$

Check $x = 4$

$4 = 8 - 2\sqrt{4}$

$4 = 8 - 2 \cdot 2$

$4 = 4$

True

$x = 4$

Check $x = 16$

$16 = 8 - 2\sqrt{16}$

$16 = 8 - 2 \cdot 4$

$16 = 0$

False

33) $4x^{2/3} - 12x^{1/3} + 9 = 0$

Let $u = x^{\frac{1}{3}}$, then $u^2 = x^{2/3}$.

$4u^2 - 12u + 9 = 0$

$(2u - 3)^2 = 0$

$2u - 3 = 0$

$2u = 3$

$u = \frac{3}{2}$

Substitute this value back into $u = x^{\frac{1}{3}}$.

$\frac{3}{2} = x^{\frac{1}{3}}$

$\left(\frac{3}{2}\right)^3 = \left(x^{\frac{1}{3}}\right)^3$

$\frac{27}{8} = x$

Check:

$4\left(\frac{27}{8}\right)^{\frac{2}{3}} - 12\left(\frac{27}{8}\right)^{\frac{1}{3}} + 9 = 0$

$4\left(\frac{3}{2}\right)^2 - 12\left(\frac{3}{2}\right) + 9 = 0$

$4\left(\frac{9}{4}\right) - 12\left(\frac{3}{2}\right) + 9 = 0$

$9 - 18 + 9 = 0$

true

$\frac{27}{8} = x$

37) $x^{-\frac{3}{2}} - \frac{1}{9}x^{-\frac{1}{2}} = 0$

$\frac{1}{x^{3/2}} - \frac{1}{9x^{\frac{1}{2}}} = 0$

Multiply by $9x^{3/2}$:

$9 - x = 0$

$9 = x$

Check:

$9^{-\frac{3}{2}} - \frac{1}{9}\left(9^{-\frac{1}{2}}\right) = 0$

$\frac{1}{9^{\frac{3}{2}}} - \frac{1}{9 \cdot 9^{\frac{1}{2}}} = 0$

$\frac{1}{3^3} - \frac{1}{9 \cdot 3} = 0$

$\frac{1}{27} - \frac{1}{27} = 0$

True

$x = 9$

41) $r = \frac{1}{2}\sqrt{\frac{A}{\pi}}$

$2r = \sqrt{\frac{A}{\pi}}$

$(2r)^2 = \left(\sqrt{\frac{A}{\pi}}\right)^2$

$4r^2 = \frac{A}{\pi}$

$4\pi r^2 = A$

When $r = 2$:

$4\pi(2)^2 = A$

$16\pi = A$

The area is 16π centimeters.

45) $V = \pi r^2 h$ is the formula for volume of a right circular cylinder.

$V = \pi r^2 h = 5$

$r^2 = \frac{5}{\pi h}$

$r = \sqrt{\frac{5}{\pi h}}$

The surface area is found in three steps.

First, calculate the area of the two circular bases, using $A = \pi r^2$.

$$A = \pi\left(\sqrt{\frac{5}{\pi h}}\right)^2 = \pi\left(\frac{5}{\pi h}\right) = \frac{5}{h}$$

There are two such areas, so $2A = 2\left(\frac{5}{h}\right) = \frac{10}{h}$

45) continued

<u>Second</u>, calculate the area of the circular side of the cylinder that is between the two circular bases. This side is actually a rectangle one of whose dimensions is the height h, and the other dimension is the circumference of the base, $C = 2\pi r$. (You can see that this side is a rectangle by imagining that the cylinder has been unrolled onto a flat surface.)

$$A = h(2\pi r)$$

$$A = 2\pi h\left(\sqrt{\frac{5}{\pi h}}\right)$$

$$A = 2\pi h \cdot \frac{\sqrt{5}}{\sqrt{\pi h}} \cdot \frac{\sqrt{\pi h}}{\sqrt{\pi h}}$$

$$A = 2\pi h \frac{\sqrt{5\pi h}}{\pi h}$$

$$A = 2\sqrt{5\pi h}$$

<u>Third</u>, add the areas found in steps one and two.

$$S = \frac{10}{h} + 2\sqrt{5\pi h}$$

49) $y = \sqrt{x}$

$y = \frac{1}{2}x$

Set the two right sides equal to each other.

$$\sqrt{x} = \frac{1}{2}x$$

$$(\sqrt{x})^2 = \left(\frac{1}{2}x\right)^2$$

$$x = \frac{1}{4}x^2$$

$$\frac{1}{4}x^2 - x = 0$$

$$x\left(\frac{1}{4}x - 1\right) = 0$$

$$x = 0 \quad \text{or} \quad \frac{1}{4}x - 1 = 0$$

$$\frac{1}{4}x = 1$$

$$x = 4$$

Substitute into the second equation to find y.

When $x = 0$, $y = \frac{1}{2}(0) = 0$

When $x = 4$, $y = \frac{1}{2}(4) = 2$.

49) continued

You can check these coordinates in the first equation.

The graphs will intersect at (0, 0) and (4, 2).

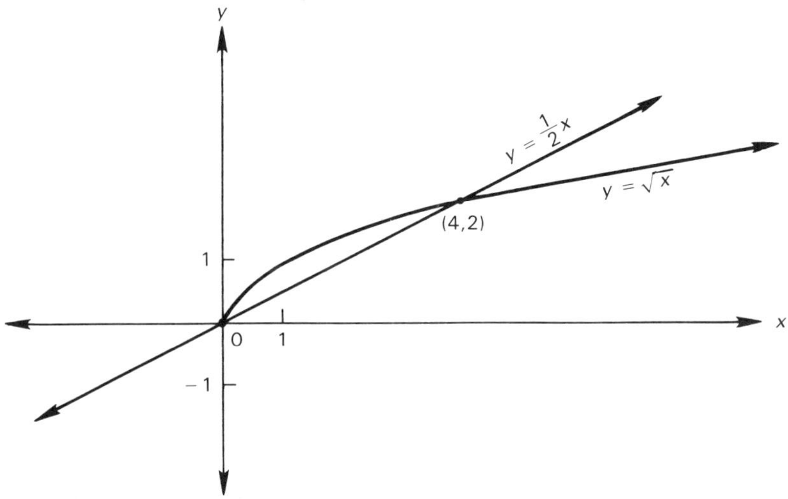

53) $\frac{1}{\sqrt{xy}}(xf + y) = f$

$xf + y = f\sqrt{xy}$

$y = f\sqrt{xy} - fx$

$y = f(\sqrt{xy} - x)$

$\frac{y}{\sqrt{xy} - x} = f$

Section 6.3 Determining the Signs of Radical Functions

1) $f(x) = (x + 4)\sqrt{x - 2}$

a) Find the domain.

$(x + 4)$ allows any real number for x

$\sqrt{x - 2}$ requires $x - 2 \geq 0$, $x \geq 2$

Domain: $x \geq 2$

1) continued

b) Determine the signs.

On the interval $x > 2$, $(x + 4) > 0$ always.

$\sqrt{x - 2} > 0$ always

$f(x) > 0$ for all $x > 2$.

c) Find the roots.

$(x + 4)\sqrt{x - 2} = 0$

$x + 4 = 0$ or $x - 2 = 0$

$\longrightarrow$ $x = -4$ $x = 2$

This x-value is <u>not</u> in the domain. 2 is the only root.

5) $f(x) = \dfrac{9 + x}{9\sqrt{x}}$

a) Find the domain.

$9 + x$ allows any real number for x.

$9\sqrt{x}$ requires $x > 0$. and its reciprocal requires $x > 0$.

Domain: $x > 0$

b) Determine the signs.

On the interval $x > 0$, both numerator and denominator will always be positive.

$f(x) > 0$ for $x > 0$.

c) Find the roots.

$\dfrac{9 + x}{9\sqrt{x}} = 0$ only when $9 + x = 0$

$x = -9$

This x-value is <u>not</u> in the domain.

No roots.

9) $f(x) = \dfrac{\sqrt{9 - x^2}}{x}$

a) Find the domain.

The denominator x requires $x \neq 0$.

$\sqrt{9 - x^2}$ requires $9 - x^2 \geq 0$, $9 \geq x^2$, $-3 \leq x \leq 3$

Domain: $-3 \leq x \leq 3$, $x \neq 0$

b) Determine the signs.

$\sqrt{9 - x^2}$ is always positive, so the sign of f is determined by the denominator x.

$f(x) < 0$ for $-3 < x < 0$.

$f(x) > 0$ for $0 < x < 3$.

9) continued

c) Find the roots.

$$\frac{\sqrt{9 - x^2}}{x} = 0 \text{ only when } \sqrt{9 - x^2} = 0$$

$$9 - x^2 = 0$$

$$9 = x^2$$

$$\pm 3 = x$$

13) $\frac{1}{2}(4 - x^2)^{-\frac{1}{2}}(-2x) = \frac{-2x}{2\sqrt{4 - x^2}} = -\frac{x}{\sqrt{4 - x^2}}$

17) $\frac{x}{(5x - 6)^{4/5}} + (5x - 6)^{\frac{1}{5}} = \frac{x + (5x - 6)^{\frac{1}{5}}(5x - 6)^{4/5}}{(5x - 6)^{4/5}}$

$$= \frac{x + (5x - 6)^{5/5}}{(5x - 6)^{4/5}} = \frac{x + 5x - 6}{(5x - 5)^{4/5}}$$

$$= \frac{6x - 6}{(5x - 6)^{4/5}} = \frac{6(x - 1)}{(5x - 6)^{4/5}}$$

21) $f(x) = x^{1/2} + \frac{1}{2}x^{-\frac{1}{2}}(x - 9) = \sqrt{x} + \frac{x - 9}{2\sqrt{x}}$

$$= \frac{\sqrt{x}(2\sqrt{x}) + (x - 9)}{2\sqrt{x}} = \frac{2x + x - 9}{2\sqrt{x}} = \frac{3x - 9}{2\sqrt{x}} = \frac{3(x - 3)}{2\sqrt{x}}$$

Determine the signs.

The denominator $2\sqrt{x}$ requires $x > 0$.

The denominator is always positive.

$(x - 3)$ will be negative on $(0, 3)$ and positive on $(3, \infty)$.

$f(x) < 0$ when $0 < x < 3$.

$f(x) > 0$ when $3 < x$.

25) $y = f(x) = \sqrt{x^2 - 9}$

$\sqrt{x^2 - 9}$ requires $x^2 - 9 \geq 0$, $x^2 \geq 9$, $x \leq -3$ or $x \geq 3$.

Domain: $x \leq -3$ or $x \geq 3$ Range: $y \geq 0$

$f(x) = f(-x)$, so the graph is symmetric around the y-axis.

If both sides of $y = \sqrt{x^2 - 9}$ are squared, the result is

25) continued

$y^2 = x^2 - 9$ or $\frac{x^2}{9} - \frac{y^2}{9} = 1$. This is a hyperbola, but the original function has only positive y-values.

x	y
3	0
4	$\sqrt{7}$
5	4

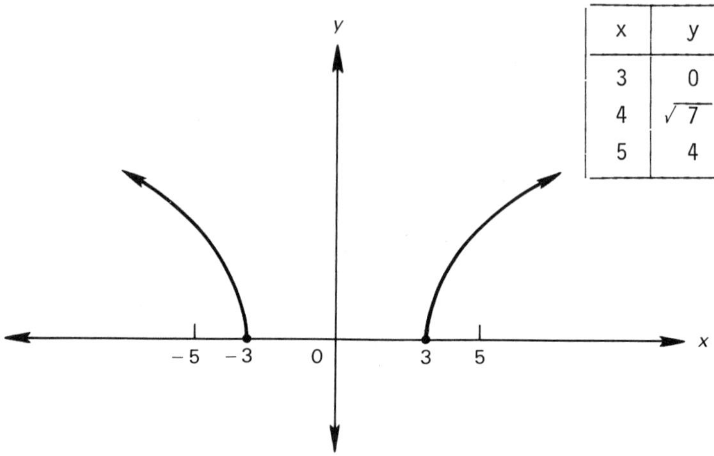

29) $y = f(x) = (x - 3)\sqrt{x}$

$(x - 3)$ allows any real number for x.

$\sqrt{x}$ requires $x \geq 0$

Domain: $x \geq 0$

$f(x) < 0$ when $0 < x < 3$.

$f(x) > 0$ when $x > 3$.

$f(x) = 0$ when $x = 3$ or $x = 0$.

x	y
0	0
1	-2
2	$-\sqrt{2}$
3	0
4	2

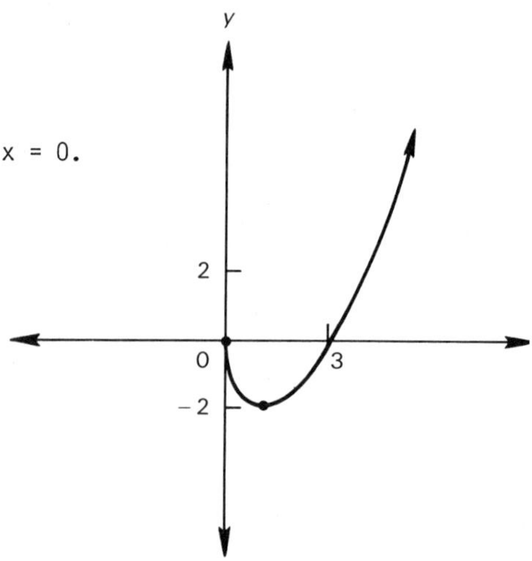

Section 6.4 Combining Functions

1) $f(x) = 2x - 3 \qquad g(x) = 3x + 2$

a) $f(1) = 2 \cdot 1 - 3 = -1 \qquad g(1) = 3 \cdot 1 + 2 = -5 \qquad f(1) + g(1) = -1 + 5 = 4$

b) $(f + g)(x) = (2x - 3) + (3x + 2) = 5x - 1$

Domain of $f(x) = 2x - 3$: all reals

Domain of $g(x) = 3x + 2$: all reals

Domain $(f + g)(x)$: all reals

c) $(f + g)(1) = 5 \cdot 1 - 1 = 4$

5) a) $g(0) = 3 \cdot 0 + 2 = 2$

$f(g(0)) = f(2) = 2 \cdot 2 - 3 = 1$

b) $(f \circ g)(x) = f(g(x)) = f(3x + 2) = 2(3x + 2) - 3 = 6x + 1$

Domain of $g(x) = 3x + 2$: all reals

Domain of $(f \circ g)(x)$: all x such that $g(x)$ is a real number, that is, all real numbers.

c) $(f \circ g)(0) = 6 \cdot 0 + 1 = 1$

9) $f(x) = x^3 - 1 \qquad g(x) = \frac{1}{x}$

a) $(f + g)(x) = (x^3 - 1) + \frac{1}{x} = x^3 + \frac{1}{x} - 1$

Domain of $f(x) = x^3 - 1$: all reals

Domain of $g(x) = \frac{1}{x}$: all $x \neq 0$

Domain of $(f + g)(x)$: all $x \neq 0$

b) $\frac{f}{g}(x) = \dfrac{x^3 - 1}{\frac{1}{x}} = x(x^3 - 1) = x^4 - x$

Domain of $\frac{f}{g}(x)$: all $x \neq 0$

c) $(f \circ g)(x) = f(g(x)) = f\left(\frac{1}{x}\right) = \left(\frac{1}{x}\right)^3 - 1 = \frac{1}{x^3} - 1$

Domain of $(f \circ g)(x)$: all $x \neq 0$

13) $f(x) = -2x + 5 \qquad g(x) = 4x - 1$

a) $(g - f)(x) = (4x - 1) - (-2x + 5) = 6x - 6$

Domain of f(x): all reals

Domain of g(x): all reals

Domain of (g - f)(x): all reals

b) $(g \cdot f)(x) = (4x - 1)(-2x + 5) = -8x^2 + 22x - 5$

Domain of $(g \cdot f)(x)$: all reals

c) $(g \circ f)(x) = g(f(x)) = g(-2x + 5) = 4(-2x + 5) - 1 = -8x + 19$

Domain of $(g \circ f)(x)$: all reals

17) $f(x) = x^2 \qquad g(x) = x - 1$

$(f \circ g)(x) = f(g(x)) = f(x - 1) = (x - 1)^2$

$(g \circ f)(x) = g(f(x)) = g(x^2) = x^2 - 1$

21) $f(x) = \sqrt{x + 1} \qquad g(x) = x^4 - 1$

$(f \circ g)(x) = f(g(x)) = f(x^4 - 1) = \sqrt{(x^4 - 1) + 1} = \sqrt{x^4} = x^2$

$(g \circ f)(x) = g(f(x)) = g(\sqrt{x + 1}) = (\sqrt{x + 1})^4 - 1 = (x + 1)^2 - 1$

$= x^2 + 2x$

25) $f(x) = \frac{1}{x} \qquad g(x) = 2x - 1 \qquad h(x) = x^{\frac{1}{3}}$

a) $(f \circ g \circ h)(x) = f(g(h(x))) = f\left(g\left(x^{\frac{1}{3}}\right)\right) = f\left(2\left(x^{\frac{1}{3}}\right) - 1\right)$

$= \frac{1}{2\sqrt[3]{x} - 1}$

b) $(g \circ f \circ h)(x) = g(f(h(x))) = g\left(f\left(x^{\frac{1}{3}}\right)\right) = g\left(\frac{1}{\sqrt[3]{x}}\right)$

$= 2\left(\frac{1}{\sqrt[3]{x}}\right) - 1 = \frac{2}{\sqrt[3]{x}} - 1$

c) $(h \circ f \circ g)(x) = h(f(g(x))) = h(f(2x - 1)) = h\left(\frac{1}{2x - 1}\right)$

$= \left(\frac{1}{2x - 1}\right)^{\frac{1}{3}} = \frac{1}{\sqrt[3]{2x - 1}}$

29) $f(x) = x^3$

$f(g(x)) = x$

$(g(x))^3 = x$

$g(x) = x^{\frac{1}{3}} = \sqrt[3]{x}$

Check:

$(f \circ g)(x) = f(g(x)) = f(\sqrt[3]{x})$

$= (\sqrt[3]{x})^3 = x$

$(g \circ f)(x) = g(f(x)) = g(x^3)$

$= \sqrt[3]{x^3} = x$

Section 6.5 Decomposition of Composite Functions (Other answers possible for numbers 1-21.)

1) $h(x) = (3x + 1)^2$

The outer function squares, so let $f(x) = x^2$, then $g(x) = 3x + 1$.

Check:

$(f \circ g)(x) = f(g(x)) = f(3x + 1) = (3x + 1)^2$

5) $h(x) = \left(\frac{x + 1}{x - 1}\right)^2$

The outer function squares, so let $f(x) = x^2$, then $g(x) = \frac{x + 1}{x - 1}$.

Check:

$(f \circ g)(x) = f(g(x)) = f\left(\frac{x + 1}{x - 1}\right) = \left(\frac{x + 1}{x - 1}\right)^2$

9) $h(x) = \sqrt{\frac{x}{x - 1}}$

The outer function takes the square root, so let $f(x) = \sqrt{x}$, then $g(x) = \frac{x}{x - 1}$.

Check:

$(f \circ g)(x) = f\left(\frac{x}{x - 1}\right) = \sqrt{\frac{x}{x - 1}}$

13) $h(x) = \frac{2}{\sqrt{4 - x^2}}$

The directions require the inner function to be either a polynomial or a rational function. If we make $g(x) = 4 - x^2$, then the outer function has to take the square root and put the expression in the denominator under 2. $f(x) = \frac{2}{\sqrt{x}}$

Check:

$(f \circ g)(x) = f(4 - x^2) = \frac{2}{\sqrt{4 - x^2}}$

17) $k(x) = \sqrt{\left(\frac{x}{x + 1}\right)^5}$

The outer function takes the square root, so let $h(x) = \sqrt{x}$.
The middle function raises to the fifth power, so let $g(x) = x^5$.

Then $f(x) = \frac{x}{x + 1}$.

Check:

$$(h \circ g \circ f)(x) = h(g(f(x))) = h\left(g\left(\frac{x}{x + 1}\right)\right) = h\left(\left(\frac{x}{x + 1}\right)^5\right)$$
$$= \sqrt{\left(\frac{x}{x + 1}\right)^5}$$

21) $k(x) = -\sqrt{(x^2 - 4x + 7)^3}$

The outer function takes the negative of the square root, so let $h(x) = -\sqrt{x}$.

The middle function raises to the third power, so let $g(x) = x^3$.

Then $f(x) = x^2 - 4x + 7$.

Check:

$$(h \circ g \circ f)(x) = h(g(f(x))) = h(g(x^2 - 4x + 7))$$
$$= h[(x^2 - 4x + 7)^3] = -\sqrt{(x^2 - 4x + 7)^3}$$

25) $(f \circ f)(x) = ((x + 1)^2 + 1)^2$

The outer and inner functions must be the same.

The outer function adds 1, then squares, so try $f(x) = (x + 1)^2$.

$(f \circ f)(x) = f\left((x + 1)^2\right) = ((x + 1)^2 + 1)^2$

Section 6.6 Inverse Functions

1) First graph $f(x) = (x - 1)^2$.

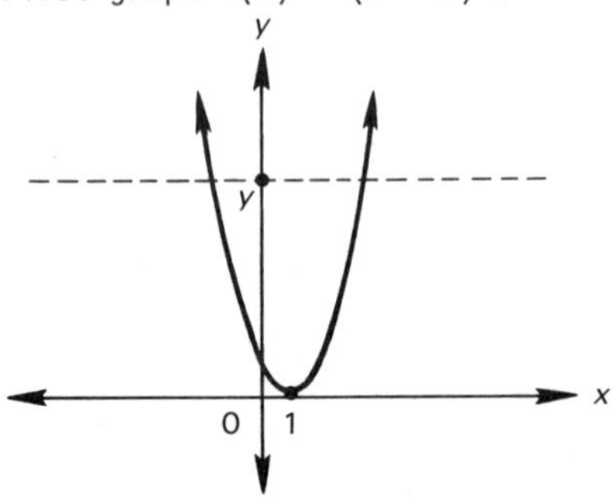

A horizontal line strikes the curve twice.
Not one-to-one.

5) First graph $f(x) = \sqrt{x} + 3$.

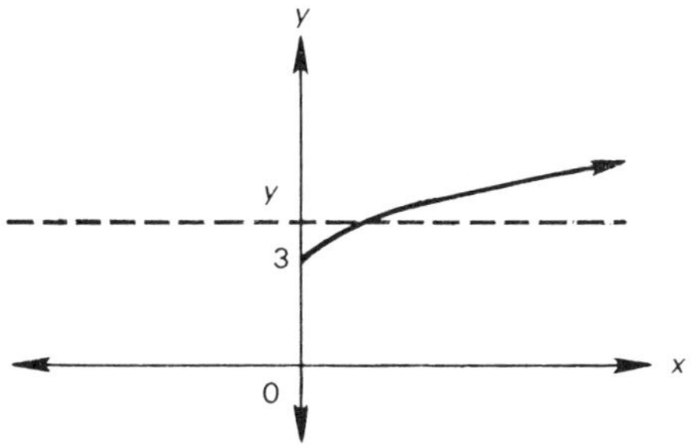

A horizontal through any range value y will intersect the curve only once. One-to-one.

9) $f(x) = (x + 1)^3$ $\qquad g(x) = \sqrt[3]{x} - 1$

$(f \circ g)(x) = f(g(x)) = f(\sqrt[3]{x} - 1) = [(\sqrt[3]{x} - 1) + 1]^3 = [\sqrt[3]{x}]^3 = x$

$(g \circ f)(x) = g(f(x)) = g((x + 1)^3) = \sqrt[3]{(x + 1)^3} - 1 = x + 1 - 1 = x$

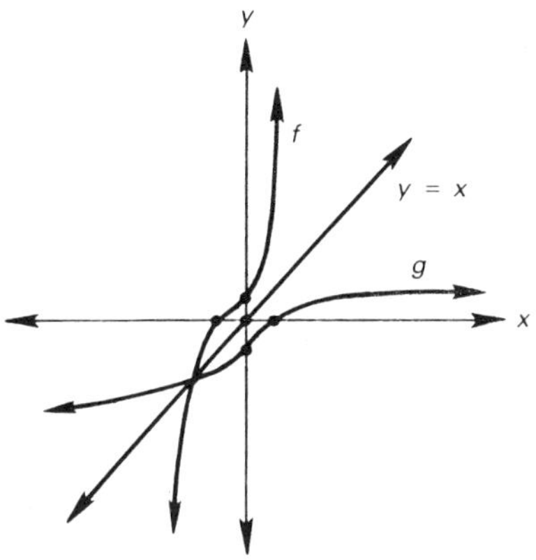

13) $y = f(x) = (x - 5)^3$

Let $y = g(x)$

$x = (y - 5)^3$

$\sqrt[3]{x} = y - 5$

$\sqrt[3]{x} + 5 = y$

$\sqrt[3]{x} + 5 = g(x)$

17) $y = f(x) = (x - 1)^5$

Let $y = g(x)$

$x = (y - 1)^5$

$\sqrt[5]{x} = y - 1$

$\sqrt[5]{x} + 1 = y$

$\sqrt[5]{x} + 1 = g(x)$

21) $f(x) = \dfrac{2}{x - 2}$

Let $y = f^{-1}(x)$

$$x = \frac{2}{y - 2}$$

$$xy - 2x = 2$$

$$xy = 2 + 2x$$

$$y = \frac{2}{x} + 2$$

$$f^{-1}(x) = \frac{2}{x} + 2$$

$$f(f^{-1}(x)) = f\left(\frac{2}{x} + 2\right) = \frac{2}{\left(\frac{2}{x} + 2\right) - 2} = \frac{2}{\frac{2}{x}} = x$$

$$f^{-1}(f(x)) = f^{-1}\left(\frac{2}{x - 2}\right) = \frac{2}{\frac{2}{x - 2}} + 2 = x - 2 + 2 = x$$

25) $f(x) = x^{-5}$

Let $y = f^{-1}(x)$

$$x = y^{-5}$$

$$x^{-\frac{1}{5}} = \left(y^{-5}\right)^{-\frac{1}{5}}$$

$$x^{-\frac{1}{5}} = y$$

$$x^{-\frac{1}{5}} = f^{-1}(x)$$

$$f(f^{-1}(x)) = f\left(x^{-\frac{1}{5}}\right) = \left(x^{-\frac{1}{5}}\right)^{-5} = x$$

$$f^{-1}(f(x)) = f^{-1}(x^{-5}) = \left(x^{-5}\right)^{-\frac{1}{5}} = x$$

29) $f(x) = \dfrac{x}{x - 1}$

$$f(f(x)) = f\left(\frac{x}{x - 1}\right) = \frac{\frac{x}{x - 1}}{\frac{x}{x - 1} - 1} = \frac{\frac{x}{x - 1}}{\frac{x - x + 1}{x - 1}} = \frac{x}{1} = x$$

33) $y = f(x) = \dfrac{1}{\sqrt{x}}$

Let $y = g(x)$.

$$x = \frac{1}{\sqrt{y}}$$

$$x^2 = \left(\frac{1}{\sqrt{y}}\right)^2$$

$$x^2 = \frac{1}{y}$$

$$\frac{1}{x^2} = y$$

$$\frac{1}{x^2} = g(x)$$

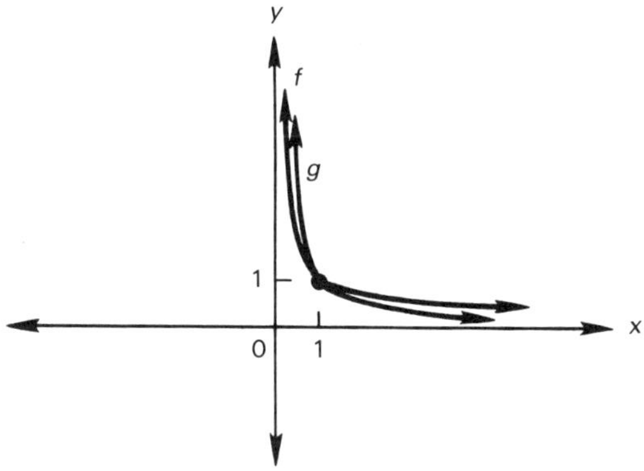

37) The graph of a linear function is a line.
If a line is going to be its own reflection through the $y = x$ line, it has to be perpendicular to $y = x$. The slope of $y = x$ is 1, so any line perpendicular to it has a slope -1.
Check:

$y = -1x + k = -x + k$ to see whether all such lines are their own inverses.

$f(f(x)) = f(-x + k) = -(-x + k) + k = x - k + k = x$.

Any line $y = mx + k$ where $m = -1$ is its own inverse.

Sample Test Questions for Chapter 6

Part I

1) $f(x) = \sqrt[3]{x + 2}$

Domain: all reals

Shift the graph of $y = \sqrt[3]{x}$ to the left 2 units.

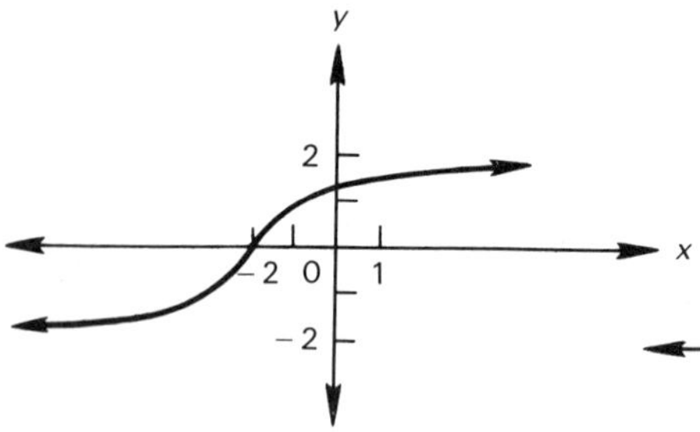

2) $g(x) = \dfrac{1}{\sqrt{x}} + 2$

Domain: $x > 0$

Asymptotes: $y = 2$, $x = 0$

Shift the graph of $\dfrac{1}{\sqrt{x}}$ up 2 units.

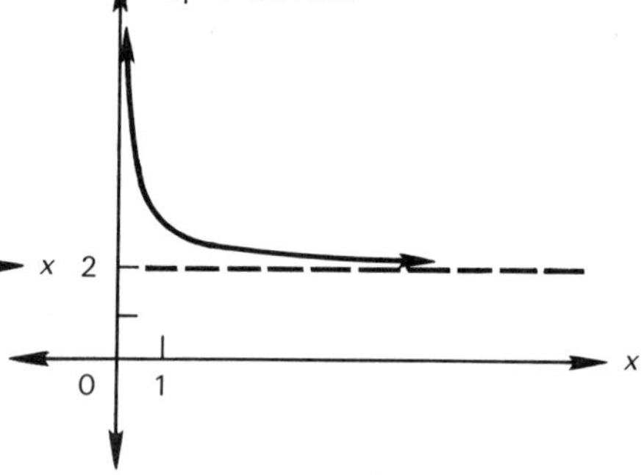

3) $f(x) = -\sqrt[3]{x - 2}$

Domain: all reals

Reflect the graph of $y = \sqrt[3]{x}$ through the x-axis and shift to the right 2 units.

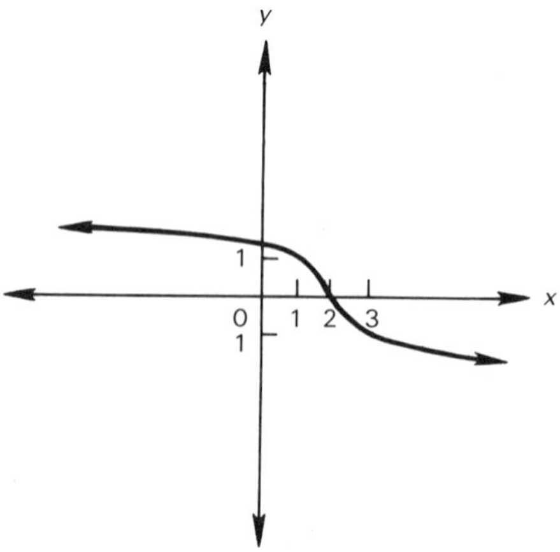

4) $g(x) = \dfrac{1}{\sqrt{x - 3}}$

Domain: $x > 3$

Asymptotes: $y = 0$, $x = 3$

Shift the graph of $y = \dfrac{1}{\sqrt{x}}$ to the right 3 units

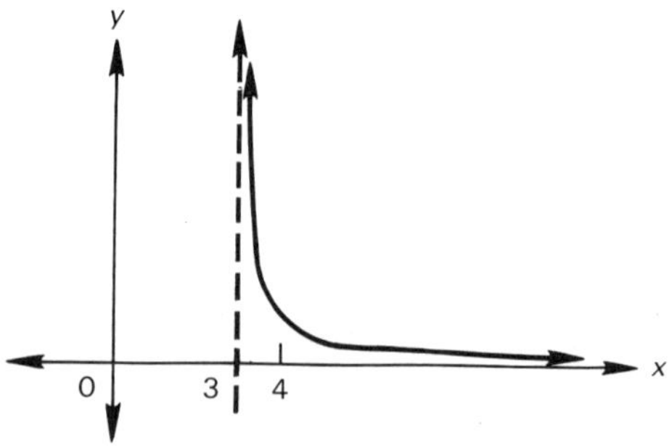

5) $\sqrt{18x + 5} - 9x = 1$

$\sqrt{18x + 5} = 9x + 1$

$(\sqrt{18x + 5})^2 = (9x + 1)^2$

$18x + 5 = 81x^2 + 18x + 1$

$81x^2 - 4 = 0$

$9x - 2 = 0$ or $(9x + 2) = 0$

$9x = 2$ $\quad$ $9x = -2$

$x = \frac{2}{9}$ $\quad$ $x = -\frac{2}{9}$

Check $\frac{2}{9}$

$\sqrt{18\left(\frac{2}{9}\right) + 5} - 9\left(\frac{2}{9}\right) = 1$

$\sqrt{9} - 2 = 1$

$3 - 2 = 1$

$1 = 1$

$x = \frac{2}{9}$

Check $-\frac{2}{9}$

$\sqrt{18\left(-\frac{2}{9}\right) + 5} - 9\left(-\frac{2}{9}\right) = 1$

$\sqrt{-4} + 2 = 1$

-4 is not in the domain of the radical function.

6) $6x^{\frac{2}{3}} + 5x^{\frac{1}{3}} - 4 = 0$

Let $u = x^{\frac{1}{3}}$, $u^2 = x^{\frac{2}{3}}$

$6u^2 + 5u - 4 = 0$

$(3u + 4)(2u - 1) = 0$

$3u + 4 = 0$ or $2u - 1 = 0$

$3u = -4$ $\quad$ $2u = 1$

$u = -\frac{4}{3}$ $\quad$ $u = \frac{1}{2}$

Substitute these values into $u = x^{\frac{1}{3}}$.

$-\frac{4}{3} = x^{\frac{1}{3}}$ $\quad$ $\frac{1}{2} = x^{\frac{1}{3}}$

$\left(-\frac{4}{3}\right)^3 = \left(x^{\frac{1}{3}}\right)^3$ $\quad$ $\left(\frac{1}{2}\right)^3 = \left(x^{\frac{1}{3}}\right)^3$

$-\frac{64}{27} = x$ $\quad$ $\frac{1}{8} = x$

Check $x = -\frac{64}{27}$

$6\left(-\frac{64}{27}\right)^{\frac{2}{3}} + 5\left(-\frac{64}{27}\right)^{\frac{1}{3}} - 4 = 0$

$6\left(-\frac{4}{3}\right)^2 + 5\left(-\frac{4}{3}\right) - 4 = 0$

$6\left(\frac{16}{9}\right) - \frac{20}{3} - 4 = 0$

$\frac{32}{3} - \frac{20}{3} - \frac{12}{3} = 0$

$0 = 0$

$x = -\frac{64}{27}$

Check $x = \frac{1}{8}$

$6\left(\frac{1}{8}\right)^{\frac{2}{3}} + 5\left(\frac{1}{8}\right)^{\frac{1}{3}} - 4 = 0$

$6\left(\frac{1}{2}\right)^2 + 5\left(\frac{1}{2}\right) - 4 = 0$

$6\left(\frac{1}{4}\right) + \frac{5}{2} - 4 = 0$

$\frac{3}{2} + \frac{5}{2} - \frac{8}{2} = 0$

$0 = 0$

$x = \frac{1}{8}$

7) $\sqrt{x - 7} + \sqrt{x + 9} = 8$

$\sqrt{x - 7} = 8 - \sqrt{x + 9}$

$(\sqrt{x - 7})^2 = (8 - \sqrt{x + 9})^2$

$x - 7 = 64 - 16\sqrt{x + 9} + (x + 9)$

$16\sqrt{x + 9} = 80$

$\sqrt{x + 9} = 5$

$(\sqrt{x + 9})^2 = 5^2$

$x + 9 = 25$

$x = 16$

Check

$\sqrt{16 - 7} + \sqrt{16 + 9} = 8$

$\sqrt{9} + \sqrt{25} = 8$

$3 + 5 = 8$

$8 = 8$

8) $(x + 1)^{1/2}(2x) + (x + 1)^{-1/2}\left(\frac{x^2}{2}\right) = 2x\sqrt{x + 1} + \frac{x^2}{2\sqrt{x + 1}}$

$= \frac{(2x\sqrt{x + 1})(2\sqrt{x + 1}) + x^2}{2\sqrt{x + 1}}$

$= \frac{4x(x + 1) + x^2}{2\sqrt{x + 1}}$

$= \frac{4x^2 + 4x + x^2}{2\sqrt{x + 1}}$

$= \frac{5x^2 + 4x}{2\sqrt{x + 1}}$

$= \frac{x(5x + 4)}{2\sqrt{x + 1}}$

9) $f(x) = \frac{\sqrt[3]{x - 4}}{x - 2}$

Domain: $x \neq 2$

Interval	$(-\infty, 2)$	$(2, 4)$	$(4, \infty)$
Sign of $\sqrt[3]{x - 4}$	-	-	+
Sign of $x - 2$	-	+	+
Sign of $f(x)$	+	-	+

$f(x) > 0$ on $(-\infty, 2)$ and $(4, \infty)$

$f(x) < 0$ on $(2, 4)$

10) $f(x) = x^{\frac{1}{2}} - \frac{1}{2}x^{-1/2}(x + 4) = \sqrt{x} - \frac{x + 4}{2\sqrt{x}} = \frac{\sqrt{x}(2\sqrt{x}) - (x + 4)}{2\sqrt{x}}$

$$= \frac{2x - x - 4}{2\sqrt{x}} = \frac{x - 4}{2\sqrt{x}}$$

Domain: $x > 0$

Because the denominator is always positive, the sign of the fraction depends on the sign of $(x - 4)$.

$f(x) > 0$ on $(4, \infty)$

$f(x) < 0$ on $(0, 4)$

11) $f(x) = \frac{1}{x^2 - 1}$ $\qquad g(x) = \sqrt{x + 2}$

$(f - g)(x) = \frac{1}{x^2 - 1} - \sqrt{x + 2}$

$x^2 - 1 \neq 0 \qquad x^2 \neq 1, \quad x \neq \pm 1$

$\sqrt{x + 2} \geq 0 \qquad x + 2 \geq 0, \quad x \geq -2$

Domain: $x \geq -2, \quad x \neq \pm 1$

12) $\frac{f}{g}(x) = \frac{1}{x^2 - 2} \div \sqrt{x + 2} = \frac{1}{(x^2 - 1)(\sqrt{x + 2})}$

$x^2 - 1 \neq 0, \quad x^2 \neq 1, \quad x \neq \pm 1$

$\sqrt{x + 2} > 0, \quad x + 2 > 0, \quad x > -2$

Domain: $x > -2, \quad x \neq \pm 1$

13) $(f \cdot g)(x) = \frac{1}{x^2 - 1} \bullet \sqrt{x + 2} = \frac{\sqrt{x + 2}}{x^2 - 1}$

$\sqrt{x + 2} \geq 0, \ x + 2 \geq 0, \quad x \geq -2$

$x^2 - 1 \neq 0, \quad x^2 \neq 1, \quad x \neq \pm 1$

Domain: $x \geq -2, \ x \neq \pm 1$

14) $f(x) = \dfrac{1}{1 - x^2}$ $g(x) = \sqrt{x}$

$(f \circ g)(x) = f(g(x)) = f(\sqrt{x}) = \dfrac{1}{1 - (\sqrt{x})^2} = \dfrac{1}{1 - x}$

Domain: all x such that g(x) is a real number $\neq 1$, that is all

$x > 0,\ x \neq 1.$

$(g \circ f)(x) = g(f(x)) = g\left(\dfrac{1}{1 - x^2}\right) = \sqrt{\dfrac{1}{1 - x^2}} = \dfrac{1}{\sqrt{1 - x^2}}$

Domain: all x such that f(x) is a real number > 0, that is

$-1 < x < 1.$

15) $h(x) = \sqrt[3]{(x - 2)^2}$

It is possible to distinguish three functions here. The outer function takes a cube root, the middle function squares, and the inner function is $(x - 2)$. The directions ask for only two functions, with the inner one a binomial.

Let $g(x) = x - 2$.

Now f(x) must <u>both</u> square <u>and</u> take a cube root, so $f(x) = x^{\frac{2}{3}}$.

Check $(f \circ g)(x) = f(x - 2) = (x - 2)^{\frac{2}{3}} = \sqrt[3]{(x - 2)^2}$.

16) $F(x) = \dfrac{1}{(2x - 1)^{3/2}}$

The outer function forms a reciprocal, so $f(x) = \dfrac{1}{x}$.

The middle function raises to the $\frac{3}{2}$ power, so $g(x) = x^{\frac{3}{2}}$.

The inner function is $h(x) = 2x - 1$.

Check:

$(f \circ g \circ h) = f(g(h(x))) = f(g(2x - 1)) = f\left((2x - 1)^{\frac{3}{2}}\right)$

$= \dfrac{1}{(2x - 1)^{\frac{3}{2}}}$

17) $f(x) = 3x - 2$

Let $y = f^{-1}(x)$

$x = 3y - 2$

$x + 2 = 3y$

$\frac{x + 2}{3} = y$

$\frac{1}{3} x + \frac{2}{3} = f^{-1}(x)$

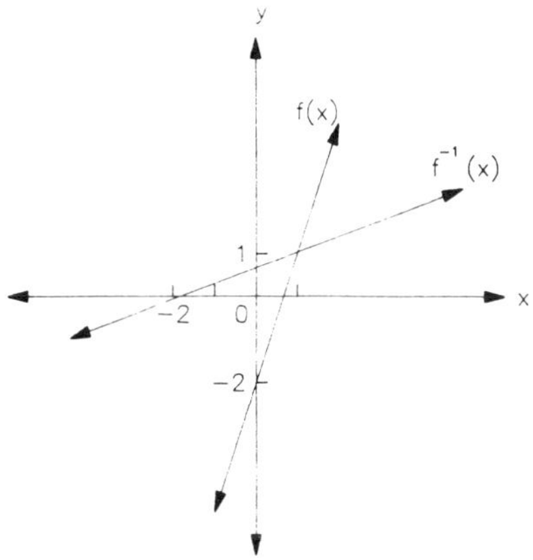

18) $y = f(x) = \sqrt[3]{x} - 1$

Let $y = g(x)$

$x = \sqrt[3]{y} - 1$

$x + 1 = \sqrt[3]{y}$

$(x + 1)^3 = (\sqrt[3]{y})^3$

$(x + 1)^3 = y$

$(x + 1)^3 = g(x)$

Check:

$(f \text{ o } g)(x) = f(g(x)) = f((x + 1)^3)$

$= \sqrt[3]{(x + 1)^3} - 1$

$= x + 1 - 1 = x$

19) $f(x) = \sqrt[3]{3x + 1} - x - 1$

Domain: all reals

x-intercepts:

$0 = \sqrt[3]{3x + 1} - x - 1$

$x + 1 = \sqrt[3]{3x + 1}$

$(x + 1)^3 = (\sqrt[3]{3x + 1})^3$

$x^3 + 3x^2 + 3x + 1 = 3x + 1$

$x^3 + 3x^2 = 0$

$x^2(x + 3) = 0$

$x^2 = 0$ or $x + 3 = 0$

$x = 0$ $\quad$ $x = -3$

Check $x = 0$

$\sqrt[3]{3 \cdot 0 + 1} - 0 - 1 = 0$

$\sqrt[3]{1} - 1 = 0$

$0 = 0$

$x = 0$

Check $x = -3$

$\sqrt[3]{3(-3) + 1} - (-3) - 1 = 0$

$\sqrt[3]{-8} + 3 - 1 = 0$

$-2 + 3 - 1 = 0$

$0 = 0$

$x = -3$

20) $y = 2\sqrt{x - 1}$

$y = \frac{1}{2}x + 1$

Set the two right sides equal to each other.

$2\sqrt{x - 1} = \frac{1}{2}x + 1$

$4\sqrt{x - 1} = x + 2$

$(4\sqrt{x - 1})^2 = (x + 2)^2$

$16(x - 1) = x^2 + 4x + 4$

$16x - 16 = x^2 + 4x + 4$

$0 = x^2 - 12x + 20$

$0 = (x - 10)(x - 2)$

$x - 10 = 0$ or $x - 2 = 0$

$x = 10$ or $x = 2$

Substitute into the second equation to find y.

When $x = 10$, $y = \frac{1}{2}(10) + 1 = 6$

When $x = 2$, $y = \frac{1}{2}(2) + 1 = 2$

You can check these coordinates in the second equation.

Solutions: (2, 2), (10, 6)

Sample Test Questions for Chapter 6

Part II

1) (c)

$$\sqrt{x^2 - 9x} + \sqrt{3x} = x$$
$$\sqrt{x^2 - 9x} = x - \sqrt{3x}$$
$$(\sqrt{x^2 - 9x})^2 = (x - \sqrt{3x})^2$$
$$x^2 - 9x = x^2 - 2x\sqrt{3x} + 3x$$
$$2x\sqrt{3x} = 12x$$
$$x\sqrt{3x} = 6x$$
$$(x\sqrt{3x})^2 = (6x)^2$$
$$x^2(3x) = 36x^2$$
$$3x^3 - 36x^2 = 0$$
$$3x^2(x - 12) = 0$$
$$x = 0 \qquad x - 12 = 0$$
$$x = 12$$

Check x = 0
$$\sqrt{0^2 - 9 \cdot 0} + \sqrt{3 \cdot 0} = 0$$
$$0 + 0 = 0$$
$$0 = 0$$
$$x = 0$$

Check x = 12
$$\sqrt{(12)^2 - 9 \cdot 12} + \sqrt{3 \cdot 12} = 12$$
$$\sqrt{36} + \sqrt{36} = 12$$
$$6 + 6 = 12$$
$$x = 12$$

2) (d)

3) (a)

4) (e)

$$y = 0 = \sqrt{2x^2 - 8x - 24}$$
$$0 = 2x^2 - 8x - 24$$
$$0 = x^2 - 4x - 12$$
$$0 = (x - 6)(x + 2)$$
$$x - 6 = 0 \quad \text{or} \quad x + 2 = 0$$
$$x = 6 \qquad x = -2$$

Check x = 6
$$\sqrt{2(6)^2 - 8 \cdot 6 - 24} = 0$$
$$\sqrt{0} = 0$$
$$0 = 0$$
$$x = 6$$

Check x = -2
$$\sqrt{2(-2)^2 - 8(-2) - 24} = 0$$
$$\sqrt{0} = 0$$
$$0 = 0$$
$$x = -2$$

5) (c)

$\sqrt[3]{x^2} - \sqrt[3]{x} - 6 = 0$

$x^{\frac{2}{3}} - x^{\frac{1}{3}} - 6 = 0$

Let $u = x^{\frac{1}{3}}$, $u^2 = x^{\frac{2}{3}}$

$u^2 - u - 6 = 0$

$(u - 3)(u + 2) = 0$

$u - 3 = 0$ or $u + 2 = 0$

$u = 3$ or $u = -2$

Substitute into $u = x^{\frac{1}{3}}$

$3 = x^{\frac{1}{3}}$ $\quad$ $-2 = x^{\frac{1}{3}}$

$3^3 = \left(x^{\frac{1}{3}}\right)^3$ $\quad$ $(-2)^3 = \left(x^{\frac{1}{3}}\right)^3$

$27 = x$ $\quad$ $-8 = x$

Check $x = 27$

$\sqrt[3]{(27)^2} - \sqrt[3]{27} - 6 = 0$

$\sqrt[3]{729} - \sqrt[3]{27} - 6 = 0$

$9 - 3 - 6 = 0$

$0 = 0$

$x = 27$

Check $x = -8$

$\sqrt[3]{(-8)^2} - \sqrt[3]{-8} - 6 = 0$

$\sqrt[3]{64} - \sqrt[3]{-8} - 6 = 0$

$4 - (-2) - 6 = 0$

$0 = 0$

$x = -8$

6) (b)

$r = \sqrt{\frac{v}{\pi h}}$

$r^2 = \frac{v}{\pi h}$

$r^2 \pi h = v$

$h = \frac{v}{r^2 \pi}$

7) (c)

$f(x) = \frac{x - 2}{\sqrt{x^2 + 4}}$

The denominator is positive for all real values of x, so the sign of the fraction is determined by $x - 2$.

$f(x) < 0$ for $x < 2$

$f(x) > 0$ for $x > 2$

8) (d)

$$f(x) = \frac{x^2}{\sqrt{1 - x^2}}$$

$(1 - x^2) > 0$, $x^2 < 1$, $-1 < x < 1$

Domain: $(-1, 1)$

9) (e)

10) (c)

$$f(x) = \frac{1}{x^2 - 1} \qquad g(x) = \sqrt{x}$$

$$\frac{f}{g}(x) = \frac{1}{x^2 - 1} \div \sqrt{x} = \frac{1}{x^2 - 1} \cdot \frac{1}{\sqrt{x}} = \frac{1}{\sqrt{x}\,(x^2 - 1)}$$

$$\frac{f}{g}(4) = \frac{1}{\sqrt{4}\,(4^2 - 1)} = \frac{1}{2(15)} = \frac{1}{30}$$

11) (d)

$$x^{\frac{1}{3}} + (x - 8)\,\frac{1}{3}\,x^{-\frac{2}{3}} = x^{\frac{1}{3}} + \frac{(x - 8)}{3x^{\frac{2}{3}}} = \frac{x^{\frac{1}{3}}(3x^{\frac{2}{3}}) + (x - 8)}{3x^{\frac{2}{3}}}$$

$$= \frac{3x + x - 8}{3x^{\frac{2}{3}}} = \frac{4x - 8}{3x^{\frac{2}{3}}} = \frac{4(x - 2)}{3x^{\frac{2}{3}}}$$

12) (d)

$$f(x) = \sqrt[3]{x} \qquad g(x) = \frac{1}{x^3 - 1}$$

$$g(f(x)) = g(\sqrt[3]{x}) = \frac{1}{(\sqrt[3]{x})^3 - 1} = \frac{1}{x - 1}$$

13) (b)

14) (b)

$f(x) = 3x - 2$

Let $z = f^{-1}(x)$

$x = 3z - 2$

$x + 2 = 3z$

$\frac{x + 2}{3} = z$

$\frac{1}{3}x + \frac{2}{3} = f^{-1}(x)$

15) (c)

CHAPTER 7

EXPONENTIAL AND LOGARITHMIC FUNCTIONS

Section 7.1 Exponential Functions

1) $f(x) = 2^x$

x	2^x
-2	$\frac{1}{4}$
-1	$\frac{1}{2}$
0	1
1	2
2	4
3	8

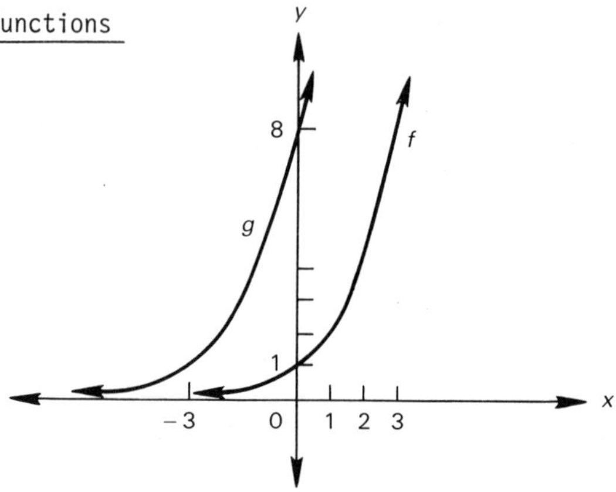

$g(x) = 2^{x+3}$ can be graphed by shifting f(x) to the left 3 units.

5) $f(x) = \left(\frac{3}{2}\right)^x$

x	$\left(\frac{3}{2}\right)^x$
-2	$\frac{4}{9}$
-1	$\frac{2}{3}$
0	1
1	$\frac{3}{2}$
2	$\frac{9}{4}$
3	$\frac{27}{8}$

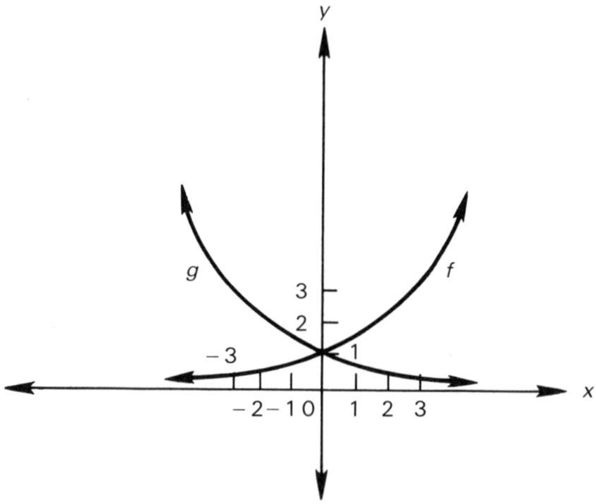

$g(x) = \left(\frac{3}{2}\right)^{-x}$ can be graphed by reflecting f(x) through the y-axis.

9) $f(x) = 2^{\frac{x}{2}}$

x	$2^{\frac{x}{2}}$
-4	$\frac{1}{4}$
-2	$\frac{1}{2}$
0	1
2	2
4	4

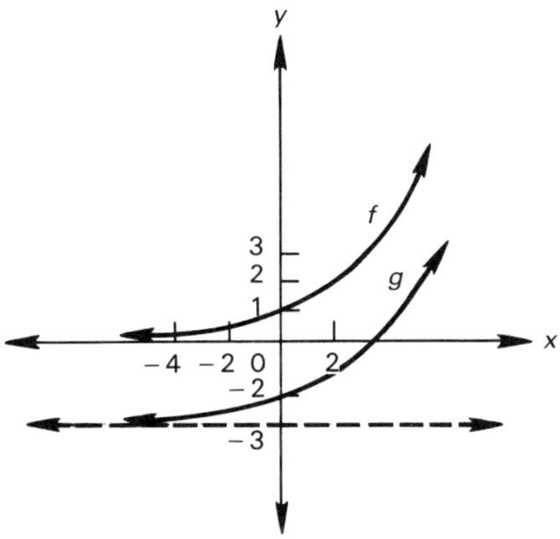

$g(x) = 2^{\frac{x}{2}} - 3$ can be graphed by shifting f(x) down 3 units.

13) $y = 2^{|x|}$

Sketch 2^x for $x \geq 0$, then reflect this part of the graph through the y-axis.

$y = -2^{|x|}$ can be graphed by reflecting $y = 2^{|x|}$ through the x-axis.

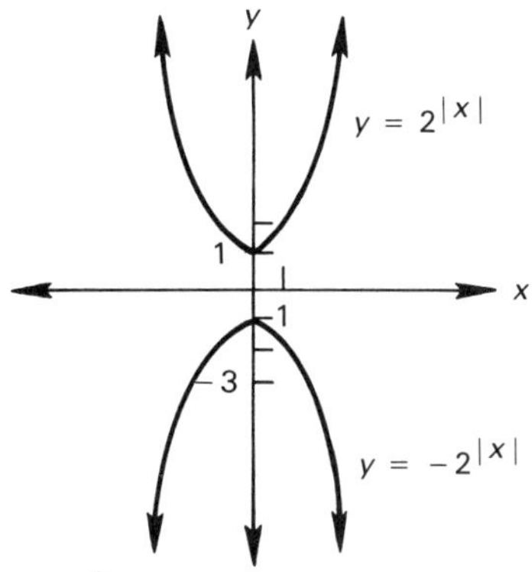

17) $2^{x^2} = 512$

$2^{x^2} = 2^9$

$x^2 = 9$

$x = \pm 3$

21) $7^{x^2+x} = 49$

$7^{x^2+x} = 7^2$

$x^2 + x = 2$

$x^2 + x - 2 = 0$

$(x + 2)(x - 1) = 0$

$x + 2 = 0$ or $x - 1 = 0$

$x = -2$ $\qquad$ $x = 1$

25) $9^x = 3$

$9^x = 9^{\frac{1}{2}}$

$x = \frac{1}{2}$

29) $\left(\frac{1}{49}\right)^x = (7^{-2})^x = 7$

$7^{-2x} = 7$

$-2x = 1$

$x = -\frac{1}{2}$

33)

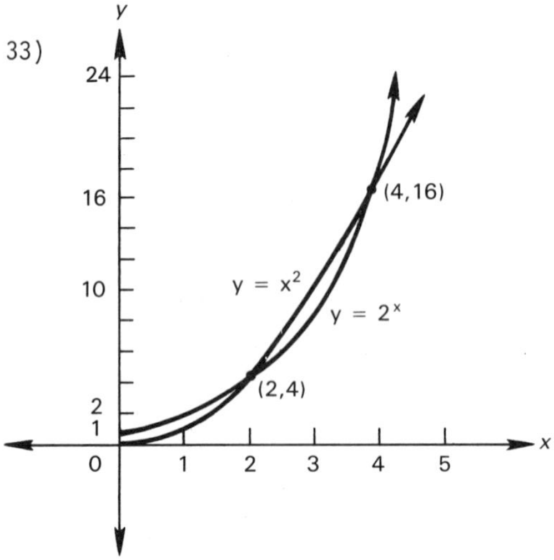

$$(5^{2x+1})(7^{2x}) = 175$$
$$5 \cdot 5^{2x} \cdot 7^{2x} = 175$$
$$5^{2x} \cdot 7^{2x} = 35$$
$$(35)^{2x} = 35$$
$$2x = 1$$
$$x = \frac{1}{2}$$

Section 7.2 Logarithmic Functions

1) $y = f(x) = 4^x \quad g(x) = \log_4 x$

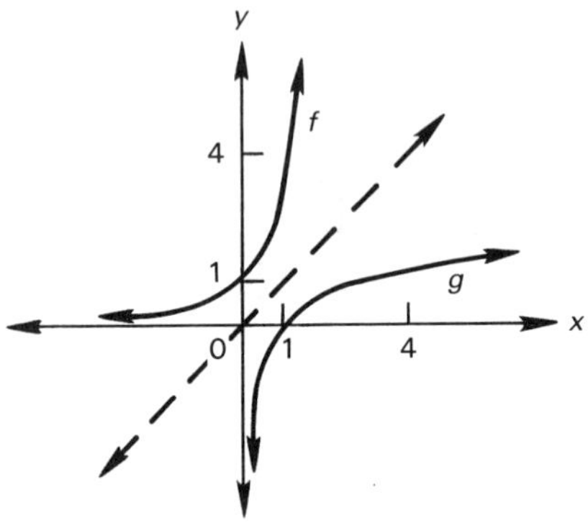

5) $g(x) = \log_3 x \quad h(x) = \log_3(x+2)$

h(x) can be graphed by shifting g(x) left 2 units.

Domain: $x + 2 > 0,\ x > -2$

Asymptote: $x = -2$

9) $f(x) = \log_{10} x$

Domain: $x > 0$

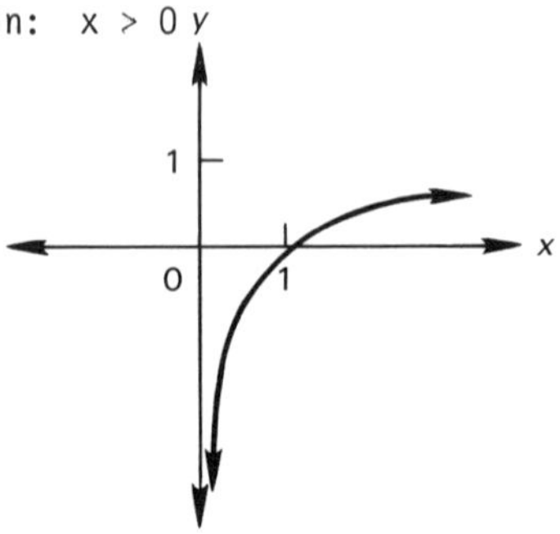

13) $f(x) = \log_{10}|x|$

Domain: all $x \neq 0$

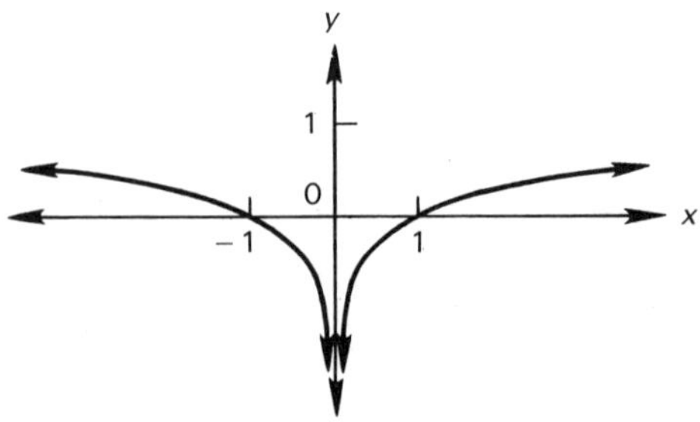

17) $\left(\frac{1}{3}\right)^{-1} = 3$

$\log_{\frac{1}{3}} 3 = -1$

21) $\log_{10} 0.0001 = -4$

$10^{-4} = 0.0001$

25) $\log_{12} \frac{1}{1728} = -3$

$12^{-3} = \frac{1}{1728}$

29) $\log_{\frac{1}{3}} 27 = y$

$\left(\frac{1}{3}\right)^{y} = 27$

$\left(\frac{1}{3}\right)^{y} = \left(\frac{1}{3}\right)^{-3}$

$y = -3$

33) $\log_b 125 = 3$

$b^3 = 125$

$(b^3)^{\frac{1}{3}} = (125)^{\frac{1}{3}}$

$b = 5$

37) $\log_{27} 3 = y$

$(3^3)^y = 3$

$3^{3y} = 3$

$3y = 1$

$y = \frac{1}{3}$

41) $\log_b \frac{1}{27} = -\frac{3}{2}$

$b^{-\frac{3}{2}} = \frac{1}{27}$

$\left(b^{-\frac{3}{2}}\right)^{-\frac{2}{3}} = \left(\frac{1}{27}\right)^{-\frac{2}{3}}$

$b = 9$

45) $\log_{0.001} 10 = y$

$(0.001)^y = 10$

$(10^{-3})^y = 10'$

$10^{-3y} = 10'$

$-3y = 1$

$y = -\frac{1}{3}$

49) $\log_{\frac{3}{4}} \left(\log_{\frac{1}{27}} \frac{1}{81}\right)$

First find $\left(\log_{\frac{1}{27}} \frac{1}{81}\right)$

$\log_{\frac{1}{27}} \frac{1}{81} = y$

$\left(\frac{1}{27}\right)^y = \frac{1}{81}$

$(3^{-3})^y = 3^{-4}$

$3^{-3y} = 3^{-4}$

$-3y = -4$

$y = \frac{4}{3}$

Substitute this value into the original expressions, and evaluate

$\log_{\frac{3}{4}} \left(\frac{4}{3}\right) = -1$

Section 7.3 The Law of Logarithms

1) $\log_b \frac{3x}{x + 1} = \log_b(3x) - \log_b(x + 1) = \log_b 3 + \log_b x - \log_b(x + 1)$

5) $\log_b \frac{1}{x^2} = \log_b 1 - \log_b x^2 = 0 - 2 \log_b x = -2 \log_b x$

9) $\frac{1}{2} \log_b(x^2 - 1) - \frac{1}{2} \log_b(x^2 + 1) = \log_b(x^2 - 1)^{\frac{1}{2}} - \log_b(x^2 + 1)^{\frac{1}{2}}$

$= \log_b\sqrt{x^2 - 1} - \log_b\sqrt{x^2 + 1}$

$= \log_b \frac{\sqrt{x^2 - 1}}{\sqrt{x^2 + 1}} = \log_b\sqrt{\frac{x^2 - 1}{x^2 + 1}}$

13) $\log_b 27 + \log_b 3 \underset{\text{(Law 1)}}{=} \log_b 27 \cdot 3 = \log_b 81$

$\log_b 243 - \log_b 3 \underset{\text{(Law 2)}}{=} \log_b \frac{243}{3} = \log_b 81$

17) a) $\log 4 = \log 2^2 = 2 \log 2 = 2(0.3010) = 0.6020$

b) $\log 8 = \log 2^3 = 3 \log 2 = 3(0.3010) = 0.9030$

c) $\log \frac{1}{2} = \log 1 - \log 2 = 0 - 0.3010 = -0.3010$

21) a) $\log \sqrt[3]{5} = \log 5^{\frac{1}{3}} = \frac{1}{3} \log 5 = \frac{1}{3} (0.6990) = 0.2330$

b) $\log \sqrt{20^3} = \log 20^{\frac{3}{2}} = \frac{3}{2} \log 20 = \frac{3}{2} \log 2^2 \cdot 5$

$= \frac{3}{2} (2 \log 2 + \log 5) = 3 \log 2 + \frac{3}{2} \log 5$

$= 3(0.3010) + \frac{3}{2} (0.6990) = 0.9030 + 1.0485 = 1.9515$

c) $\log \sqrt{900} = \log 30 = \log 2 \cdot 3 \cdot 5 = \log 2 + \log 3 + \log 5$

$= 0.3010 + 0.4771 + 0.6990 = 1.4771$

25) $\log_{10}5 - \log_{10}x = 2$

$$\log_{10}\frac{5}{x} = 2$$
$$10^2 = \frac{5}{x}$$
$$100x = 5$$
$$x = \frac{1}{20}$$

Check:

$$\log_{10}5 - \log_{10}\frac{1}{20} = \log_{10}\left(\frac{5}{\frac{1}{20}}\right)$$
$$= \log_{10}100 = 2$$

29) $\log_{16}x + \log_{16}(x - 4) = \frac{5}{4}$ (Note: $x > 4$.)

$$\log_{16}x(x - 4) = \frac{5}{4}$$
$$\log_{16}(x^2 - 4x) = \frac{5}{4}$$
$$16^{\frac{5}{4}} = x^2 - 4x$$
$$32 = x^2 - 4x$$
$$0 = x^2 - 4x - 32$$
$$0 = (x - 8)(x + 4)$$

$x - 8 = 0$ or $x + 4 = 0$

$x = 8$ $\quad$ $x = -4$

Rejected because $-4 < 4$.

Check:

$$\log_{16}8 + \log_{16}(8 - 4) = \frac{5}{4}$$
$$\log_{16}8 \cdot 4 = \frac{5}{4}$$
$$\log_{16}32 = \frac{5}{4}$$
$$16^{\frac{5}{4}} = 32$$
$$32 = 32$$

33) $\log_{\frac{1}{7}} x + \log_{\frac{1}{7}}(5x - 28) = -2$

Note: $5x - 28 > 0$, $5x > 28$, $x > \frac{28}{5}$

$$\log_{\frac{1}{7}} x(5x - 28) = -2$$

$$\log_{\frac{1}{7}}(5x^2 - 28x) = -2$$

$$\left(\frac{1}{7}\right)^{-2} = 5x^2 - 28x$$

$$49 = 5x^2 - 28x$$

$$0 = 5x^2 - 28x - 49$$

$$0 = (x - 7)(5x + 7)$$

$$x - 7 = 0 \quad \text{or} \quad 5x + 7 = 0$$

$$x = 7 \qquad\qquad x = -\frac{7}{5}$$

Rejected because $-\frac{7}{5} < \frac{28}{5}$.

Check:

$$\log_{\frac{1}{7}} 7 + \log_{\frac{1}{7}}(5 \cdot 7 - 28) = -2$$

$$\log_{\frac{1}{7}} 7(7) = -2$$

$$\log_{\frac{1}{7}} 49 = -2$$

$$\left(\frac{1}{7}\right)^{-2} = 49$$

$$49 = 49$$

37) $2 \log_{25} x - \log_{25}(25 - 4x) = \frac{1}{2}$

Note: $25 - 4x > 0$, $25 > 4x$, $x < \frac{25}{4}$ and $x > 0$, so $0 < x < \frac{25}{4}$.

$$\log_{25} x^2 - \log_{25}(25 - 4x) = \frac{1}{2}$$

$$\log_{25} \frac{x^2}{25 - 4x} = \frac{1}{2}$$

$$25^{\frac{1}{2}} = \frac{x^2}{25 - 4x}$$

$$5 = \frac{x^2}{25 - 4x}$$

$$125 - 20x = x^2$$

37) continued

$$x^2 + 20x - 125 = 0$$
$$(x + 25)(x - 5) = 0$$
$$x + 25 = 0 \quad \text{or} \quad x - 5 = 0$$
$$x = -25 \qquad\qquad x = 5$$

Rejected because 25 < 0

Check:

$$2 \log_{25}5 - \log_{25}(25 - 4 \cdot 5) = \frac{1}{2}$$
$$\log_{25}(5^2) - \log_{25}5 = \frac{1}{2}$$
$$\log_{25} \frac{25}{5} = \frac{1}{2}$$
$$\log_{25}5 = \frac{1}{2}$$
$$25^{\frac{1}{2}} = 5$$
$$5 = 5$$

41) $(x + 2)\log_b b^x = x$

$$\log_b b^x = \frac{x}{x + 2}$$
$$x = \frac{x}{x + 2}$$
$$x^2 + 2x = x$$
$$x^2 + x = 0$$
$$x(x + 1) = 0$$
$$x = 0 \quad \text{or} \quad x + 1 = 0$$
$$x = -1$$

Check $x = 0$

$$(0 + 2)\log_b b^0 = 0$$
$$2 \cdot 0 = 0$$
$$0 = 0$$

Check $x = -1$

$$(-1 + 2)\log_b b^{-1} = -1$$
$$1(-1) = -1$$
$$-1 = -1$$

45) $B^{\log_B N} = N$

$$\log_b B^{\log_B N} = \log_b N$$
$$(\log_B N)(\log_b B) = \log_b N \qquad \text{(By Law 3)}$$
$$\log_B N = \frac{\log_b N}{\log_b B}$$

Section 7.4 The Base e

1) $y = e^x$

$y = e^{x-2}$ can be graphed by shifting $y = e^x$ to the right 2 units.

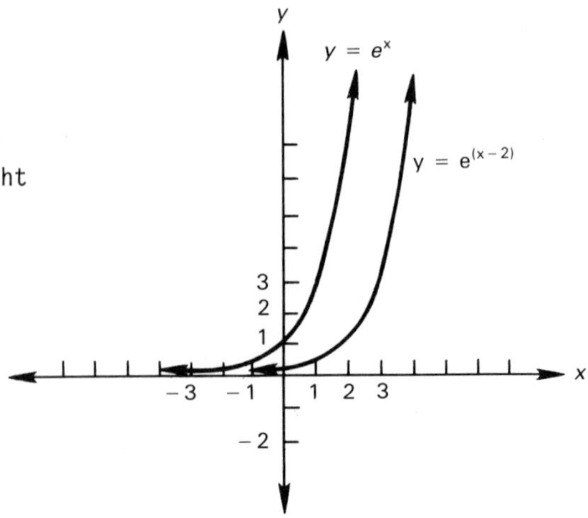

5) $y = \ell n\ x$

$y = \ell n(-x)$ can be graphed by reflecting $y = \ell n\ x$ through the y-axis.

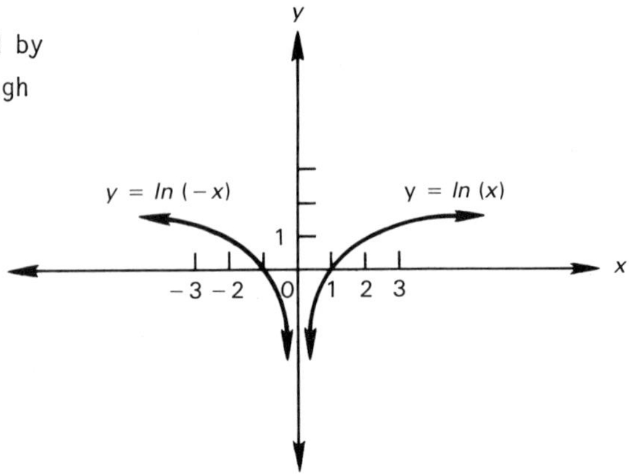

9) $f(x) = -e^{-x}$

This is a double reflection. First reflect e^x through the y-axis to get $y = e^{-x}$, then reflect $y = e^{-x}$ through the x-axis to get $y = -e^{-x}$.

$g(x) = 1 - e^{-x} = -e^{-x} + 1$.

Shift the graph of $y = -e^{-x}$ up 1 unit.

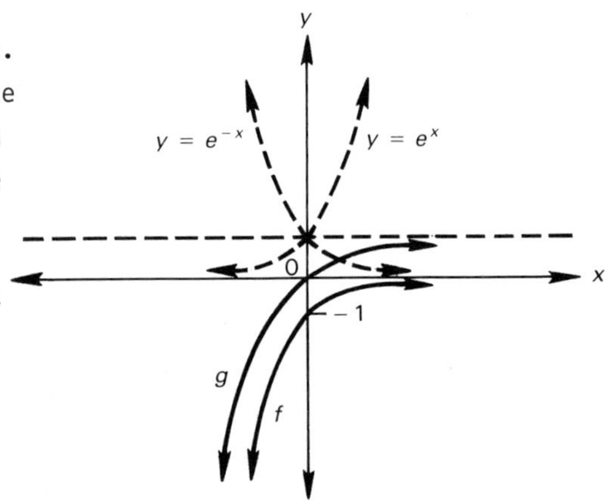

13) $f(x) = \ell n\ ex = \ell n\ e + \ell n\ x = 1 + \ell n\ x$

Shift the graph of $y = \ell n\ x$ up 1 unit.

17) $f(x) = \ell n(x^2 - 1) - \ell n(x + 1) = \ell n \frac{x^2 - 1}{x + 1} = \ell n \frac{(x + 1)(x - 1)}{(x+1)}$

$= \ell n(x - 1)$

Shift the graph of $y = \ell n\ x$ to the right 1 unit.

21) $f(x) = \ell n(2x - 1)$

$2x - 1 > 0$

$2x > 1$

$x > \frac{1}{2}$

25) $\ell n\ f(x) = \ell n \frac{5x}{x^2 - 4} = \ell n\ 5x - \ell n\ (x^2 - 4)$

$= \ell n\ 5 + \ell n\ x - \ell n\ (x + 2)(x - 2)$

$= \ell n\ 5 + \ell n\ x - [\ell n(x + 2) + \ell n(x - 2)]$

$= \ell n\ 5 + \ell n\ x - \ell n(x + 2) - \ell n(x - 2)$

29) $\ell n\ f(x) = \ell n \sqrt{x^3(x + 1)} = \ell n(x^3(x + 1))^{\frac{1}{2}}$

$= \frac{1}{2}\ \ell n\ x^3\ (x + 1) = \frac{1}{2}\ (\ell n\ x^3 + \ell n\ (x + 1)) = \frac{1}{2}(3\ \ell n\ x + \ell n(x + 1))$

$= \frac{3}{2} \ell n\ x + \frac{1}{2} \ell n(x + 1)$

33) $3\ \ell n(x + 1) + 3\ \ell n(x - 1) = \ell n(x + 1)^3 + \ell n(x - 1)^3$

$= \ell n(x + 1)^3(x - 1)^3 = \ell n(x^2 - 1)^3$

37) $e^{\ell n \sqrt{x}} = \sqrt{x}$

41) $\ell n \frac{e^x}{e^{x-1}} = \ell n\ e^x - \ell n\ e^{x-1} = x - (x - 1) = x - x + 1 = 1$

45) $e^{\ell n(1-x)} = 2x$

$1 - x = 2x$

$1 = 3x$

$\frac{1}{3} = x$

49) $\ell n\ e^{\sqrt{x+1}} = 3$

$\sqrt{x + 1} = 3$

$(\sqrt{x + 1})^2 = 3^2$

$x + 1 = 9$

$x = 8$

53) $\ln x = \frac{1}{2}\ln 4 + \frac{2}{3}\ln 8$

$\ln x = \ln 4^{\frac{1}{2}} + \ln 8^{\frac{2}{3}}$

$\ln x = \ln 2 + \ln 4$

$\ln x = \ln (2 \cdot 4)$

$\ln x = \ln 8$

$x = 8$

57) $f(n) = \left(1 + \frac{1}{n}\right)^n$

$f(2) = \left(1 + \frac{1}{2}\right)^2 = (1.5)^2 = 2.2500$

$f(10) = \left(1 + \frac{1}{10}\right)^{10} = (1.1)^{10} \approx 2.5937$

$f(100) = \left(1 + \frac{1}{100}\right)^{100} = (1.01)^{100} \approx 2.7048$

$f(500) = \left(1 + \frac{1}{500}\right)^{500} = (1.002)^{500} \approx 2.7156$

$f(1000) = \left(1 + \frac{1}{1000}\right)^{1000} = (1.001)^{1000} \approx 2.7169$

$f(5000) = \left(1 + \frac{1}{5000}\right)^{5000} = (1.0002)^{5000} \approx 2.7180$

$f(10{,}000) = \left(1 + \frac{1}{10000}\right)^{10000} = (1.0001)^{10000} \approx 2.7181$

61) $h(x) = \ln \frac{x}{x+1}$

The outer function takes the natural log, so let $f(x) = \ln x$.

Then $g(x) = \frac{x}{x+1}$

$(f \circ g)(x) = f(g(x)) = f\left(\frac{x}{x+1}\right) = \ln \frac{x}{x+1}$

65) $F(x) = e^{(3x-1)^2}$

The outer function exponentiates on base e. ("Exponentiate" means to use the expression as an exponent on some base.)

So let $f(x) = e^x$.

The middle function squares, so let $g(x) = x^2$.

Then $h(x) = 3x - 1$.

$(f \circ g \circ h)(x) = f(g(h(x))) = f(g(3x-1)) = f((3x-1)^2) = e^{(3x-1)^2}$.

69) $f(x) = e^{2x} - 2xe^{2x} = e^{2x}(1 - 2x)$

$e^{2x} > 0$ for all x, so the sign of f(x) will be determined by the sign of $(1 - 2x)$.

$1 - 2x = 0$ when $x = \frac{1}{2}$

$f(x) > 0$ when $1 - 2x > 0$, on $(-\infty, \frac{1}{2})$.

$f(x) < 0$ when $1 - 2x < 0$, on $(\frac{1}{2}, \infty)$.

73) $$\ell n\left(\frac{x}{4} - \frac{\sqrt{x^2 - 4}}{4}\right) = \ell n\left(\frac{x - \sqrt{x^2 - 4}}{4}\right)$$

$$= \ell n\left(\frac{x - \sqrt{x^2 - 4}}{4} \cdot \frac{x + \sqrt{x^2 - 4}}{x + \sqrt{x^2 - 4}}\right)$$

$$= \ell n\left(\frac{x^2 - (x^2 - 4)}{4\ (x + \sqrt{x^2 - 4})}\right) = \ell n\left(\frac{4}{4\ (x + \sqrt{x^2 - 4})}\right)$$

$$= \ell n\left(\frac{1}{x + \sqrt{x^2 - 4}}\right) = \ell n\left(x + \sqrt{x^2 - 4}\right)^{-1}$$

$$= -\ \ell n\left(x + \sqrt{x^2 - 4}\right)$$

Section 7.5 Exponential Growth and Decay

1) $\frac{\ell n\ 6}{\ell n\ 2} = \frac{1.792}{0.693} = 2.586$

5) $\frac{\ell n\ 15}{\ell n\ 3} = \frac{\ell n\ (1.5 \times 10)}{\ell n\ 3} = \frac{\ell n\ 1.5 + \ell n\ 10}{\ell n\ 3} = \frac{0.405 + 2.303}{1.099} = \frac{2.708}{1.099} = 2.464$

9) $y = Ae^{kx}$

$y = 100e^{0.75(4)}$

$y = 100e^{3}$

$y = 100(20.1)$

$y = 2010$

13) $5000 = 50e^{2k}$

$100 = e^{2k}$

$\ell n\ 100 = 2k$

$\frac{\ell n\ 100}{2} = k$

17) $y = 10{,}000e^{0.6(7)}$

$y = 10{,}000e^{4.2}$

$y = 10{,}000(66.7)$

$y = 667{,}000$

21) $S = S_0e^{-0.04t}$

S_0 is 50. If half of it decays, then S = 25. Solve for t.

$$25 = 50e^{-0.04t}$$

$$\frac{1}{2} = e^{-0.04t}$$

$$\ell n\ \frac{1}{2} = -0.04t$$

$$-25(-\ \ell n\ 2) = t$$

$$-25(-0.693) = t$$

$$17.33 = t$$

17.33 years

25) Use the formula $y = Ae^{kx}$. To find the half-life, k must be known. Use the given information to find k.

$$4 = 5e^{k(30)}$$

$$0.8 = e^{k(30)}$$

$$\ell n\ 0.8 = 30k$$

$$-\ \frac{0.223}{30} = k$$

$$-0.00743 = k$$

Use this value of k to solve for x and get the half-life.

$$\frac{1}{2} = e^{-0.00743x}$$

$$\ell n\ 0.5 = -0.00743x$$

$$\frac{-0.693}{-0.00743} = x$$

$$93.3 = x$$

93.3 seconds

29) Use the formula $y = Ae^{(\ell n\ 0.5/5750)x}$

$y = \frac{3}{5}A$

$$\frac{3}{5}A = Ae^{(\ell n 0.5/5750)x}$$

$$\frac{3}{5} = e^{(\ell n\ 0.5/5750)x}$$

$$\ell n\ 0.6 = \frac{\ell n\ 0.5}{5750}\ x$$

$$-0.511 = \frac{-0.693}{5750}\ x$$

$$\frac{(-0.511)(5750)}{-0.693} = x$$

$$4239.9 = x$$

4200 years

33) Use the formula $A = P\left(1 + \frac{r}{n}\right)^{nt}$

$P = 10{,}000 \quad t = 5 \quad r = 0.09$

a) $n = 4$

$$A = 10{,}000\left(1 + \frac{0.09}{4}\right)^{4(5)}$$
$$A = 10{,}000(1.0225)^{20}$$
$$A = 10{,}000(1.5605)$$
$$A = \$15{,}605$$

b) $n = 12$

$$A = 10{,}000\left(1 + \frac{0.09}{12}\right)^{12(5)}$$
$$A = 10{,}000(1.0075)^{60}$$
$$A = 10{,}000(1.5657)$$
$$A = \$15{,}657$$

c) $n = 52$

$$A = 10{,}000\left(1 + \frac{0.09}{52}\right)^{52(5)}$$
$$A = 10{,}000(1.0017308)^{260}$$
$$A = 10{,}000(1.5677)$$
$$A = \$15{,}677$$

d) $n = 365$

$$A = 10{,}000\left(1 + \frac{0.09}{365}\right)^{365(5)}$$
$$A = 10{,}000(1.0002465)^{1825}$$
$$A = 10{,}000(1.5682)$$
$$A = \$15{,}682$$

e) Use the formula $A = Pe^{rt}$

$$A = 10000e^{0.09(5)}$$
$$A = 10000e^{0.45}$$
$$A = 10000(1.5683)$$
$$A = \$15{,}683$$

37) Use the formula $A = Pe^{rt}$

$P = 5000 \quad A = 10000 \quad r = 0.09$

Solve for t.

$$10{,}000 = 5000e^{0.09t}$$
$$2 = e^{0.09t}$$
$$\ell n\ 2 = 0.09t$$
$$\frac{\ell n\ 2}{0.09} = t$$
$$\frac{0.693}{0.09} = t$$
$$7.7 = t$$

7.7 years

To find the doubling time at 12%, change to $r = 0.12$.

$$2 = e^{0.12t}$$

$\frac{\ell n\ 2}{0.12} = t \qquad \frac{0.693}{0.12} = t \qquad 5.8 = t \qquad$ 5.8 years

41) Use the formula $A = P\left(1 + \frac{r}{n}\right)^{nt}$

$P = 4{,}000$ $\quad$ $A = 8000$ $\quad$ $r = 0.08$ $\quad$ $n = 12$

Solve for t.

$$8000 = 4000\left(1 + \frac{0.08}{12}\right)^{12t}$$

$$2 = (1.00\overline{6})^{12t}$$

$$\ln 2 = 12t \ln(1.00\overline{6})$$

$$\frac{\ln 2}{12 \ln(1.00\overline{6})} = t$$

$$\frac{0.693}{0.080} = t$$

$$8.7 = t$$

8.7 years

45) Use the formula $A = P\left(1 + \frac{r}{n}\right)^{nt}$.

$A = 20{,}000$ $\quad$ $r = 0.12$ $\quad$ $n = 12$ $\quad$ $t = 5$

Solve for P.

$$20{,}000 = P\left(1 + \frac{0.12}{12}\right)^{12(5)}$$

$$20{,}000 = P(1.01)^{60}$$

$$20{,}000 = P(1.8167)$$

$$\frac{20{,}000}{1.8167} = P$$

$$\$11{,}009 = P$$

Section 7.6 Scientific Notation

1) $4{,}680 = 4.68 \times 10^3$

5) $7{,}583{,}000 = 7.583 \times 10^6$

9) $0.000000555 = 5.55 \times 10^{-7}$

13) $7.89 \times 10^4 = 78900$

17) $1.74 \times 10^{-1} = 0.174$

21) $9.06 \times 10^{-2} = 0.0906$

25) $\dfrac{10^1 \times 10^2 \times 10^3 \times 10^4}{10^{10}} = \dfrac{10^{10}}{10^{10}} = 10^0$

29) $\dfrac{1}{0.0005} = \dfrac{1}{5 \times 10^{-4}} = \dfrac{1}{5} \times \dfrac{1}{10^{-4}} = \dfrac{1}{5} \times 10^4 = 0.2 \times 10^4 = 2000$

33) $\frac{(0.000025)}{(0.0625)(0.02)} = \frac{2.5 \times 10^{-5}}{(6.25 \times 10^{-2})(2 \times 10^{-2})} = \frac{2.5}{(6.25)(2)} \times \frac{10^{-5}}{10^{-2}10^{-2}}$

$= 0.2 \times 10^{-1} = 0.02$

37) $\sqrt{1,440,000} = (1,440,000)^{\frac{1}{2}} = (1.44 \times 10^6)^{\frac{1}{2}} = (1.44)^{\frac{1}{2}} \times (10^6)^{\frac{1}{2}}$

$= 01.2 \times 10^3 = 1200$

41) $\frac{(1,728,000)^{1/3}}{(0.06)(400)^2} = \frac{(1.728 \times 10^6)^{1/3}}{(6 \times 10^{-2})(4 \times 10^2)^2} = \frac{(1.728)^{\frac{1}{3}}}{6(4)^2} \times \frac{(10^6)^{\frac{1}{3}}}{(10^{-2})(10^2)^2}$

$= \frac{1.2}{96} \times \frac{10^2}{(10^{-2})(10^4)} = 0.0125 \times 10^0 = 0.0125$

Section 7.7 Common Logarithms and Applications

1) $\log 457 = \log (4.57 \times 10^2) = \log 4.57 + \log 10^2 = 0.6599 + 2 = 2.6599$

5) $\log 72.9 = \log (7.29 \times 10^1) = \log 7.29 + \log 10^1 = 0.8627 + 1 = 1.8627$

9) $\log N = 1.8331 = .8331 + 1$

$N = 6.81 \times 10^1$

$N = 68.1$

13) $P = (512)(84,000)$

$\log P = \log(512) + \log(84,000) = \log(5.12 \times 10^2) + \log(8.4 \times 10^4)$

$= 2.7093 + 4.9243 = 7.6336$

$P = 4.3 \times 10^7$

$P = 43,000,000$

17) $Q = \frac{(0.421)(81.7)}{(368)(750)}$

Add the logs of the numerator:

$\log 0.421 = 9.6243 - 10$

$\log 81.7 = 1.9122$

$11.5365 - 10 = \log N$

17) continued

Add the logs of the denominator:

$$\begin{aligned} \log 368 &= 2.5658 \\ \log 750 &= \underline{2.8751} \\ & \quad 5.4409 = \log D \end{aligned}$$

Find the log of the quotient with log N - log D.

$$\begin{array}{r} 11.5365 - 10 \\ \underline{-\ 5.4409 \qquad} \\ 6.0956 - 10 = \log Q \\ 1.25 \times 10^{-4} = Q \\ 0.000125 = Q \end{array}$$

21) $R = \sqrt{\frac{28.3}{(579)(621)}} = \left(\frac{28.3}{(579)(621)}\right)^{\frac{1}{2}}$

$$\log R = \frac{1}{2} \log \left(\frac{28.3}{(579)(621)} \right)$$

Find the log of the numerator:

$$\log 28.3 = 1.4518 = \log N$$

Add the logs of the denominator:

$$\begin{aligned} \log 579 &= 2.7627 \\ \log 621 &= \underline{2.7931} \\ & \quad 5.5558 = \log D \end{aligned}$$

Find the log of the quotient with log N - log D.

$$\begin{array}{r} 1.4518 = 11.4518 - 10 \\ \underline{-\ 5.5558 \qquad} \\ 5.8960 - 10 \end{array}$$

$$\log R = \frac{1}{2} (5.8960 - 10) = 2.9480 - 5$$

$$R = 8.87 \times 10^{-3}$$

$$R = 0.00887$$

25) Cost per gallon equals total cost divided by the number of gallons.

$$C = \frac{16.93}{14}$$

$$\log C = \log 16.93 - \log 14 - 1.2279 - 1.1461 = 0.0818$$

$$C = \$1.21 \text{ per gallon}$$

29) $P = \dfrac{A}{(1 + r)^t}$

$A = \$8440 \qquad r = 0.128 \qquad t = 6$

$P = \dfrac{8440}{(1 + 0.128)^6} = \dfrac{8440}{(1.128)^6}$

$\log P = \log 8440 - \log (1.128)^6 = \log 8440 - 6 \log 1.128$

$= 3.9263 - 6(0.0531) = 3.9263 - 0.3186 = 3.6077$

$P = 4.05 \times 10^3$

$P = \$4050$

33) $V = \frac{4}{3} \pi r^2$

$V = \frac{4}{3} (3.14)(12)^3$

$\log V = \log(1.33) + \log 3.14 + \log (12)^3 = 0.1239 + \log 3.14 + 3 \log 12$

$= 0.1239 + 0.4969 + 3(1.0792) = 0.1239 + 0.4969 + 3.2376$

$= 3.8584$

$V = 7.22 \times 10^3$

$V = 7220$ cubic centimeters

37) $1 \leq x < 10$

$\log 1 \leq \log x < \log 10$

$0 \leq \log x < 1$

Sample Test Questions for Chapter 7

Part I

1) (i) a (ii) c (iii) e (iv) h (v) d

(vi) b

2) (a) $81^x = 9$

$81^x = 81^{\frac{1}{2}}$

$x = \frac{1}{2}$

(b) $e^{\ln x} = 9$

$x = 9$

3) $\frac{1}{2^{x+1}} = 64$

$2^{-(x+1)} = 64$

$2^{-(x+1)} = 2^6$

$-x - 1 = 6$

$x = -7$

4) (a) $\log_5 125 = 3$

$5^3 = 125$

(b) $\log_9 \frac{1}{81} = -2$

$9^{-2} = \frac{1}{81}$

5) (a) $16^{\frac{3}{4}} = 8$

$\log_{16} 8 = \frac{3}{4}$

(b) $7^{-2} = \frac{1}{49}$

$\log_7 \frac{1}{49} = -2$

6) (a) $\log_b \frac{27}{8} = -3$

$\left(b^{-3}\right) = \frac{27}{8}$

$(b^{-3})^{-\frac{1}{3}} = \left(\frac{27}{8}\right)^{-\frac{1}{3}}$

$b = \frac{1}{\left(\frac{27}{8}\right)^{\frac{1}{3}}}$

$b = \frac{2}{3}$

(b) $\log_{10} 0.01$

$10^x = 0.01$

$10^x = 10^{-2}$

$x = -2$

7) $y = f(x) = 2^x - 4$

Domain: all reals

Asymptote: $y = -4$

8) $y = f(x) = \log_3(x + 4)$

Domain: $x + 4 > 0$, $x > -4$

Asymptote: $x = -4$

9) $\frac{1}{3}\log_b x - 2\log_b(x+2) = \log_b x^{\frac{1}{3}} - \log_b(x+2)^2 = \log_b \frac{x^{\frac{1}{3}}}{(x+2)^2}$

10) $\log_{25}x^2 - \log_{25}(2x-5) = \frac{1}{2}$

$$\log_{25}\frac{x^2}{2x-5} = \frac{1}{2}$$

$$25^{\frac{1}{2}} = \frac{x^2}{2x-5}$$

$$5 = \frac{x^2}{2x-5}$$

$$10x - 25 = x^2$$

$$0 = x^2 - 10x + 25$$

$$0 = (x-5)^2$$

$$x - 5 = 0$$

$$x = 5$$

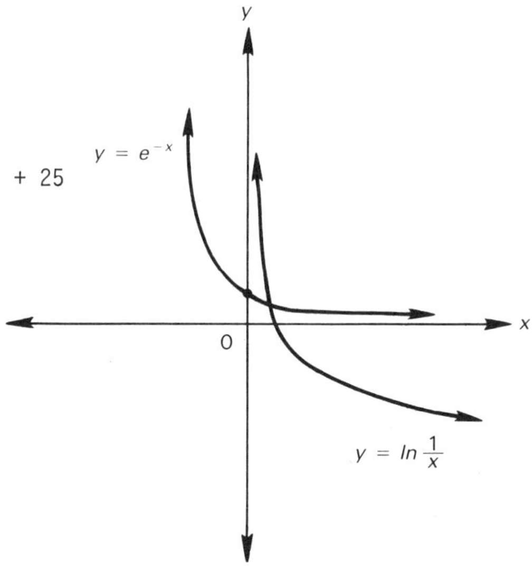

11) $y = e^{-x}$

$g(x) = x = e^{-y}$

$\ell n\ x = -y$

$-\ell n\ x = y$

$\ell n \frac{1}{x} = y$

Also, $y = e^{-x} = \left(\frac{1}{e}\right)^x$.

Thus $y = \ell n_{\frac{1}{e}} x$ is another form

of the inverse.

12) $\ell n \frac{x^3}{(x+1)\sqrt{x^2+2}} = \ell n\ x^3 - [\ell n(x+1) + \ell n(x^2+2)^{\frac{1}{2}}]$

$= 3\ \ell n\ x - \ell n(x+1) - \frac{1}{2}\ell n\ (x^2+2)$

13) $4^{2x} = 5$

$\ell n\ 4^{2x} = \ell n\ 5$

$2x\ \ell n\ 4 = \ell n\ 5$

$x = \frac{\ell n\ 5}{2\ \ell n\ 4}$

14) $y = Ae^{-0.04t}$

$A = 50 \qquad y = 25$

Solve for t.

$25 = 50e^{-0.04t}$

$\frac{1}{2} = e^{-0.04t}$

$\ell n \frac{1}{2} = -0.04t$

$-\ell n \frac{1}{2} = \frac{1}{25}t$

$\ell n \left(\frac{1}{2}\right)^{-1} = \frac{1}{25}t$

$25 \ell n\ 2 = t$

15) $A = P\left(1 + \frac{r}{n}\right)^{nt}$

$A = 2000\left(1 + \frac{0.08}{4}\right)^{4(6)}$

$A = 2000(1.02)^{24}$

16) $A = Pe^{rt}$

$A = 6000e^{0.07(10)}$

$A = 6000e^{0.7}$

$A = \$12083$

(\$12060 using Table II)

17) (a) $42{,}900 = 4.29 \times 10^4$

(b) $3.25 \times 10^{-3} = 0.00325$

18) $$\frac{\sqrt{144{,}000{,}000}}{(2000)^2(0.0005)} = \frac{(1.44 \times 10^8)^{\frac{1}{2}}}{(2 \times 10^3)^2(5 \times 10^{-4})} = \frac{(1.44)^{\frac{1}{2}}}{2^2 \cdot 5} \times \frac{(10^8)^{\frac{1}{2}}}{(10^3)^2(10^{-4})}$$

$$= \frac{1.2}{20} \times \frac{10^4(10^4)}{10^6} = 0.06 \times 10^2 = 6$$

19) $$N = \frac{(2430)^2}{(0.842)\sqrt{27.9}}$$

$$\log N = \log \frac{(2430)^2}{(0.842)\sqrt{27.9}}$$

Find the log of the numerator:

$\log(2430)^2 = 2 \log 2430 = 2(3.3856) = 6.7712$

Add the logs of the denominator:

$\log (0.842) = \qquad 9.9253 - 10$

$\log \sqrt{27.9} = \frac{1}{2} \log 27.9 = \frac{1}{2}(1.4456) = \underline{0.7228}$

$10.6481 - 10$

$.6481$

19) continued

Find the log of the quotient by subtracting the log of the denominator from the log of the numerator.

$\log N = 6.7712 - .6481 = 6.1231$

$$N = 1.33 \times 10^6$$
$$N = 1{,}330{,}000$$

20) $\ell n\, 2 - \ell n(1 - x) = 1 - \ell n(x + 1)$

$$\ell n \frac{2}{1 - x} = 1 - \ell n(x + 1)$$

$$\ell n \frac{2}{1 - x} + \ell n(x + 1) = 1$$

$$\ell n \left[\frac{2}{1 - x} \cdot (x + 1)\right] = 1$$

$$e^1 = \frac{2x + 2}{1 - x}$$

$$e(1 - x) = 2x + 2$$
$$e - ex = 2x + 2$$
$$e - 2 = 2x + ex$$
$$e - 2 = x(2 + e)$$
$$\frac{e - 2}{e + 2} = x$$

Sample Test Questions for Chapter 7

Part II

1) (d)

2) (a)

$b^{x^2-2x} = 1$

$b^{x^2-2x} = b^0$

$x^2 - 2x = 0$

$x(x - 2) = 0$

$x = 0 \quad \text{or} \quad x = 2$

3) (c)

4) (e)

$x + 3 > 0, \; x > -3$

5) (a)

$\log_3 x + \log_3(2x + 51) = 4$

$\log_3 x(2x + 51) = 4$

$x(2x + 51) = 3^4$

$2x^2 + 51x = 81$

$2x^2 + 51x - 81 = 0$

$(2x - 3)(x + 27) = 0$

$2x - 3 = 0 \quad \text{or} \quad x + 27 = 0$

$2x = 3 \qquad x = -27$

$x = \frac{3}{2}$ — $x = -27$: Not allowed because $x > 0$.

6) (c)

$\log_b \frac{1}{64} = -\frac{3}{2}$

$b^{-\frac{3}{2}} = \frac{1}{64}$

$\left(b^{-\frac{3}{2}}\right)^{-\frac{2}{3}} = \left(\frac{1}{64}\right)^{-\frac{2}{3}}$

$b = 16$

7) (e) False; $2 \log_b x = \log_b x^2$

I. True

II. False; $\log_b A + \log_b B = \log_b AB$

III. False; $\log_b A - \log_b B = \log_b \frac{A}{B}$

8) (d)

9) (c)

10) (b)

$$\ell n\ (x + 1) = 1 + \ell n\ x$$
$$\ell n\ (x + 1) - \ell n\ x = 1$$
$$\ell n \frac{(x + 1)}{x} = 1$$
$$e^{1} = \frac{x + 1}{x}$$
$$ex = x + 1$$
$$ex - x = 1$$
$$x(e - 1) = 1$$
$$x = \frac{1}{e - 1}$$

11) (a)

$$9^{x-1} = 4$$
$$\ell n\ 9^{x-1} = \ell n\ 4$$
$$(x - 1)\ \ell n\ 9 = \ell n\ 4$$
$$x - 1 = \frac{\ell n\ 4}{\ell n\ 9}$$
$$x = \frac{\ell n\ 4}{\ell n\ 9} + 1$$

12) (b)

$$30 = 40e^{8k}$$
$$\frac{3}{4} = e^{8k}$$
$$\ell n \frac{3}{4} = \ell n\ e^{8k}$$
$$\ell n \frac{3}{4} = 8k\ \ell n\ e$$
$$\ell n\ 0.75 = 8k$$
$$\frac{\ell n\ 0.75}{8} = k$$
$$y = 40e^{kx}$$
$$y = 40e^{\left(\frac{\ell n\ 0.75}{8}\right) x}$$
$$y = 40e^{\frac{x\ \ell n\ 0.75}{8}}$$

13) (c) $A_t = P\left(1 + \frac{r}{n}\right)^{nt}$

$$A_5 = 3000\left(1 + \frac{0.084}{12}\right)^{12.5}$$
$$A = 3000(1.007)^{60}$$

14) (b)

$$\frac{(0.00054)(9200)}{370} = \frac{(5.4 \times 10^{-4})(9.2 \times 10^{3})}{(3.7) \times 10^{2}} = \frac{(5.4)(9.2)}{3.7} \times 10^{-4+3-2}$$
$$= \frac{(5.4)(9.2)}{3.7} \times 10^{-3}$$

15) (d)

CHAPTER 8
TRIGONOMETRY

Section 8.1 Trigonometric Ratios

1) $\sin\theta = \dfrac{\text{opposite}}{\text{hypotenuse}} = \dfrac{8}{17}$ $\qquad \cot\theta = \dfrac{\text{adjacent}}{\text{opposite}} = \dfrac{15}{8}$

$\cos\theta = \dfrac{\text{adjacent}}{\text{hypotenuse}} = \dfrac{15}{17}$ $\qquad \sec\theta = \dfrac{\text{hypotenuse}}{\text{adjacent}} = \dfrac{17}{15}$

$\tan\theta = \dfrac{\text{opposite}}{\text{adjacent}} = \dfrac{8}{15}$ $\qquad \csc\theta = \dfrac{\text{hypotenuse}}{\text{opposite}} = \dfrac{17}{8}$

5) $\sin\theta = \dfrac{\text{opposite}}{\text{hypotenuse}} = \dfrac{12t}{13t} = \dfrac{12}{13}$ $\qquad \cot\theta = \dfrac{\text{adjacent}}{\text{opposite}} = \dfrac{5}{12}$

$\cos\theta = \dfrac{\text{adjacent}}{\text{hypotenuse}} = \dfrac{5t}{13t} = \dfrac{5}{13}$ $\qquad \sec\theta = \dfrac{\text{hypotenuse}}{\text{adjacent}} = \dfrac{13}{5}$

$\tan\theta = \dfrac{\text{opposite}}{\text{adjacent}} = \dfrac{12t}{5t} = \dfrac{12}{5}$ $\qquad \csc\theta = \dfrac{\text{hypotenuse}}{\text{opposite}} = \dfrac{13}{12}$

9) $\sqrt{3} = \tan\theta$

$60^\circ = \theta$

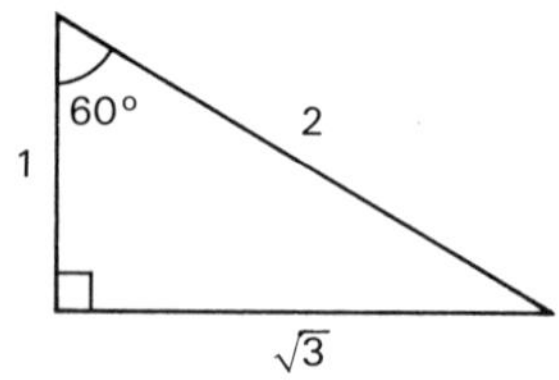

13) $a = 6, \quad b = 8$

$a^2 + b^2 = 6^2 + 8^2 = 100 = c^2$

$10 = c$

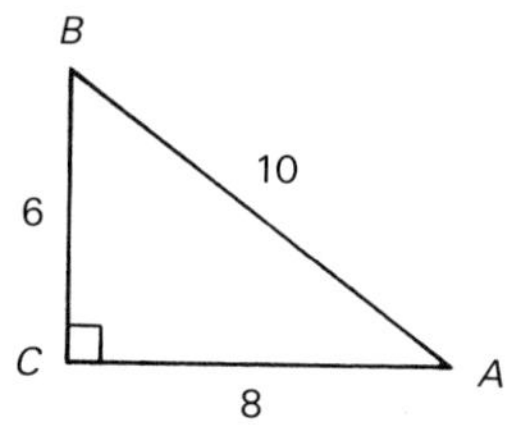

$\sin A = \dfrac{3}{5} = \cos B$ $\qquad \cot A = \dfrac{4}{3} = \tan B$

$\cos A = \dfrac{4}{5} = \sin B$ $\qquad \sec A = \dfrac{5}{4} = \csc B$

$\tan A = \dfrac{3}{4} = \cot B$ $\qquad \csc A = \dfrac{5}{3} = \sec B$

17) $a = 1 \qquad b = 2$

$a^2 + b^2 = 1^2 + 2^2 = 5 = c^2$

$\sqrt{5} = c$

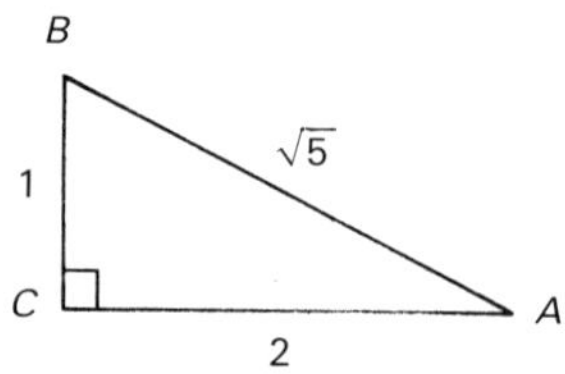

$\sin A = \dfrac{1}{\sqrt{5}} = \cos B$ $\qquad \cot A = 2 = \tan B$

$\cos A = \dfrac{2}{\sqrt{5}} = \sin B$ $\qquad \sec A = \dfrac{\sqrt{5}}{2} = \csc B$

17) continued

$\tan A = \frac{1}{2} = \cot B$ $\qquad$ $\csc A = \sqrt{5} = \sec B$

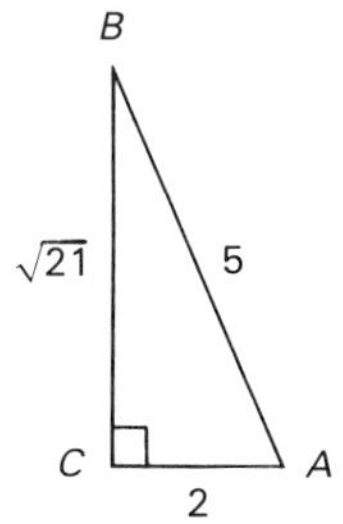

21) $b = 2 \qquad c = 5$

$c^2 = b^2 + a^2$

$25 = 4 + a^2$

$21 = a^2$

$\sqrt{21} = a$

$\sin A = \frac{\sqrt{21}}{5} = \cos B$ $\qquad$ $\cot A = \frac{2}{\sqrt{21}} = \tan B$

$\cos A = \frac{2}{5} = \sin B$ $\qquad$ $\sec A = \frac{5}{2} = \csc B$

$\tan A = \frac{\sqrt{21}}{2} = \cot B$ $\qquad$ $\csc A = \frac{5}{\sqrt{21}} = \sec B$

25) $(\tan X)(\cot X) = \frac{x}{y} \cdot \frac{y}{x} = 1$

29) $\sin^2 X + \cos^2 X = \left(\frac{x}{z}\right)^2 + \left(\frac{y}{z}\right)^2 = \frac{x^2}{z^2} + \frac{y^2}{z^2} = \frac{x^2 + y^2}{z^2}$

$= \frac{z^2}{z^2}$ (By the Pythagorean theorem)

$= 1$

33) $\sin A = \frac{3}{4} = \frac{a}{c} \qquad a = 3 \qquad c = 4$

$c^2 = a^2 + b^2$

$16 = 9 + b^2$

$7 = b^2$

$\sqrt{7} = b$

$\cos A = \frac{b}{c} = \frac{\sqrt{7}}{4}$

$\tan A = \frac{a}{b} = \frac{3}{\sqrt{7}}$

$\cot A = \frac{b}{a} = \frac{\sqrt{7}}{3}$

$\sec A = \frac{c}{b} = \frac{4}{\sqrt{7}}$

$\csc A = \frac{c}{a} = \frac{4}{3}$

37) $\sec B = \dfrac{\sqrt{11}}{\sqrt{2}} = \dfrac{c}{a}$ $c = \sqrt{11}$, $a = \sqrt{2}$

$$c^2 = a^2 + b^2$$
$$11 = 2 + b^2$$
$$9 = b^2$$
$$3 = b$$

$$\sin B = \frac{b}{c} = \frac{3}{\sqrt{11}}$$

$$\cos B = \frac{a}{c} = \frac{\sqrt{2}}{\sqrt{11}}$$

$$\tan B = \frac{b}{a} = \frac{3}{\sqrt{2}}$$

$$\cot B = \frac{a}{b} = \frac{\sqrt{2}}{3}$$

$$\csc B = \frac{c}{b} = \frac{\sqrt{11}}{3}$$

41) $a = x$ $c = 4$

$$a^2 + b^2 + c^2$$
$$x^2 + b^2 = 16$$
$$b^2 = 16 - x^2$$
$$b = \sqrt{16 - x^2}$$

$$\sin A = \frac{a}{c} = \frac{x}{4}$$

$$\cos A = \frac{b}{c} = \frac{\sqrt{16 - x^2}}{4}$$

$$\frac{4\sin^2 A}{\cos A} = \frac{4\left(\frac{x}{4}\right)^2}{\frac{\sqrt{16 - x^2}}{4}} = \frac{16\left(\frac{x^2}{16}\right)}{\sqrt{16 - x^2}} = \frac{x^2}{\sqrt{16 - x^2}}$$

45) Since the sum $x^2 + 16$ is involved the perpendicular sides of the triangle must have measures x and 4 so that the hypotenuse will have measure $\sqrt{x^2 + 16}$.

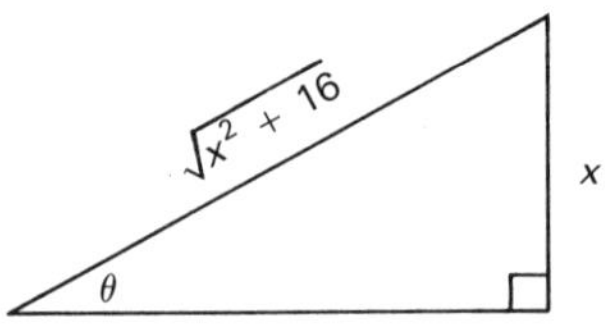

$$\tan\theta = \frac{x}{4}$$

$$\cos\theta = \frac{4}{\sqrt{x^2 + 16}}$$

$$(\tan^2\theta)(\cos^2\theta) = \left(\frac{x}{4}\right)^2\left(\frac{4}{\sqrt{x^2 + 16}}\right)^2 = \frac{x^2}{16} \cdot \frac{16}{x^2 + 16} = \frac{x^2}{x^2 + 16}$$

49) The chord has length 2y. Therefore each leg of the isosceles right triangle measures 2y.

$$A = \frac{1}{2}\text{ bh} = \frac{1}{2}(2y)(2y) = 2y^2$$

Also $x^2 + y^2 = 9$

$$y^2 = 9 - x^2$$

Substitute $(9 - x^2)$ for y^2 in the area equation.

$$A = 2(9 - x)^2$$

Section 8.2 Angle Measure

1) $45 \cdot \frac{\pi}{180} = \frac{\pi}{4}$

5) $270 \cdot \frac{\pi}{180} = \frac{3\pi}{2}$

9) $225 \cdot \frac{\pi}{180} = \frac{5\pi}{4}$

13) $330 \cdot \frac{\pi}{180} = \frac{11\pi}{6}$

17) $100 \cdot \frac{\pi}{180} = \frac{5\pi}{9}$

$100(0.0175) = 1.75$

21) $\pi \cdot \frac{180}{\pi} = 180^{\circ}$

25) $\frac{5\pi}{9} \cdot \frac{180}{\pi} = 100^{\circ}$

29) $\frac{5\pi}{4} \cdot \frac{180}{\pi} = 225^{\circ}$

33) $\frac{5\pi}{18} \cdot \frac{180}{\pi} = 50^{\circ}$

37) $3 \cdot \frac{180}{\pi} = \left(\frac{540}{\pi}\right)^{\circ}$

$3(57.296) = 171.9^{\circ}$

41) $\frac{\pi}{6} \cdot \frac{180}{\pi} = 30^{\circ}$

$\tan\frac{\pi}{6} = \tan 30^{\circ} = \frac{1}{\sqrt{3}} = \frac{\sqrt{3}}{3}$

45) $\theta = \frac{\frac{\pi}{2}}{1} = \frac{\pi}{2}$

49) $30 \cdot \frac{\pi}{180} = \frac{\pi}{6}$

$A = \frac{1}{2}(12)^2\left(\frac{\pi}{6}\right)$

$A = 12\pi \text{ cm}^2$

53) $315 \cdot \frac{\pi}{180} = \frac{21\pi}{12}$

$A = \frac{1}{2}(12)^2\left(\frac{21\pi}{12}\right)$

$A = 126\pi \text{ cm}^2$

57) The area formula requires r.

Use $\theta = \frac{s}{r}$ and the given information to find r.

$$45 \cdot \frac{\pi}{180} = \frac{\pi}{4} = \theta \qquad s = \frac{\pi}{2}$$

$$\theta = \frac{s}{r}$$

$$\frac{\pi}{4} = \frac{\frac{\pi}{2}}{r}$$

$$r = \frac{\frac{\pi}{2}}{\frac{\pi}{4}} = 2$$

$$A = \frac{1}{2}(2)^2\left(\frac{\pi}{4}\right) = \frac{\pi}{2} \text{ cm}^2 \approx 1.6 \text{ cm}^2$$

61) The surface area of the cup is the same as the area of the sector.

$$270 \cdot \frac{\pi}{180} = \frac{3\pi}{2} = \theta \qquad r = 8$$

$$A = \frac{1}{2}(8)^2\left(\frac{3\pi}{2}\right) = 48\pi \approx 150.7 \text{ sq. in.}$$

65) a) Calculations to find the velocity require the arc length, which in this case is one complete orbit. The radius of the circular orbit is the earth's radius of 4000 miles plus the 300 miles of altitude above the earth, a total of 4300 miles.

$$s = 2\pi r = 2(3.14)(4300) = 27{,}004$$

A complete orbit takes 1.75 hr.

$$v = \frac{s}{t} = \frac{27004}{1.75} \approx 15431 \text{ mph}$$

There are 5280 feet in a mile and 3600 seconds in an hour.

$$\frac{15431(5280)}{1(3600)} = \frac{81475680}{3600} \approx 22632 \text{ ft/sec}$$

Note: If the approximation $\pi = 3.14$ is replaced by $\pi = 3.1415927$ as is used in some calculators, the answers in part (a) become 15439 mph and 22644 ft/sec.

b) The satelite completes a full circle of 2π radians in $\frac{7}{4}$ hours.

$$w = \frac{2\pi}{\frac{7}{4}} = \frac{8\pi}{7} \text{ rad/hr.}$$

69) a) Calculations for the speed require the circumference.

$$C = \pi D = 22\pi \text{ inches}$$

The bike travels 240 revolutions per minute and 22π inches each revolution.

$$240 \cdot 22\pi = 5280\pi \text{ inches/minute} = 440\pi \text{ ft/min.}$$

b) There are 5280 feet in a mile and the bike is going 440π feet per minute. Use $T = \frac{D}{R}$

$\frac{5280}{440\pi} = 3.8$ minutes (approximately)

Section 8.3 Right Triangle Trigonometry

1) $\tan \frac{\pi}{6} = \frac{x}{25}$

$\frac{1}{\sqrt{3}} = \frac{x}{25}$

$x\sqrt{3} = 25$

$x = \frac{25}{\sqrt{3}}$

5) $\sin \frac{\pi}{3} = \frac{x}{40}$

$\frac{\sqrt{3}}{2} = \frac{x}{40}$

$2x = 40\sqrt{3}$

$x = 20\sqrt{3}$

9) $\sin \theta = \frac{5}{5\sqrt{2}}$

$\sin \theta = \frac{1}{\sqrt{2}}$

$\theta = \frac{\pi}{4}$

13) $\sin 42^\circ = 0.6691$

17) $\sec 89^\circ = 57.30$

21) $\cot 44^\circ 50' = 1.006$

25) $\tan \theta = 6.314$

$\theta = 81^\circ$

$\theta = 1.4137$ radians

29) $\cos \theta = 0.9940$

(use the nearest entry, 0.9939)

$\theta = 6^\circ 20'$

$\theta = 0.1105$ radians

33) $B = 40^o20'$ $\quad$ a = 20 is the side adjacent to B

b is the side opposite B

$$\tan B = \frac{b}{a}$$

$$\tan 42^o20' = \frac{b}{20}$$

$$0.9110 = \frac{b}{20}$$

$$18.2 = b$$

37) B = ? $\quad$ b = 9 is the side opposite B

c = 25 is the hypotenuse

$$\sin B = \frac{b}{c}$$

$$\sin B = \frac{9}{25}$$

$$\sin B = 0.36$$

(The nearest entry is 0.3611)

$$B = 21^o10'$$

41)

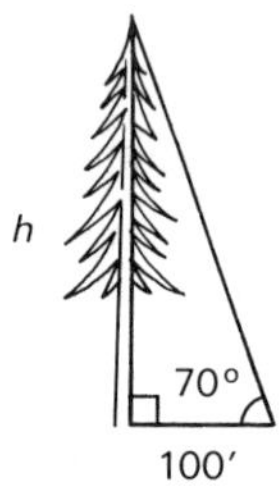

Let h represent the height of the tree.

$$\tan 70^o = \frac{h}{100}$$

$$2.747 = \frac{h}{100}$$

$$274.7 = h$$

The tree is 274.7 feet tall.

45)

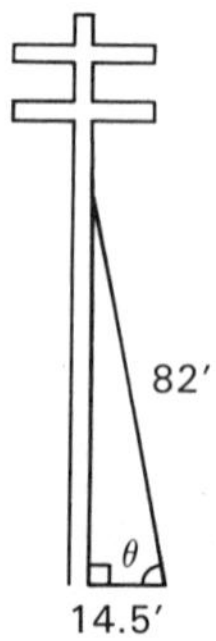

$$\cos \theta = \frac{14.5}{82}$$

$$\cos \theta = 0.1768$$

(The closest entry is 0.1765.)

$$\theta = 79^o50'$$

The cable makes a $79^o50'$ angle with the ground.

49)

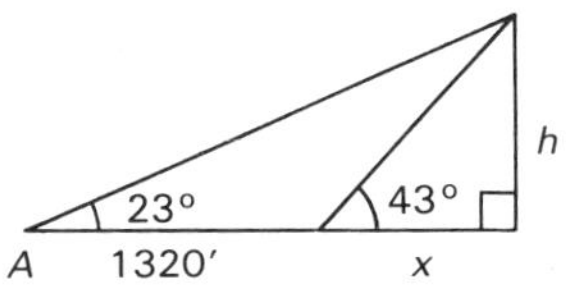

Let h represent the height of the mountain.
Let x represent the base of the right triangle with 43^0 angle.
Now use the cotangent of 23^0 and of 43^0 to solve for h as follows.

$$\cot 23^0 = \frac{1320 + x}{h} = \frac{1320}{h} + \frac{x}{h}$$

$$= \frac{1320}{h} + \cot 43^0, \left(\cot 43^0 = \frac{x}{h}\right)$$

Then

$$2.356 = \frac{1320}{h} + 1.072$$

$$1.284 = \frac{1320}{h}$$

$$h = \frac{1320}{1.284} = 1028$$

The mountain is 1028 ft. high.

Section 8.4 Extending the Definitions

1) a) 73^0 is in quadrant I.

$\emptyset = 73^0$

b) The coterminal angle is

$73^0 - 360^0 = -287^0$.

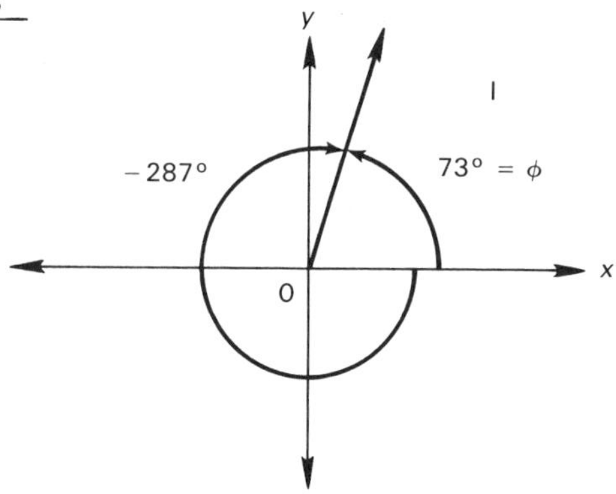

5) a) $\frac{5\pi}{4}$ is in quadrant III.

$\emptyset = \frac{\pi}{4}$

b) The coterminal angle is

$\frac{5\pi}{4} - 2\pi = -\frac{3\pi}{4}$

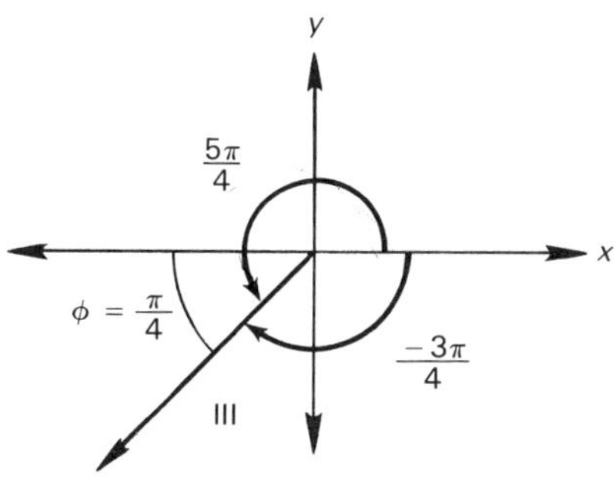

9) a) $-850^{\circ} + 2(360^{\circ}) = -130^{\circ}$

-850° is in quadrant III.

$\emptyset = 50^{\circ}$

b) The coterminal angle is

$-130^{\circ} + 360^{\circ} = 230^{\circ}$.

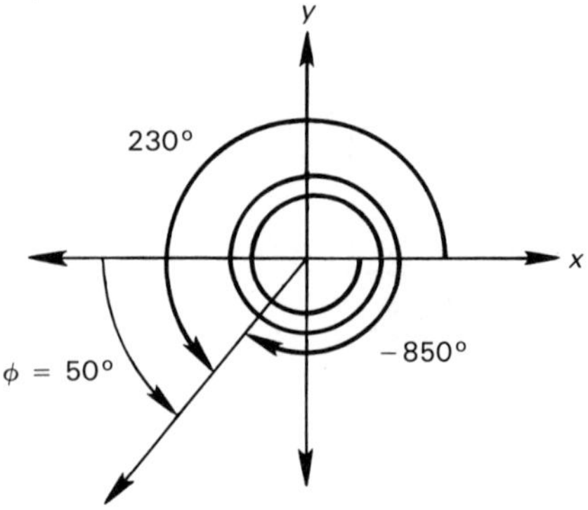

13) $\theta = \frac{2\pi}{3}$

The reference angle is $\frac{\pi}{3}$.

The hypotenuse is 1.

$\sin\frac{\pi}{3} = y$

$\frac{\sqrt{3}}{2} = y$

$-\cos\frac{\pi}{3} = x$

$-\frac{1}{2} = x$

$P = \left(-\frac{1}{2}, \frac{\sqrt{3}}{2}\right)$

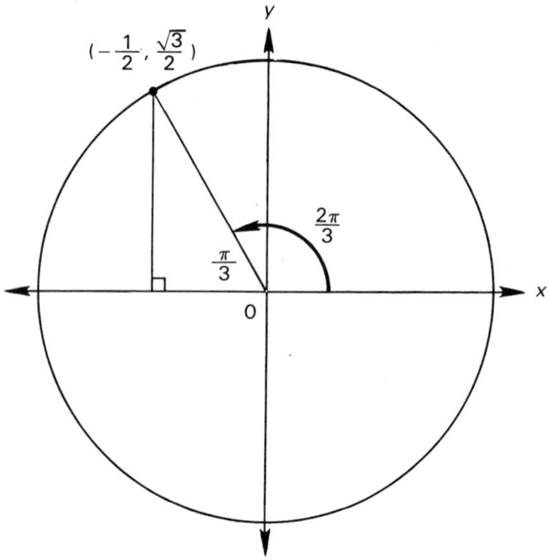

$\sin\frac{2\pi}{3} = y = \frac{\sqrt{3}}{2}$

$\cos\frac{2\pi}{3} = x = -\frac{1}{2}$

$\tan\frac{2\pi}{3} = \frac{y}{x} = -\sqrt{3}$

$\cot\frac{2\pi}{3} = \frac{x}{y} = -\frac{1}{\sqrt{3}} = -\frac{\sqrt{3}}{3}$

$\sec\frac{2\pi}{3} = \frac{1}{x} = -2$

$\csc\frac{2\pi}{3} = \frac{1}{y} = \frac{2}{\sqrt{3}} = \frac{2\sqrt{3}}{3}$

17) $\theta = 405^{\circ}$

The reference angle is 45°.

$\sin 45^{\circ} = y$

$\frac{\sqrt{2}}{2} = y$

$\cos 45^{\circ} = x$

$\frac{\sqrt{2}}{2} = x$

$P = \left(\frac{\sqrt{2}}{2}, \frac{\sqrt{2}}{2}\right)$

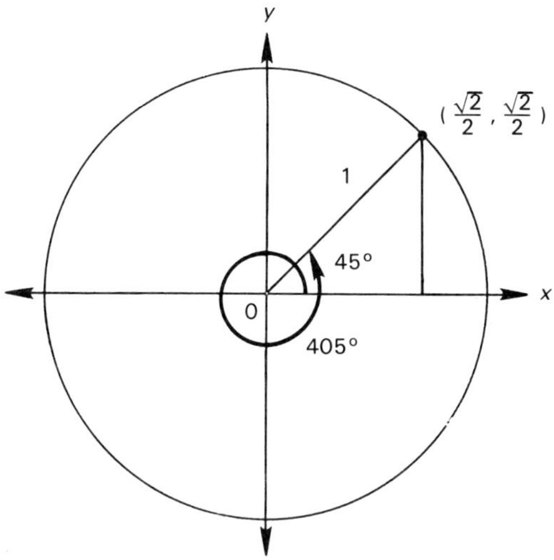

$\sin 405^{\circ} = y = \frac{\sqrt{2}}{2}$

$\cos 405^{\circ} = x = \frac{\sqrt{2}}{2}$

$\tan 405^{\circ} = \frac{y}{x} = 1$

$\cot 405^{\circ} = \frac{x}{y} = 1$

$\sec 405^{\circ} = \frac{1}{x} = \frac{2}{\sqrt{2}} = \sqrt{2}$

$\csc 405^{\circ} = \frac{1}{y} = \frac{2}{\sqrt{2}} = \sqrt{2}$

21) $\theta = -\frac{7\pi}{2}$

$P(0, 1)$

$\sin -\left(\frac{7\pi}{2}\right) = y = 1$

$\cos -\left(\frac{7\pi}{2}\right) = x = 0$

$\tan -\left(\frac{7\pi}{2}\right) = \frac{y}{x} = \frac{1}{0}$ undefined

$\cot -\left(\frac{7\pi}{2}\right) = \frac{x}{y} = 0$

$\sec -\left(\frac{7\pi}{2}\right) = \frac{1}{x} = \frac{1}{0}$ undefined

$\csc -\left(\frac{7\pi}{2}\right) = \frac{1}{y} = \frac{1}{1} = 1$

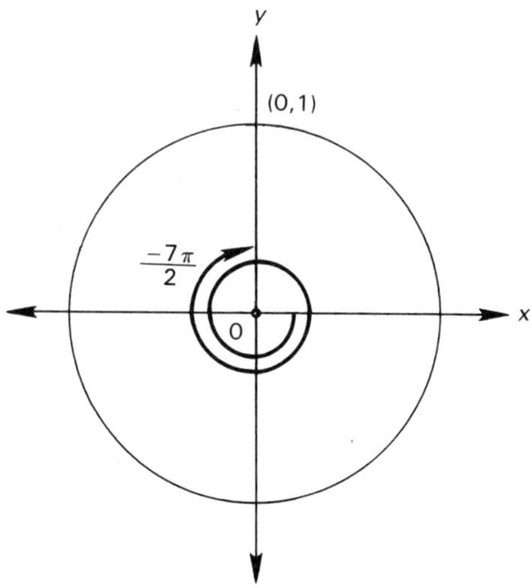

25) $\sin 261^\circ$

261° in quadrant III, sine is negative

reference angle is 81°

$\sin 261^\circ = -\sin 81^\circ = -0.9877$

29) $\cot\left(\frac{7\pi}{2}\right)$

coterminal with $\frac{3\pi}{2}$

$\cot \frac{7\pi}{2} = \cot \frac{3\pi}{2} = \frac{0}{-1} = 0$

33) $\tan 8\pi$

coterminal with 0

$\tan 8\pi = \tan 0 = \frac{0}{1} = 0$

37) $\cot 368^\circ$

$368^\circ - 360 = 8^\circ$

368° in quadrant I, cotangent is positive

reference angle is 8°

$\cot 368^\circ = \cot 8^\circ = 7.115$

41) $\cot 1200^\circ$

$1200 - 3(360) = 120^\circ$

1200° in quadrant II, cotangent is negative

reference angle 60°

$\cot 1200^\circ = -\cot 60^\circ = -\frac{1}{\sqrt{3}} = -\frac{\sqrt{3}}{3}$

45) $\cos(-200^\circ 50')$

$-200^\circ 50'$ in quadrant II, cosine is negative

reference angle $20^\circ 50'$

$\cos -200^\circ 50' = -\cos 20^\circ 50' = -0.9346$

49) $\csc(-480^{o})$

$-480^{o} + 360^{o} = -120^{o}$

-120° in quadrant III, cosecant is negative

reference angle 60°

$\csc(-120^{o}) = -\csc 60^{o} = -\dfrac{2}{\sqrt{3}} = -\dfrac{2\sqrt{3}}{3}$

53) a) $P\left(\dfrac{2}{3}, \dfrac{\sqrt{5}}{3}\right)$ is on the unit circle

if a right triangle with legs $\frac{2}{3}$ and $\frac{\sqrt{5}}{3}$ has hypotenuse 1.

$$\left(\frac{2}{3}\right)^2 + \left(\frac{\sqrt{5}}{3}\right)^2 = \frac{4}{9} + \frac{5}{9} = \frac{9}{9} = 1$$

b)

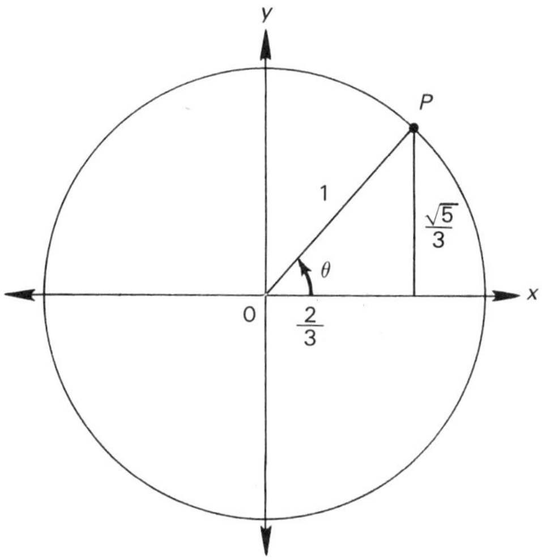

c) $\sin\theta = y = \dfrac{\sqrt{5}}{3}$

$\cos\theta = x = \dfrac{2}{3}$

$\tan\theta = \dfrac{y}{x} = \dfrac{\sqrt{5}}{2}$

$\cot\theta = \dfrac{x}{y} = \dfrac{2}{\sqrt{5}} = \dfrac{2\sqrt{5}}{5}$

$\sec\theta = \dfrac{1}{x} = \dfrac{3}{2}$

$\csc\theta = \dfrac{1}{y} = \dfrac{3}{\sqrt{5}} = \dfrac{3\sqrt{5}}{5}$

d) $\cos\theta = 0.\overline{6}$

In the cosine column, the slosest entry is 0.6670.

$\theta = 48^{o}10'$

57) $\sin\theta = 1$

$\theta = \frac{\pi}{2}$

61) $\sin\theta = \frac{1}{2}$

$\theta = \frac{\pi}{6}$ or $\theta = \frac{5\pi}{6}$

65) $\tan\theta = 1$ $\sin = -\frac{\sqrt{2}}{2}$

The reference angle is $\frac{\pi}{4}$, and θ is in quadrant III where tangent is positive, sine negative.

$\cos\theta = -\frac{\sqrt{2}}{2}$ $\sec\theta = -\sqrt{2}$

$\cot\theta = 1$ $\csc\theta = -\sqrt{2}$

69) The terminal side of θ is in quadrant II since $\frac{\pi}{2} < \theta < \pi$. Also, since $\sin\theta = \frac{1}{3}$ we may use a reference triangle in which $y = 1$, hypotenuse = 3, and $x = -\sqrt{9 - 1} = -2\sqrt{2}$.
Then

$$\cos\theta = \frac{-2\sqrt{2}}{3} = -\frac{2\sqrt{2}}{3}$$

73) By definition of tangent, for $\triangle OBA$

$\tan\theta = \frac{AB}{OA} = \frac{AB}{1} = AB$

$\sec\theta = \frac{OB}{OA} = \frac{OB}{1} = OB$

Section 8.5 Identities

1) $\frac{1}{\csc\theta} = \sin\theta$

5) $\sin\theta\cot\theta = \cos\theta$

9) $\tan^2(3^\circ20') - \sec^2(3^\circ20') = -1$
because $\tan^2\theta = \sec^2\theta = -1$ is true for all θ not coterminal with $\frac{\pi}{2}, \frac{3\pi}{2}$

13) $\cos^2(\tan^2\theta + 1) = \cos^2\theta(\sec^2\theta) = \cos^2\theta\left(\frac{1}{\cos^2\theta}\right) = 1$

17) $\frac{1}{\tan\theta}$ is defined for all θ except quadrantal angles.

21) $\dfrac{\tan\theta}{\cot\theta} = \dfrac{\dfrac{\sin\theta}{\cos\theta}}{\dfrac{\cos\theta}{\sin\theta}} = \dfrac{\sin\theta}{\cos\theta} \cdot \dfrac{\sin\theta}{\cos\theta} = \dfrac{\sin^2\theta}{\cos^2\theta}$, θ not a quadrantal angle

25) $\dfrac{\sec\theta + \csc\theta}{\cos\theta + \sin\theta} = \dfrac{\dfrac{1}{\cos\theta} + \dfrac{1}{\sin\theta}}{\cos\theta + \sin\theta} = \dfrac{\dfrac{\sin\theta + \cos\theta}{\sin\theta\cos\theta}}{\cos\theta + \sin\theta}$

$$= \frac{\cancel{\sin\theta + \cos\theta}}{\sin\theta\cos\theta} \cdot \frac{1}{\cancel{\cos\theta + \sin\theta}} = \frac{1}{\sin\theta\cos\theta} ,$$

θ not a quadrantal angle.

29) $\dfrac{\cot\theta - \sin\theta}{\csc\theta} = \dfrac{\dfrac{\cos\theta}{\sin\theta} - \sin\theta}{\dfrac{1}{\sin\theta}} = \dfrac{\cos\theta - \sin^2\theta}{\sin\theta} \cdot \dfrac{\sin\theta}{1}$

$$= \cos\theta - \sin^2\theta = \cos\theta - (1 - \cos^2\theta)$$

$$= \cos^2\theta + \cos\theta = 1$$

Note: A single identity can be verified correctly in several different ways. The methods shown below and in the answer key at the back of the textbook will show you some approaches. You may find others, equally correct, and easier for you.

33) $(\tan\theta - 1)^2 = \tan^2\theta - 2\tan\theta + 1 = (\tan^2\theta + 1) - 2\tan\theta$

$$= \sec^2\theta - 2\tan\theta$$

0 not coterminal with $\frac{\pi}{2}$, $\frac{3\pi}{2}$

37)

$\dfrac{\cot\theta - 1}{1 - \tan\theta}$	$\dfrac{\csc\theta}{\sec\theta}$
$\dfrac{\dfrac{\cos\theta}{\sin\theta} - 1}{1 - \dfrac{\sin\theta}{\cos\theta}}$	$\dfrac{\dfrac{1}{\sin\theta}}{\dfrac{1}{\cos\theta}}$
$\dfrac{\dfrac{\cos\theta - \sin\theta}{\sin\theta}}{\dfrac{\cos\theta - \sin\theta}{\cos\theta}}$	$\dfrac{\cos\theta}{\sin\theta}$
$\dfrac{\cos\theta - \sin\theta}{\sin\theta} \cdot \dfrac{\cos\theta}{\cos\theta - \sin\theta}$	
$\dfrac{\cos\theta}{\sin\theta}$	$\dfrac{\cos\theta}{\sin\theta}$

37) continued

From line 1, $\tan\theta \neq 1$, so θ not coterminal with $\frac{\pi}{4}$ or $\frac{5\pi}{4}$.

From line 2, $\sin\theta \neq 0$ and $\cos\theta \neq 0$, so θ not quadrantal.

41) $(\sec\theta + \tan\theta)(1 - \sin\theta) = \left(\frac{1}{\cos\theta} + \frac{\sin\theta}{\cos\theta}\right)(1 - \sin\theta)$

$$= \left(\frac{1 + \sin\theta}{\cos\theta}\right)\left(\frac{1 - \sin\theta}{1}\right)$$

$$= \frac{1 - \sin^2\theta}{\cos\theta} = \frac{\cos^2\theta}{\cos\theta} = \cos\theta$$

45) $\tan^2\theta - \sin^2\theta = \frac{\sin^2\theta}{\cos^2\theta} - \sin^2\theta = \sin^2\theta\left(\frac{1}{\cos^2\theta} - 1\right) = \sin^2\theta(\sec^2\theta - 1)$

$$= \sin^2\theta\tan^2\theta$$

49) $\frac{1}{1 + \cos\theta} + \frac{1}{1 - \cos\theta} = \frac{1 - \cos\theta}{1 - \cos^2\theta} + \frac{1 + \cos\theta}{1 - \cos^2\theta} = \frac{2}{1 - \cos^2\theta}$

$$= \frac{2}{\sin^2\theta} = 2\csc^2\theta$$

53) $\frac{\sec\theta + \csc\theta}{\sec\theta - \csc\theta} = \frac{\frac{1}{\cos\theta} + \frac{1}{\sin\theta}}{\frac{1}{\cos\theta} - \frac{1}{\sin\theta}} = \frac{\frac{\sin\theta + \cos\theta}{\cos\theta\sin\theta}}{\frac{\sin\theta - \cos\theta}{\cos\theta\sin\theta}}$

$$= \frac{\sin\theta + \cos\theta}{\sin\theta - \cos\theta}$$

57) $(\sin^2\theta + \cos^2\theta)^5 = 1^5 = 1$

61) $\frac{\tan^2\theta + 1}{\tan^2\theta} = \frac{\sec^2\theta}{\tan^2\theta} = \frac{\frac{1}{\cos^2\theta}}{\frac{\sin^2\theta}{\cos^2\theta}} = \frac{1}{\sin^2\theta} = \csc^2\theta$

65) $\sec^3\theta + \frac{\sin^2\theta}{\cos^3\theta} = \frac{1}{\cos^3\theta} + \frac{\sin^2\theta}{\cos^3\theta} = \frac{1 + \sin^2\theta}{\cos^3\theta}$

$$= \frac{(\cos^2\theta + \sin^2\theta) + \sin^2\theta}{\cos^3\theta} = \frac{\cos^2\theta + 2\sin^2\theta}{\cos^3\theta}$$

69) $\dfrac{\sec\theta}{\tan\theta - \sin\theta} = \dfrac{\dfrac{1}{\cos\theta}}{\dfrac{\sin\theta}{\cos\theta} - \sin\theta} = \dfrac{\dfrac{1}{\cos\theta}}{\dfrac{\sin\theta - \sin\theta\cos\theta}{\cos\theta}}$

$$= \frac{1}{\sin\theta(1 - \cos\theta)} = \frac{1}{\sin\theta(1 - \cos\theta)} \cdot \frac{1 + \cos\theta}{1 + \cos\theta}$$

$$= \frac{1 + \cos\theta}{\sin\theta(1 - \cos^2\theta)} = \frac{1 + \cos\theta}{\sin^3\theta}$$

73) $-\ln|\sec x - \tan x| = \ln|\sec x - \tan x|^{-1} = \ln \dfrac{1}{|\sec x - \tan x|}$

$$= \ln\left|\frac{1}{\sec x - \tan x}\right| = \ln\left|\frac{\sec x + \tan x}{\sec^2 x - \tan^2 x}\right|$$

$$= \ln\left|\frac{\sec x + \tan x}{1}\right| = \ln|\sec x + \tan x|$$

77) $\sin\frac{\pi}{4} = \frac{\sqrt{2}}{2} \neq -\frac{\sqrt{2}}{2} = -\sqrt{\frac{1}{2}} = -\sqrt{1 - \cos^2\frac{\pi}{4}}$

81) a) $\ln(\sec\theta + \tan\theta) = \ln\left(\sqrt{\tan^2\theta + 1} + \tan\theta\right)$

$$= \ln\left(\sqrt{\left(\frac{x}{3}\right)^2 + 1} + \frac{x}{3}\right)$$

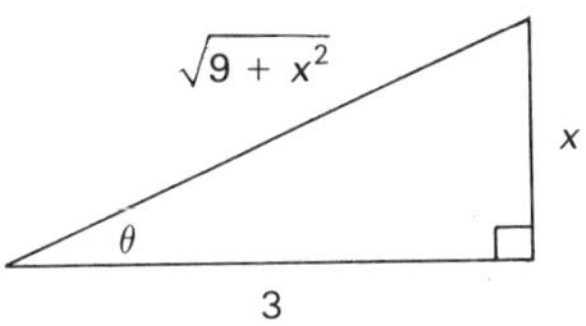

$$= \ln\left(\sqrt{\frac{x^2 + 9}{9}} + \frac{x}{3}\right)$$

$$= \ln\left(\frac{\sqrt{x^2 + 9}}{3} + \frac{x}{3}\right)$$

$$= \ln\frac{\sqrt{x^2 + 9} + x}{3}$$

b) $F(4) - F(0) = \ln\dfrac{\sqrt{4^2 + 9} + 4}{3} - \ln\dfrac{\sqrt{0 + 9} + 0}{3}$

$$= \ln\frac{\sqrt{25} + 4}{3} - \ln(1) = \ln 3$$

85)

$$\frac{\sec^2 x - 2}{(1 + \tan x)^2}$$

$$\frac{(\tan^2 x + 1) - 2}{(\tan x + 1)^2}$$

$$\frac{\tan^2 x - 1}{(\tan x + 1)^2}$$

$$\frac{\cancel{(\tan x + 1)}(\tan x - 1)}{\cancel{(\tan x + 1)}(\tan x + 1)}$$

$$\frac{\tan x - 1}{\tan x + 1}$$

$$\frac{1 - \cot x}{1 + \cot x}$$

$$\frac{1 - \frac{1}{\tan x}}{1 + \frac{1}{\tan x}}$$

$$\frac{\frac{\tan x - 1}{\tan x}}{\frac{\tan x + 1}{\tan x}}$$

$$\frac{\tan x - 1}{\tan x + 1}$$

$$\frac{\tan x - 1}{\tan x + 1}$$

Therefore the given equation is an identity.

Sample Test Questions for Chapter 8

Part I

1. a) $\tan A = \frac{30}{16} = \frac{15}{8}$

b) $\sin B = \frac{16}{34} = \frac{8}{17}$

c) $\sec A = \frac{34}{16} = \frac{17}{8}$

2. $\sec A = \frac{3}{2} = \frac{\text{hypotenuse}}{\text{adjacent}} = \frac{c}{b}$

$b = 2 \quad c = 3$

$a^2 + 2^2 = 3^2$ by the Pythagorean theorem

$a^2 + 4 = 9$

$a^2 = 5$

$a^2 = \sqrt{5}$

$\sin A = \frac{a}{c} = \frac{\sqrt{5}}{3}$

$\cos A = \frac{b}{c} = \frac{2}{3}$

$\tan A = \frac{a}{b} = \frac{\sqrt{5}}{2}$

3. $c^2 = x^2 + 3^2$ by the Pythagorean theorem

$c = \sqrt{x^2 + 9}$

$\sin A = \frac{x}{\sqrt{x^2 + 9}}$

$\sec A = \frac{\sqrt{x^2 + 9}}{3}$

$$3(\sin^2 A)(\sec A) = 3\left(\frac{x}{\sqrt{x^2 + 9}}\right)^2 \left(\frac{\sqrt{x^2 + 9}}{3}\right)$$

$$= 3\left(\frac{x^2}{x^2 + 9}\right)\left(\frac{\sqrt{x^2 + 9}}{3}\right) = \frac{x^2\sqrt{x^2 + 9}}{x^2 + 9} \cdot \frac{\sqrt{x^2 + 9}}{\sqrt{x^2 + 9}}$$

$$= \frac{x^2 \cancel{(x^2 + 9)}}{\cancel{(x^2 + 9)}\sqrt{x^2 + 9}} = \frac{x^2}{\sqrt{x^2 + 9}}$$

4. $s = r\theta$

$r = 10 \qquad 120^o = 120 \cdot \frac{\pi}{180} = \frac{2\pi}{3}$ radians

$s = 10\left(\frac{2\pi}{3}\right) = \frac{20\pi}{3}$

5. a) $\sin\frac{\pi}{3} = \frac{\sqrt{3}}{2}$

b) $\cot\frac{3\pi}{2} = 0$

c) $\sin(-135^o)$

-135° in quadrant III; sine is negative;

reference angle is 45°

$\sin(-135^o) = -\sin 45^o = -\frac{\sqrt{2}}{2}$

6. a) $\tan 510^o$

$510^o - 360^o = 150^o$

150° in quadrant II; tangent is negative;

reference angle is 30°

$\tan 510^o = -\tan 30^o = -\frac{\sqrt{3}}{3}$

b) $\cos 3\pi$

$3\pi - 2\pi = \pi$

$\cos 3\pi = \cos\pi = -1$

c) $\csc\left(-\frac{7\pi}{6}\right)$

$-\frac{7\pi}{6}$ in II; csc positive;

reference angle is $\frac{\pi}{6}$

$\csc\left(-\frac{7\pi}{6}\right) = \csc\frac{\pi}{6} = 2$

7. $20^2 + 20^2 = c^2$

$800 = c^2$

$\sqrt{800} = c$

$20\sqrt{2} = c$

8.

Let h represent the height of the building.

$$\tan 75^o = \frac{h}{100}$$

$$3.732 = \frac{h}{100}$$

$$373.2 = h$$

The building is about 373 feet high.

9. $\cot^2\theta - \cos^2\theta = \dfrac{\cos^2\theta}{\sin^2\theta} - \cos^2\theta = \cos^2\theta\left(\dfrac{1}{\sin^2\theta} - 1\right)$

$$= \cos^2\theta(\csc^2\theta - 1) = \cos^2\theta \cot^2\theta$$

10. $(\sec\theta - \tan\theta)^2 = \left(\dfrac{1}{\cos\theta} - \dfrac{\sin\theta}{\cos\theta}\right)^2 = \dfrac{(1 - \sin\theta)^2}{\cos^2\theta}$

$$= \frac{(1 - \sin\theta)^2}{1 - \sin^2\theta} = \frac{(1 - \sin\theta)^2}{(1 - \sin\theta)(1 + \sin\theta)}$$

$$= \frac{1 - \sin\theta}{1 + \sin\theta}$$

11. $\dfrac{\tan\theta - \sin\theta}{\csc\theta - \cot\theta} = \dfrac{(\tan\theta - \sin\theta)(\csc\theta + \cot\theta)}{(\csc\theta - \cot\theta)(\csc\theta + \cot\theta)}$

$$= \frac{\tan\theta\csc\theta - \sin\theta\csc\theta + \tan\theta\cot\theta - \sin\theta\cot\theta}{\csc^2\theta - \cot^2\theta}$$

$$= \frac{\tan\theta\csc\theta - 1 + 1 - \sin\theta\cot\theta}{1}$$

$$= \frac{\sin\theta}{\cos\theta}\cdot\frac{1}{\sin\theta} - \sin\theta\cdot\frac{\cos\theta}{\sin\theta} = \frac{1}{\cos\theta} - \cos\theta$$

$$= \sec\theta - \cos\theta$$

12. Pick any point, P, on the line $y = 2x$. The coordinates are (x, y). Drop a perpendicular to the x-axis from P. The resulting right triangle has legs of measure x and y. The hypotenuse has measure

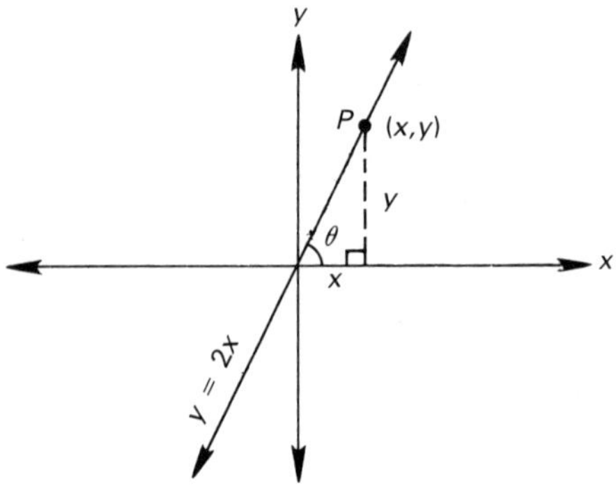

$$c = \sqrt{x^2 + y^2}$$

Substitute 2x for y to obtain

$$c = \sqrt{x^2 + (2x)^2} = \sqrt{5x^2} = x\sqrt{5} \quad (\text{since } x > 0)$$

$$\cos \theta = \frac{x}{x\sqrt{5}} = \frac{1}{\sqrt{5}} = \frac{\sqrt{5}}{5}$$

13. The shaded area is the difference between the area of the triangle and the area of the sector. The side adjacent to the 30° angle is the base of the triangle and the radius of the arc. The measure of this side is therefore needed for both area formulas.
Let x represent the length of the side.

$$\tan 30^o = \frac{2}{x}$$

$$\frac{1}{\sqrt{3}} = \frac{2}{x}$$

$$2\sqrt{3} = x$$

Area of the triangle: $A_t = \frac{1}{2} bh$

$$A_t = \frac{1}{2}(2\sqrt{3})(2) = 2\sqrt{3}$$

Area of the sector: $A_s = \frac{1}{2} r^2\theta$

θ must be in radians. $30^o = \frac{\pi}{6}$ radians

$$A_s = \frac{1}{2}(2\sqrt{3})^2\left(\frac{\pi}{6}\right) = \frac{\pi}{12}(4 \cdot 3) = \pi$$

Area of the shaded region:

$$A = A_t - A_s = 2\sqrt{3} - \pi \text{ square units.}$$

14. $\sin \theta = -\frac{\sqrt{3}}{2}$ $\tan \theta = \sqrt{3}$

θ is in quadrant III where tangent is positive; sine negative.

reference angle is 60^{0}

$\theta = 240^{0}$

$\sec 240^{0} = -\sec 60^{0} = -2$

15. a) The wheels are making 246 complete revolutions per minute.

$246 \cdot 2\pi = 492\pi$ rad/min.

b) A point on the circumference of the wheel travels 246 times the circumference in one minute.

$C = \pi D = 26\pi$

$246(26\pi) = 6396\pi$ inches per minute

$\frac{6396\pi}{12} = 533\pi$ feet per minute

c) $30(533\pi) = 15{,}990\pi$ feet in 30 minutes

$\frac{30(533\pi)}{5280} = \frac{30(533\pi)}{30(176)} = \frac{533\pi}{176} = 9.51$ miles in 30 minutes

Sample Test Questions for Chapter 8

Part II

1. (b)

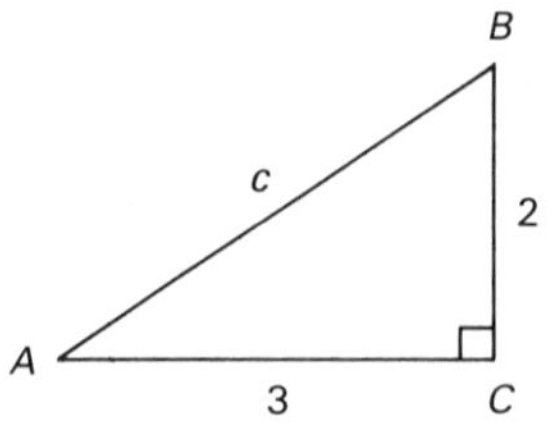

$c = \sqrt{2^2 + 3^2} = \sqrt{13}$

$\sec B = \frac{\sqrt{13}}{2}$

2. (c)

$\sin 30^o + \cos 60^o - \tan 45^o = \frac{1}{2} + \frac{1}{2} - 1 = 0$

3. (a)

$\frac{11\pi}{6} \cdot \frac{180}{\pi} = 330^o$

4. (c)

Area $= \frac{1}{2} r^2\theta - \frac{1}{2}(6^2)\frac{\pi}{6} = 3\pi$

5. (b)

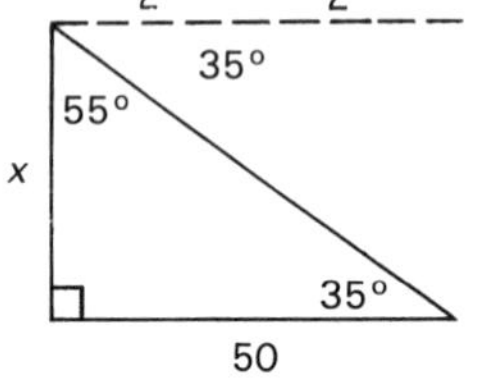

6. (a)

$\tan\left(-\frac{3\pi}{4}\right) = \frac{-1}{-1} = 1$

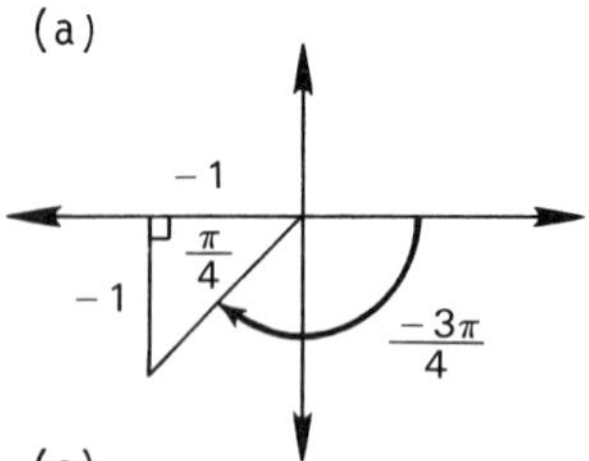

7. (e)

$\cos 330^o = \frac{\sqrt{3}}{2}$

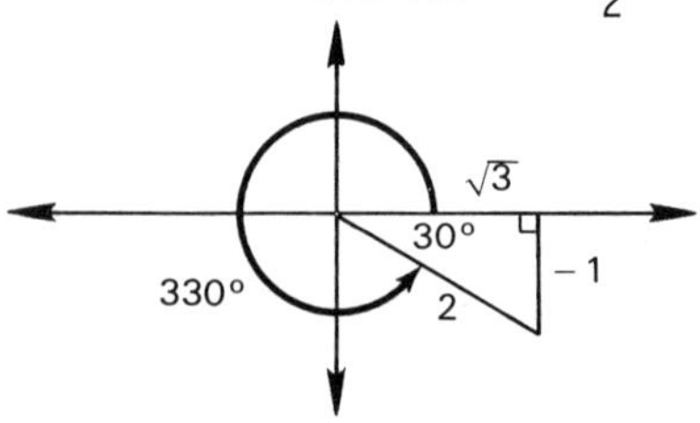

8. (d)

9. (b)

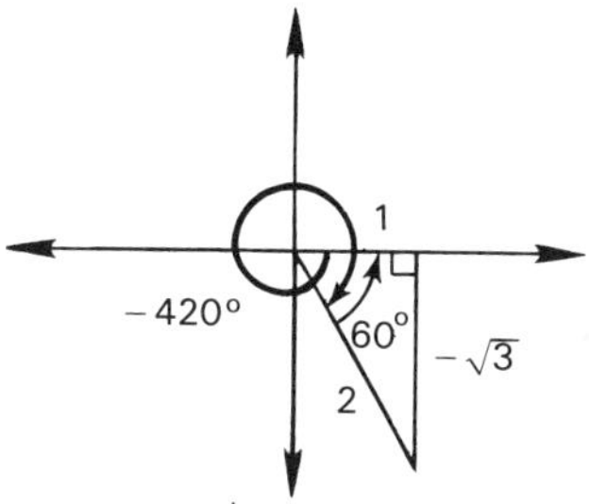

10. (d)

11. (a)

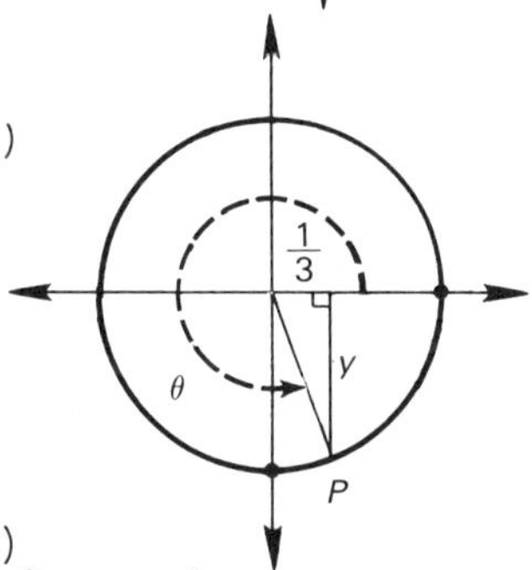

$$y = -\sqrt{1 - \frac{1}{9}} = -\sqrt{\frac{8}{9}} = -\frac{2\sqrt{2}}{3}$$

$$\tan\theta = \frac{y}{\frac{1}{3}} = \frac{\frac{-2\sqrt{2}}{3}}{\frac{1}{3}} = -2\sqrt{2}$$

12. (d)

$$\sec^2\theta - \tan^2\theta = 1 + \tan^2\theta - \tan^2\theta = 1$$

$$\csc^2\theta - \cot^2\theta = 1 + \cot^2\theta - \cot^2\theta = 1$$

$$(\cos\theta)(\tan\theta) = (\cos\theta)\frac{\sin\theta}{\cos\theta} = \sin\theta$$

13. (c)

$$\sin^2\frac{\pi}{4} - \cos^2\frac{\pi}{4} = \left(\frac{1}{\sqrt{2}}\right)^2 - \left(\frac{1}{\sqrt{2}}\right)^2 = \frac{1}{2} - \frac{1}{2} = 0 \neq 1$$

14. (c)

In one minute the angular rotation is $\theta = (280)(60) = 16800$ radians.

Then, $s = r\theta = \frac{3}{2}(16800) = 25200$ ft.

15. (b)

$$\sec^2\theta + 2\sec\theta\tan\theta + \tan^2\theta = (\sec\theta + \tan\theta)^2 = \left(\frac{1}{\cos\theta} + \frac{\sin\theta}{\cos\theta}\right)^2$$

$$= \left(\frac{1 + \sin\theta}{\cos\theta}\right)^2 = \frac{(1 + \sin\theta)^2}{\cos^2\theta}$$

$$= \frac{(1 + \sin\theta)^2}{1 - \sin^2\theta} = \frac{(1 + \sin\theta)^2}{(1 + \sin\theta)(1 - \sin\theta)}$$

$$= \frac{1 + \sin\theta}{1 - \sin\theta}$$

CHAPTER 9

THE CIRCULAR FUNCTIONS

Section 9.1 The Sine and Cosine Functions

1)

x	$-\pi$	$-\frac{5\pi}{6}$	$-\frac{2\pi}{3}$	$-\frac{\pi}{2}$	$-\frac{\pi}{3}$	$-\frac{\pi}{6}$	0
y = sin x	0	$-\frac{1}{2}$	$-\frac{\sqrt{3}}{2}$	-1	$-\frac{\sqrt{3}}{2}$	$-\frac{1}{2}$	0

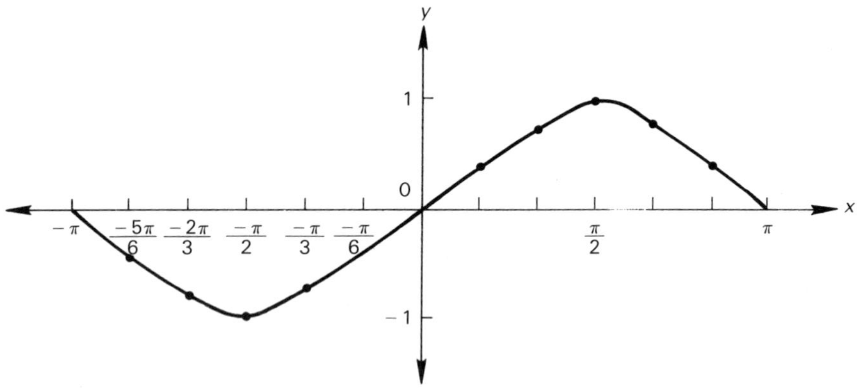

5) Shift the graph of f(x) to the right $\frac{\pi}{3}$ units.

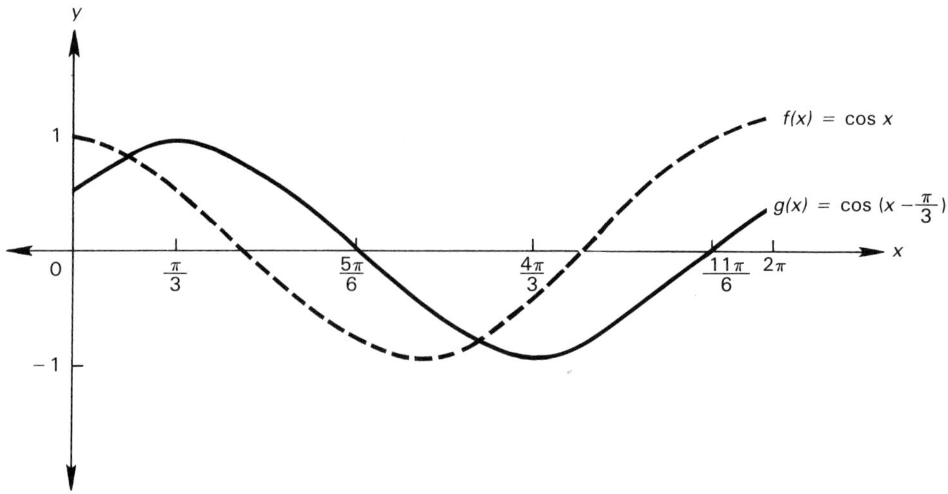

9) $y = |\sin x|$

Graph $y = \sin x$, then take the portions below the x-axis and reflect them through the x-axis.

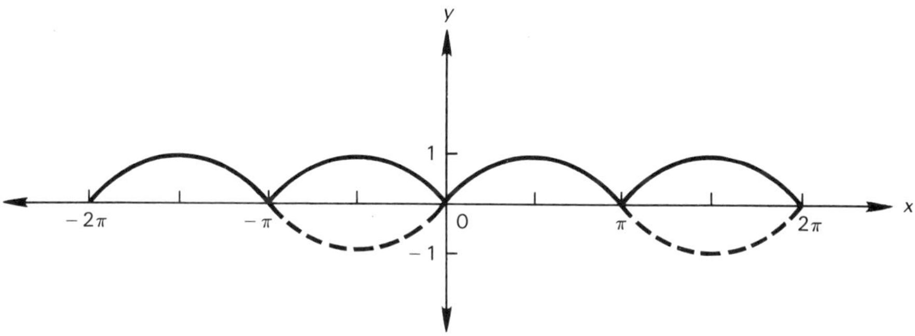

13) $y = \cos 2x$ Amplitude: $|a| = |1| = 1$

Period: $p = \dfrac{2\pi}{|b|} = \dfrac{2\pi}{|2|} = \pi$

x	y
0	1
$\frac{p}{4} = \frac{\pi}{4}$	0
$\frac{p}{2} = \frac{\pi}{2}$	-1
$\frac{3p}{4} = \frac{3\pi}{4}$	0
$p = \pi$	1

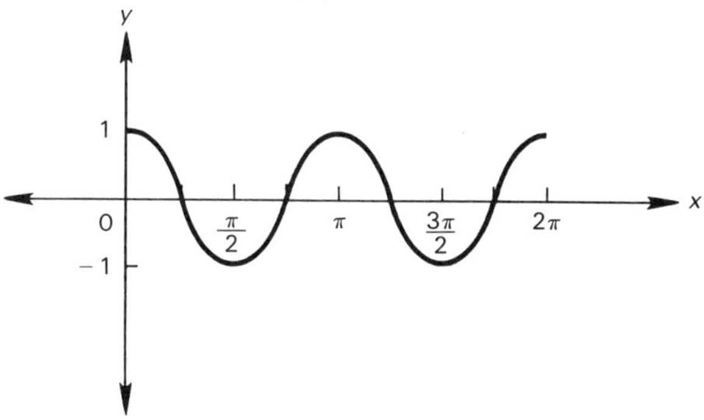

17) $y = -\cos \frac{1}{2} x$

Amplitude: $|a| = |-1| = 1$

Period: $p = \frac{2\pi}{|b|} = \frac{2\pi}{\left|\frac{1}{2}\right|} = 4\pi$

x	y
0	-1
$\frac{p}{4} = \pi$	0
$\frac{p}{2} = 2\pi$	1
$\frac{3p}{4} = 3\pi$	0
$p = 4\pi$	-1

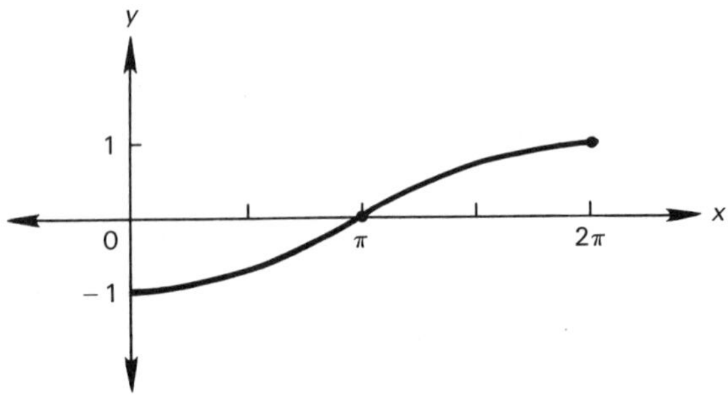

21) $y = 2 \sin \pi x$

Period: $p = \frac{2\pi}{|b|} = \frac{2\pi}{|\pi|} = 2$

x	y
0	0
$\frac{p}{4} = \frac{1}{2}$	2
$\frac{p}{2} = 1$	0
$\frac{3p}{4} = \frac{3}{2}$	-2
$p = 2$	0

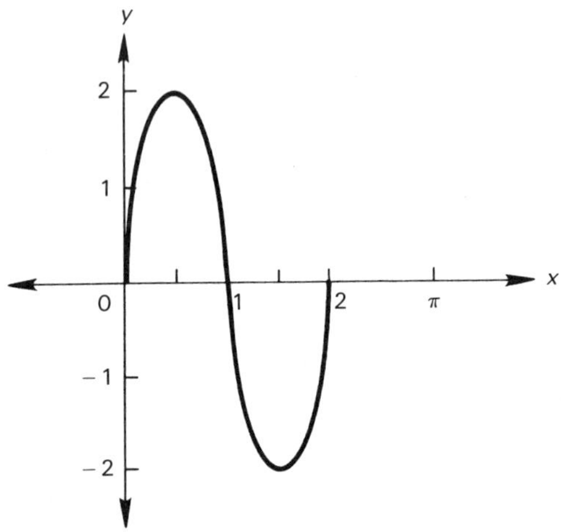

25) $a = -3$ $\quad p = 4\pi$

$$\frac{2\pi}{|b|} = 4\pi$$

$$|b| = \frac{1}{2}$$

$y = -3 \cos \frac{1}{2} x$

29) $u = -2 \sin(2x + \pi) = -2 \sin 2\left(x + \frac{\pi}{2}\right)$

Amplitude: $|a| = |-2| = 2$

Period: $p = \frac{2\pi}{|b|} = \frac{2\pi}{|-2|} = \pi$

Phase shift: $-\frac{\pi}{2}$ $\left(\frac{\pi}{2} \text{ to the left}\right)$

33) $y = \sin(4x - \pi) = \sin 4\left(x - \frac{\pi}{4}\right)$

Amplitude: $|a| = |1| = 1$

Period: $p = \frac{2\pi}{|b|} = \frac{2\pi}{|4|} = \frac{\pi}{2}$

Phase shift: $+\frac{\pi}{4}$ $\left(\frac{\pi}{4} \text{ to the right}\right)$

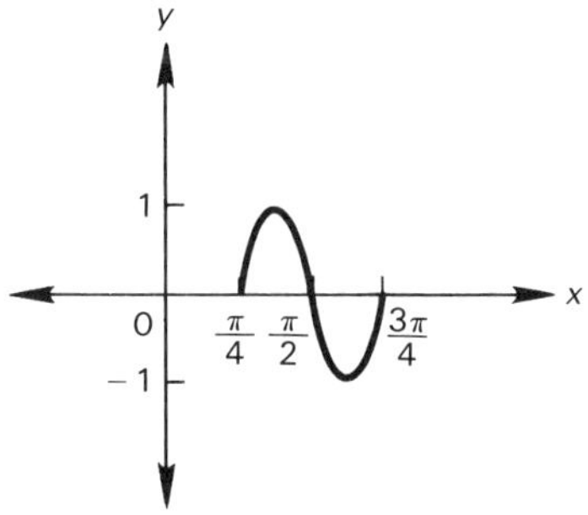

	x	y
0	$0 + \frac{\pi}{4} = \frac{\pi}{4}$	0
$\frac{p}{4}$	$\frac{\pi}{8} + \frac{\pi}{4} = \frac{3\pi}{8}$	1
$\frac{p}{2}$	$\frac{\pi}{4} + \frac{\pi}{4} = \frac{\pi}{2}$	0
$\frac{3p}{4}$	$\frac{3\pi}{8} + \frac{\pi}{4} = \frac{5\pi}{8}$	-1
p	$\frac{\pi}{2} + \frac{\pi}{4} = \frac{3\pi}{4}$	0

37) $y = -\frac{5}{2} \sin\left(\frac{x}{2} + \frac{\pi}{4}\right) = -\frac{5}{2} \sin \frac{1}{2}\left(x + \frac{\pi}{2}\right)$

Amplitude: $|a| = \left|-\frac{5}{2}\right| = \frac{5}{2}$

Period: $\frac{2\pi}{|b|} = \frac{2\pi}{\left|\frac{1}{2}\right|} = 4\pi$

Phase shift: $-\frac{\pi}{2}$ $\left(\frac{\pi}{2} \text{ to the left}\right)$

37) continued

	x	y
0	$0 - \frac{\pi}{2} = -\frac{\pi}{2}$	0
$\frac{p}{4}$	$\pi - \frac{\pi}{2} = \frac{\pi}{2}$	$-\frac{5}{2}$
$\frac{p}{2}$	$2\pi - \frac{\pi}{2} = \frac{3\pi}{2}$	0
$\frac{3p}{4}$	$3\pi - \frac{\pi}{2} = \frac{5\pi}{2}$	$\frac{5}{2}$
p	$4\pi - \frac{\pi}{2} = \frac{7\pi}{2}$	0

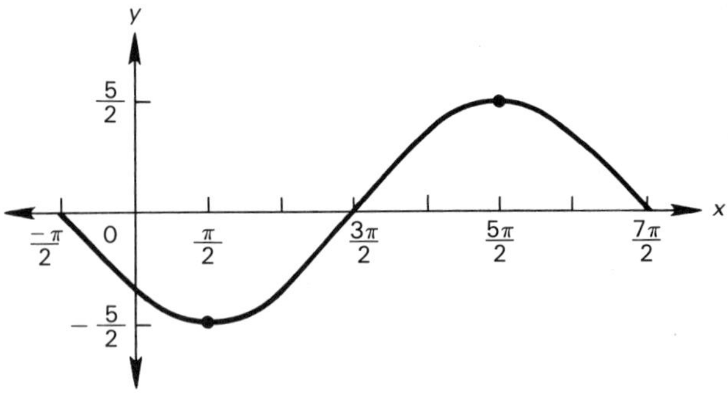

41) $f(x) = \sin x - \cos x$

x	sin x	cos x	sin x - cos x
0	0	1	-1
$\frac{\pi}{4}$	$\frac{\sqrt{2}}{2}$	$\frac{\sqrt{2}}{2}$	0
$\frac{\pi}{2}$	1	0	1
$\frac{3\pi}{4}$	$\frac{\sqrt{2}}{2}$	$-\frac{\sqrt{2}}{2}$	$\sqrt{2}$
π	0	-1	1
$\frac{5\pi}{4}$	$-\frac{\sqrt{2}}{2}$	$-\frac{\sqrt{2}}{2}$	0
$\frac{3\pi}{2}$	-1	0	-1
$\frac{7\pi}{4}$	$-\frac{\sqrt{2}}{2}$	$\frac{\sqrt{2}}{2}$	$-\sqrt{2}$
2π	0	1	-1

41) continued

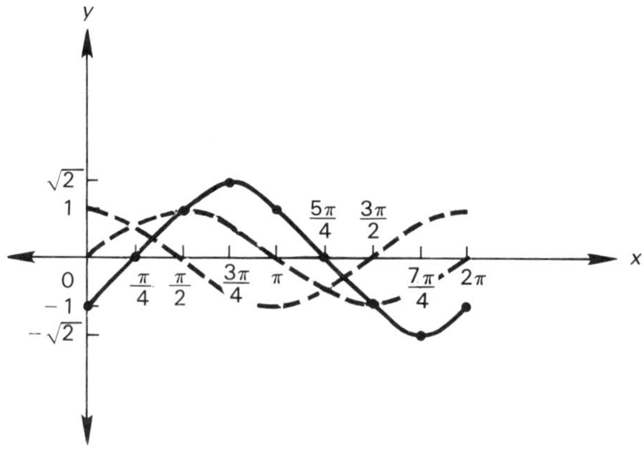

45) $f(x) = |\sin x| + |\cos x|$

x	$\|\sin x\|$	$\|\cos x\|$	$\|\sin x\| + \|\cos x\|$
0	0	1	1
$\frac{\pi}{4}$	$\frac{\sqrt{2}}{2}$	$\frac{\sqrt{2}}{2}$	$\sqrt{2}$
$\frac{\pi}{2}$	1	0	1
$\frac{3\pi}{4}$	$\frac{\sqrt{2}}{2}$	$\frac{\sqrt{2}}{2}$	$\sqrt{2}$
π	0	1	1
$\frac{5\pi}{4}$	$\frac{\sqrt{2}}{2}$	$\frac{\sqrt{2}}{2}$	$\sqrt{2}$
$\frac{3\pi}{2}$	1	0	1
$\frac{7\pi}{4}$	$\frac{\sqrt{2}}{2}$	$\frac{\sqrt{2}}{2}$	$\sqrt{2}$
2π	0	1	1

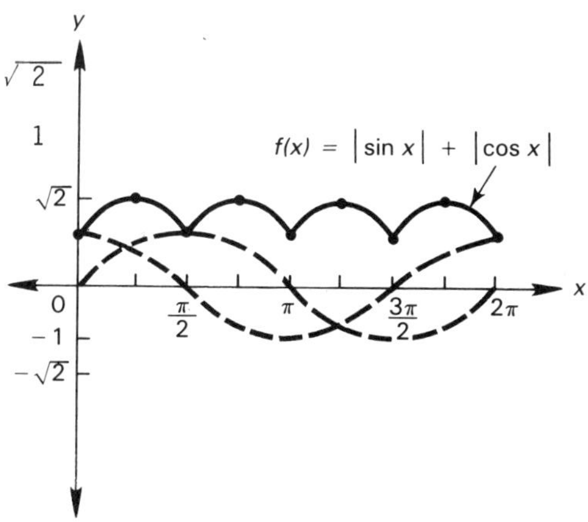

49) $h(x) = \cos(5x^2)$

Let $g(x) = 5x^2$, Let $f(x) = \cos x$

$(f \circ g)(x) = f(5x^2) = \cos(5x^2)$

53) $f(x + 2p) = f((x + p) + p)$

$= f(x + p)$ by definition of p

$= f(x)$ by definition of p

Section 9.2 Graphing Other Circular Functions

1)

x	-1.4	-1.3	$-\frac{\pi}{3}$	$-\frac{\pi}{4}$	$-\frac{\pi}{6}$	0
y = tan x	-5.8	-3.6	$-\sqrt{3}$	-1	$-\frac{\sqrt{3}}{3}$	0

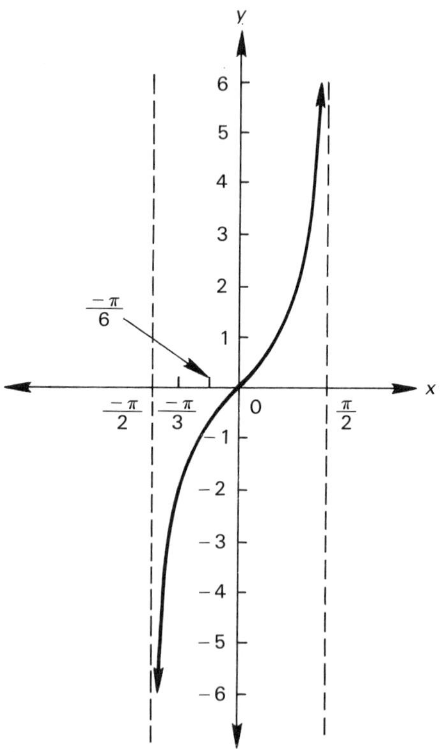

5) Shift g(x) to the left $\frac{\pi}{2}$ units.

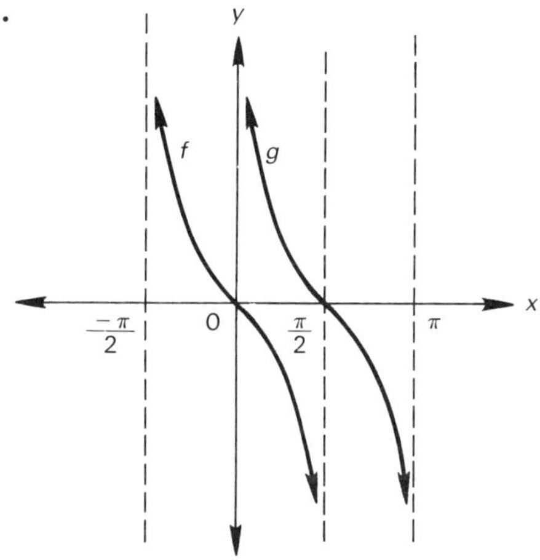

9) $y = \cot 3x$

Perid: $= = \frac{\pi}{|b|} = \frac{\pi}{3}$

Asymptotes: The function $y = \cot x$ has an asymptote at 0 and another every π units, where $\pi = p$. $y = \cot 3x$ is also undefined when $x = 0$, so the first asymptote is at $x = 0$ and the others occur every p units, with $p = \frac{\pi}{3}$; $x = k\left(\frac{\pi}{3}\right)$, k any integer, are the asymptotes.

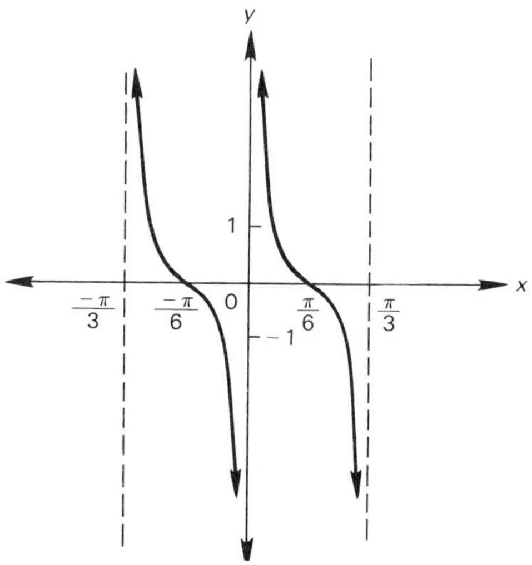

13) $y = \sec 4x$

Period: $p = \frac{2\pi}{|b|} = \frac{2\pi}{4} = \frac{\pi}{2}$

Asymptotes: The function $y = \sec x$ crosses the y-axis at $y = 1$. It has an asymptote at $\frac{\pi}{2}$, where $\frac{\pi}{2} = \frac{p}{4}$, or

$p = 2\pi$, the period of $y = \sec x$. There is an asymptote again every $\pm\frac{p}{2} = \pm\frac{2\pi}{2} = \pm\pi$.

$y = \sec 4x$ will have an asymptote at $\frac{p}{4} = \frac{\frac{\pi}{2}}{4} = \frac{\pi}{8}$

and every $\pm\frac{p}{2} = \pm\frac{\pi}{4}$ units; $x = \frac{\pi}{8} + k\left(\frac{\pi}{4}\right)$, k any integer, are the asymptotes.

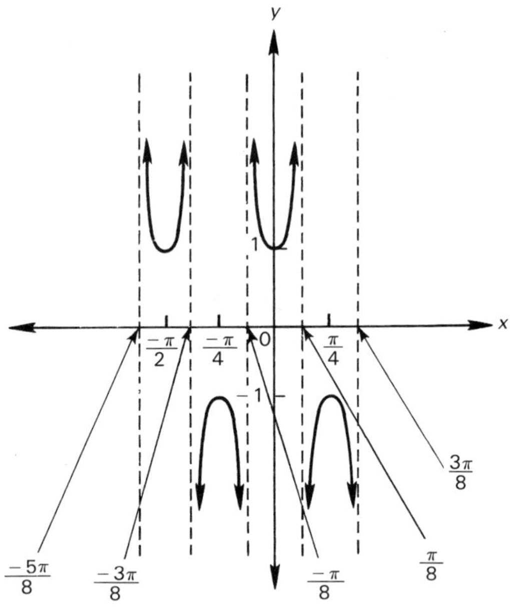

17) $y = 2 \sec \frac{3x}{2} = 2 \sec \frac{3}{2}x$

Period: $p = \frac{2\pi}{|b|} = \frac{2\pi}{\frac{3}{2}} = \frac{4\pi}{3}$

Asymptotes: (See solution for exercise 13).

$x = \frac{p}{4} + k\left(\frac{p}{2}\right)$

$x = \frac{\pi}{3} + k\left(\frac{2\pi}{3}\right)$, k any integer.

Range values will be twice those for $y = \sec x$.

17) continued

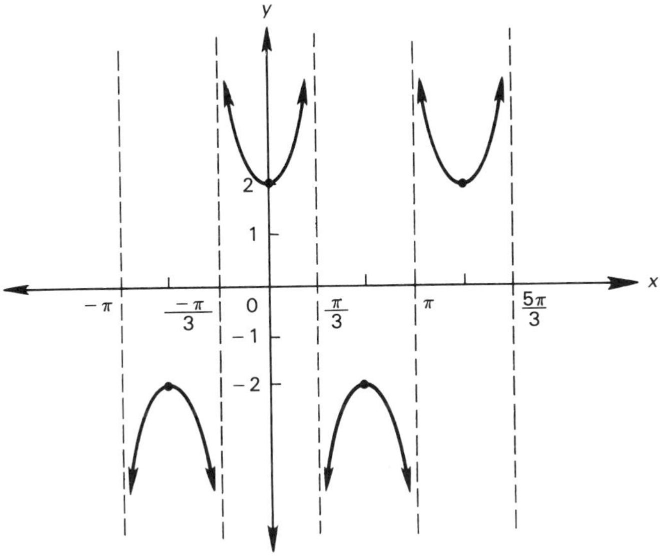

21) $y = -\csc\left(\frac{x}{2} - \frac{\pi}{4}\right) \quad = -\csc\frac{1}{2}\left(x - \frac{\pi}{2}\right)$

Period: $p = \frac{2\pi}{|b|} = \frac{2\pi}{\frac{1}{2}} = 4\pi$

Phase shift: $\frac{\pi}{2}$

Asymptote: (The reasoning is similar to that for secant in exercise 13.)

$x = \frac{\pi}{2} + k\left(\frac{p}{2}\right)$

$x = \frac{\pi}{2} + k(2\pi)$, k any integer

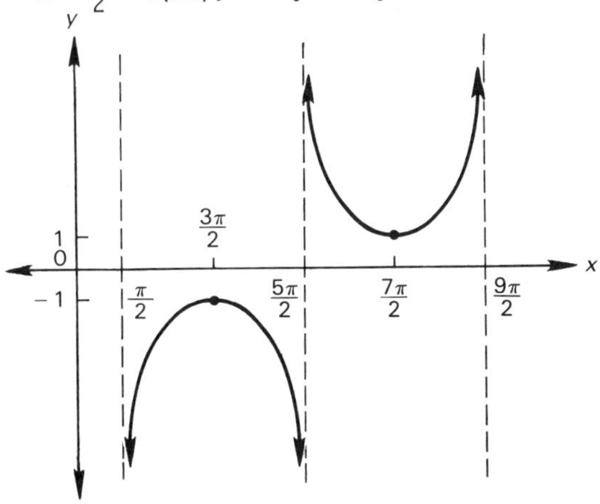

25) $f(x) = x^2$ $\qquad g(x) = \tan x$

$(f \circ g)(x) = f(g(x)) = f(\tan x) = \tan^2 x$

$(g \circ f)(x) = \ (f(x)) = g(x^2) = \tan(x^2)$

29) $F(x) = \sqrt{\tan(2x + 1)}$

The outer function takes the square root, so let $f(x) = \sqrt{x}$.

The middle function gives the tangent, so let $g(x) = \tan x$.

Then $h(x) = 2x + 1$

$(f \circ g \circ h)(x) = f(g(h(x))) = f(g(2x + 1)) = f(\tan(2x + 1))$

$= \sqrt{\tan(2x + 1)}$.

33)

h	1	0.5	0.25	0.1	0.01
$\frac{\sin h}{h}$	0.841471	0.958851	0.989616	0.998334	0.999983

$\frac{\sin h}{h}$ appears to be approaching 1 as h approaches 0.

Section 9.3 The Addition and Subtraction Formulas

1) $\cos 22^\circ \cos 38^\circ - \sin 22^\circ \sin 38^\circ = \cos(22^\circ + 38^\circ)$

$= \cos 60^\circ = \frac{1}{2}$

5) $\theta = 75^\circ = 30^\circ + 45^\circ$

$\sin(30^\circ + 45^\circ) = \sin 30^\circ \cos 45^\circ + \cos 30^\circ \sin 45^\circ$

$$= \left(\frac{1}{2}\right)\left(\frac{\sqrt{2}}{2}\right) + \left(\frac{\sqrt{3}}{2}\right)\left(\frac{\sqrt{2}}{2}\right)$$

$$= \frac{\sqrt{2}}{4} + \frac{\sqrt{6}}{4} = \frac{1}{4}(\sqrt{2} + \sqrt{6})$$

$\cos(30^\circ + 45^\circ) = \cos 30^\circ \cos 45^\circ - \sin 30^\circ \sin 45^\circ$

$$= \left(\frac{\sqrt{3}}{2}\right)\left(\frac{\sqrt{2}}{2}\right) - \left(\frac{1}{2}\right)\left(\frac{\sqrt{2}}{2}\right) = \frac{\sqrt{6}}{4} - \frac{\sqrt{2}}{4}$$

$$= \frac{1}{4}(\sqrt{6} - \sqrt{2})$$

5) continued

$$\tan(30^\circ + 45^\circ) = \frac{\tan 30^\circ + \tan 45^\circ}{1 - \tan 30^\circ \tan 45^\circ} = \frac{\frac{1}{\sqrt{3}} + 1}{1 - \frac{1}{\sqrt{3}} \cdot 1}$$

$$= \frac{\frac{1 + \sqrt{3}}{\sqrt{3}}}{\frac{\sqrt{3} - 1}{\sqrt{3}}} = \frac{\sqrt{3} + 1}{\sqrt{3} - 1}$$

9) $\theta = 165^\circ = 135^\circ + 30^\circ$

$$\sin(135^\circ + 30^\circ) = \sin 135^\circ \cos 30^\circ + \cos 135^\circ \sin 30^\circ$$

$$= \left(\frac{\sqrt{2}}{2}\right)\left(\frac{\sqrt{3}}{2}\right) + \left(-\frac{\sqrt{2}}{2}\right)\left(\frac{1}{2}\right)$$

$$= \frac{\sqrt{6}}{4} - \frac{\sqrt{2}}{4} = \frac{1}{4}(\sqrt{6} - \sqrt{2})$$

$$\cos(135^\circ + 30^\circ) = \cos 135^\circ \cos 30^\circ - \sin 135^\circ \sin 30^\circ$$

$$= \left(-\frac{\sqrt{2}}{2}\right)\left(\frac{\sqrt{3}}{2}\right) - \left(\frac{\sqrt{2}}{2}\right)\left(\frac{1}{2}\right)$$

$$= -\frac{\sqrt{6}}{4} - \frac{\sqrt{2}}{4} = -\frac{1}{4}(\sqrt{6} + \sqrt{2})$$

$$\tan(135^\circ + 30^\circ) = \frac{\tan 135^\circ + \tan 30^\circ}{1 - \tan 135^\circ \tan 30^\circ}$$

$$= \frac{(-1) + \frac{1}{\sqrt{3}}}{1 - (-1)\left(\frac{1}{\sqrt{3}}\right)} = \frac{-\sqrt{3} + 1}{\sqrt{3} + 1}$$

13) $\sin(\pi - \theta) = \sin \pi \cos \theta - \cos \pi \sin \theta = 0 \cdot \cos \theta - (-1)\sin \theta$
$= \sin \theta$

17) $\cos(90^\circ - \theta) = \cos 90^\circ \cos \theta + \sin 90^\circ \sin \theta = 0 \cdot \cos \theta + 1 \cdot \sin \theta$
$= \sin \theta$

21) $\sin\alpha = -\frac{1}{3}$ $\qquad$ $\cos\beta = \frac{2}{5}$

$a = -1 \qquad c = 3 \qquad b = 2 \qquad c = 5$

$(-1)^2 + b^2 = 3^2 \qquad a^2 + 2^2 = 5^2$

$b^2 = 8 \qquad a^2 = 21$

$b = -2\sqrt{2}$ $\begin{pmatrix} b < 0 \text{ in} \\ \text{quad. III} \end{pmatrix}$ $a = -\sqrt{21}$ (a < 0 in quad IV)

$\cos\alpha = \frac{-2\sqrt{2}}{3} \qquad \sin\beta = \frac{-\sqrt{21}}{5}$

$$\sin(\alpha + \beta) = \sin\alpha\cos\beta + \cos\alpha\sin\beta$$

$$= \left(-\frac{1}{3}\right)\left(\frac{2}{5}\right) + \left(\frac{-2\sqrt{2}}{3}\right)\left(\frac{-\sqrt{21}}{5}\right)$$

$$= \frac{-2}{15} + \frac{2\sqrt{42}}{15} = \frac{2}{15}(\sqrt{42} - 1)$$

$$\cos(\alpha + \beta) = \cos\alpha\cos\beta - \sin\alpha\sin\beta$$

$$= \left(\frac{-2\sqrt{2}}{3}\right)\left(\frac{2}{5}\right) - \left(-\frac{1}{3}\right)\left(\frac{-\sqrt{21}}{5}\right)$$

$$= -\frac{4\sqrt{2}}{15} - \frac{\sqrt{21}}{15} = -\frac{1}{15}(4\sqrt{2} + \sqrt{21})$$

25) $\cos\left(\theta - \frac{\pi}{4}\right) = \cos\theta\cos\frac{\pi}{4} + \sin\theta\sin\frac{\pi}{4}$

$$= (\cos\theta)\left(\frac{\sqrt{2}}{2}\right) + (\sin\theta)\left(\frac{\sqrt{2}}{2}\right)$$

$$= \frac{\sqrt{2}}{2}(\cos\theta + \sin\theta)$$

29) $\dfrac{\sin\left(\theta + \frac{\pi}{2}\right)}{\cos\left(\theta + \frac{\pi}{2}\right)} = \dfrac{\cos\theta}{-\sin\theta} = -\cot\theta$

33) $\sin(\alpha + \beta) - \sin(\alpha - \beta) = (\sin\alpha\cos\beta + \cos\alpha\sin\beta)$

$- (\sin\alpha\cos\beta - \cos\alpha\sin\beta) = \sin\alpha\cos\beta - \sin\alpha\cos\beta$

$+ \cos\alpha\sin\beta + \cos\alpha\sin\beta = 2\cos\alpha\sin\beta$

37) $2\sin\left(\frac{\pi}{4} - \theta\right)\sin\left(\frac{\pi}{4} + \theta\right) =$

$$2\left[\sin\frac{\pi}{4}\cos\theta - \cos\frac{\pi}{4}\sin\theta\right]\left[\sin\frac{\pi}{4}\cos\theta + \cos\frac{\pi}{4}\sin\theta\right]$$

$$= 2\left(\frac{1}{\sqrt{2}}\cos\theta - \frac{1}{\sqrt{2}}\sin\theta\right)\left(\frac{1}{\sqrt{2}}\cos\theta + \frac{1}{\sqrt{2}}\sin\theta\right)$$

$$= 2\left(\frac{\cos^2\theta}{2} - \frac{\sin^2\theta}{2}\right) = \cos^2\theta - \sin^2\theta$$

41) $\sin(\alpha - \beta) = \sin(\alpha + (-\beta)) = \sin\alpha\cos(-\beta) + \cos\alpha\sin(-\beta)$

$$= \sin\alpha\cos\beta + (\cos\alpha)(-\sin\beta) = \sin\alpha\cos\beta - \cos\alpha\sin\beta$$

45) $\sin\theta = -\cos\left(\theta + \frac{\pi}{2}\right)$

Shift $y = \cos\theta$ to the left $\frac{\pi}{2}$ units and reflect it through the θ-axis.

49) From the figure, find $\tan\alpha$ and $\tan(\alpha + \beta)$.

$$\tan\alpha = \frac{2}{8} = \frac{1}{4} \qquad \tan(\alpha + \beta) = \frac{3 + 2}{8} = \frac{5}{8}$$

Now use the addition formula, substituting these values.

$$\tan(\alpha + \beta) = \frac{\tan\alpha + \tan\beta}{1 - \tan\alpha\tan\beta}$$

$$\frac{5}{8} = \frac{\frac{1}{4} + \tan\beta}{1 - \frac{1}{4}\tan\beta}$$

$$\frac{5}{8}\left(1 - \frac{1}{4}\tan\beta\right) = \frac{1}{4} + \tan\beta$$

$$\frac{5}{8} - \frac{5}{32}\tan\beta = \frac{1}{4} + \tan\beta$$

$$\frac{3}{8} = \frac{37}{32}\tan\beta$$

$$\frac{12}{37} = \tan\beta$$

Section 9.4 The Double- and Half-angle Forumulas

1) $1 - \sin^2 \frac{7\pi}{12} = \cos 2\left(\frac{7\pi}{12}\right) = \cos \frac{7\pi}{6} = -\frac{\sqrt{3}}{2}$

5) $\cos 75^\circ = \cos \frac{150^\circ}{2} = \sqrt{\frac{1 + \cos 150^\circ}{2}} = \sqrt{\frac{1 - \frac{\sqrt{3}}{2}}{2}}$

$= \sqrt{\frac{2 - \sqrt{3}}{4}} = \frac{1}{2}\sqrt{2 - \sqrt{3}}$

9) a) $\tan 15^\circ = \tan \frac{30^\circ}{2} = \frac{1 - \cos 30^\circ}{\sin 30^\circ} = \frac{1 - \frac{\sqrt{3}}{2}}{\frac{1}{2}} = 2 - \sqrt{3}$

b) $\tan 15^\circ = 0.2679$ (Table 5)

$2 - \sqrt{3} = 2 - 1.7321 = 0.2679$

13) If $\tan \theta = -\frac{15}{8}$ and θ is obtuse, then θ is in II.

The side opposite θ (in standard form) is 15 and the side adjacent is -8.

hypotenuse $= \sqrt{(15)^2 + (-8)^2} = \sqrt{289} = 17$

$\sin \theta = \frac{15}{17} \qquad \cos \theta = \frac{-8}{17}$

a) $\sin 2\theta = 2 \sin \theta \cos \theta = 2\left(\frac{15}{17}\right)\left(\frac{-8}{17}\right) = -\frac{240}{289}$

b) $\cos 2\theta = \cos^2\theta - \sin^2\theta = \left(\frac{-8}{17}\right)^2 - \left(\frac{15}{17}\right)^2 = \frac{64}{289} - \frac{225}{289} = -\frac{161}{289}$

c) $\tan 2\theta = \frac{2 \tan \theta}{1 - \tan^2\theta} = \frac{2\left(-\frac{15}{8}\right)}{1 - \left(\frac{-15}{8}\right)^2} = \frac{-\frac{15}{4}}{1 - \frac{225}{64}} = \frac{-\frac{15}{4}}{-\frac{161}{64}} = \frac{240}{161}$

17) $9 \sin 2\theta = 9(2 \sin \theta \cos \theta) = 18\left(\frac{x}{3}\right)\left(\frac{\sqrt{9 - x^2}}{3}\right) = 2x\sqrt{9 - x^2}$

21) From the sketch: $\tan\theta = \frac{h}{150}$, $\tan 2\theta = \frac{h}{50}$

$$\tan 2\theta = \frac{2\tan\theta}{1-\tan^2\theta}$$

$$\frac{h}{50} = \frac{2\left(\frac{h}{150}\right)}{1-\left(\frac{h}{150}\right)^2}$$

$$\frac{h}{50} = \frac{\frac{h}{75}}{1-\frac{h^2}{22500}}$$

$$\frac{h}{50} = \frac{h}{75}\cdot\frac{22500}{22500-h^2}$$

$$\frac{h}{50} = \frac{300h}{22500-h^2}$$

$$15000h = h(22500-h^2)$$

$$15000 = 22500 - h^2 \qquad (h \neq 0)$$

$$h^2 = 7500$$

$$h = 50\sqrt{3}\ \text{ft.}$$

25) $$\sec^2\frac{x}{2} = \frac{1}{\cos^2\frac{x}{2}} = \frac{1}{\frac{1+\cos 2\left(\frac{x}{2}\right)}{2}} = \frac{2}{1+\cos x}$$

29) $$\cot\frac{x}{2} = \frac{1}{\tan\frac{x}{2}} = \frac{1}{\frac{\sin x}{1+\cos x}} = \frac{1+\cos x}{\sin x} = \frac{1}{\sin x} + \frac{\cos x}{\sin x}$$

$$= \csc x + \cot x$$

33) $$\tan 3x = \tan(2x+x) = \frac{\tan(2x)+\tan x}{1-\tan(2x)\tan x}$$

$$= \frac{\frac{2\tan x}{1-\tan^2 x} + \tan x}{1-\left(\frac{2\tan x}{1-\tan^2 x}\right)\tan x} = \frac{\frac{2\tan x + \tan x - \tan^3 x}{1-\tan^2 x}}{\frac{1-\tan^2 x - 2\tan^2 x}{1-\tan^2 x}}$$

$$= \frac{3\tan x - \tan^3 x}{1-3\tan^2 x}$$

37) $\csc 2x - \cot 2x = \dfrac{1}{\sin 2x} - \dfrac{\cos 2x}{\sin 2x} = \dfrac{1 - \cos 2x}{\sin 2x} = \tan \dfrac{2x}{2} = \tan x$

41) $\cos 4x = \cos 2(2x) = 2\cos^2(2x) - 1 = 2[2\cos^2 x - 1]^2 - 1$

$= 2[4\cos^4 x - 4\cos^2 x + 1] - 1 = 8\cos^4 x - 8\cos^2 x + 1$

45) a)

$$\begin{array}{ll} \cos\alpha\cos\beta - \sin\alpha\sin\beta & = \cos(\alpha+\beta) \\ \underline{\cos\alpha\cos\beta + \sin\alpha\sin\beta} & \underline{= \cos(\alpha-\beta)} \\ 2\cos\alpha\cos\beta & = \cos(\alpha+\beta) + \cos(\alpha-\beta) \\ \cos\alpha\cos\beta & = \frac{1}{2}[\cos(\alpha+\beta) + \cos(\alpha-\beta)] \end{array}$$

b) Subtract:

$$\begin{array}{ll} \cos\alpha\cos\beta + \sin\alpha\sin\beta & = \cos(\alpha-\beta) \\ \underline{\cos\alpha\cos\beta - \sin\alpha\sin\beta} & \underline{= \cos(\alpha+\beta)} \\ 2\sin\alpha\sin\beta & = \cos(\alpha-\beta) - \cos(\alpha+\beta) \\ \sin\alpha\sin\beta & = \frac{1}{2}[\cos(\alpha-\beta) - \cos(\alpha+\beta)] \end{array}$$

Add:

$$\begin{array}{ll} \sin\alpha\cos\beta + \cos\alpha\sin\beta & = \sin(\alpha+\beta) \\ \underline{\sin\alpha\cos\beta - \cos\alpha\sin\beta} & \underline{= \sin(\alpha-\beta)} \\ 2\sin\alpha\cos\beta & = \sin(\alpha+\beta) + \sin(\alpha-\beta) \\ \sin\alpha\cos\beta & = \frac{1}{2}[\sin(\alpha+\beta) + \sin(\alpha-\beta)] \end{array}$$

Subtract:

$$\begin{array}{ll} \sin\alpha\cos\beta + \cos\alpha\sin\beta & = \sin(\alpha+\beta) \\ \underline{\sin\alpha\cos\beta - \cos\alpha\sin\beta} & \underline{= \sin(\alpha-\beta)} \\ 2\cos\alpha\sin\beta & = \sin(\alpha+\beta) - \sin(\alpha-\beta) \\ \cos\alpha\sin\beta & = \frac{1}{2}[\sin(\alpha+\beta) - \sin(\alpha-\beta)] \end{array}$$

49) a) (i) If $\frac{\pi}{2} < \theta < \pi$, then $\frac{\pi}{4} < \frac{\theta}{2} < \frac{\pi}{2}$. Both θ and $\frac{\theta}{2}$ are in I where sine and tangent are both positive.

(ii) If $\pi < \theta < \frac{3\pi}{2}$, then $\frac{\pi}{2} < \frac{\theta}{2} < \frac{3\pi}{4}$. θ is in III where sine is negative and $\frac{\theta}{2}$ is in II where tangent is negative.

(iii) If $\frac{3\pi}{2} < \theta < 2\pi$, then $\frac{3\pi}{4} < \frac{\theta}{2} < \pi$. θ is in IV where sine is negative, and $\frac{\theta}{2}$ is in II where tangent is negative.

49) continued

b) $\tan^2 \frac{\theta}{2} = \frac{1 - \cos\theta}{1 + \cos\theta} = \frac{1 - \cos\theta}{1 + \cos\theta} \cdot \frac{1 + \cos\theta}{1 + \cos\theta}$

$$= \frac{1 - \cos^2\theta}{(1 + \cos\theta)^2} = \frac{\sin^2\theta}{(1 + \cos\theta)^2}$$

$$\left|\tan \frac{\theta}{2}\right| = \left|\frac{\sin\theta}{1 + \cos\theta}\right|$$

From part (a) $\tan \frac{\theta}{2}$ and $\sin\theta$ have the same sign, and

$(1 + \cos\theta) > 0$. Therefore:

$$\tan \frac{\theta}{2} = \frac{\sin\theta}{1 + \cos\theta}$$

c) $\angle Q$ is inscribed in the circle and therefore has a measure of

$\frac{1}{2}$ (arc AP) $= \frac{\theta}{2}$

$\tan Q = \tan \frac{\theta}{2} = \frac{RP}{QR} = \frac{RP}{1 + OR} = \frac{\sin\theta}{1 + \cos\theta}$

Section 9.5 Trigonometric Equations

1) $\cos x = 1$; on $[0, 2\pi)$ we have $x = 0$;
$x = 0 + 2k\pi = 2k\pi$, k any integer

5) $\sec x = -1$; on $[0, 2\pi)$ we have $x = \pi$;
$x = \pi + 2k\pi$, k any integer

9) $\frac{1}{2} \sin^2 x = 1$

$\sin^2 x = 2$

$\sin x = \pm\sqrt{2}$

But $-1 \leq \sin\theta \leq 1$ and $\sqrt{2} > 1$,

$\therefore$ No solutions.

13) $2 \cos(x + 1) = -2$

$\cos(x + 1) = -1$; on $[0, 2\pi)$ we have $(x + 1) = \pi$

$x = \pi - 1$

$x = \pi + 2k\pi - 1$, k any integer

17) $\sqrt{3}\csc\theta = 2$

$$\csc\theta = \frac{2}{\sqrt{3}}$$

$$\sin\theta = \frac{\sqrt{3}}{2}$$

$$\theta = 60^{\circ},\ 120^{\circ}$$

21) $\sin^2\theta - \cos^2\theta = 1$

$$\sin^2\theta - (1 - \sin^2\theta) = 1$$

$$2\sin^2\theta - 1 = 1$$

$$2\sin^2\theta = 2$$

$$\sin^2\theta = 1$$

$$\sin\theta = \pm 1$$

$$\theta = 90^{\circ},\ 270^{\circ}$$

25) $-1 + \tan\frac{3\theta}{2} = 0$

$$\tan\frac{3\theta}{2} = 1$$

If $0^{\circ} \le \theta < 360^{\circ}$, then $0^{\circ} \le \frac{3\theta}{2} < 540^{\circ}$.

$$\frac{3\theta}{2} = 45^{\circ},\ 225^{\circ},\ 405^{\circ}$$

$$3\theta = 90^{\circ},\ 450^{\circ},\ 810^{\circ}$$

$$\theta = 30^{\circ},\ 150^{\circ},\ 270^{\circ}$$

29) $(\cos x - 1)(2\sin x + 1) = 0$

$\cos x - 1 = 0$ or $2\sin x + 1 = 0$

$\cos x = 1$ $\qquad$ $\sin x = -\frac{1}{2}$

$x = 0$ $\qquad$ $x = \frac{7\pi}{6},\ \frac{11\pi}{6}$

33) $2\cos^2 x - \sqrt{3}\cos x = 0$

$$\cos x(2\cos x - \sqrt{3}) = 0$$

$\cos x = 0$ or $2\cos x - \sqrt{3} = 0$

$x = \frac{\pi}{2},\ \frac{3\pi}{2}$ $\qquad$ $\cos x = \frac{\sqrt{3}}{2}$

$$x = \frac{\pi}{6},\ \frac{11\pi}{6}$$

37) $\sin 2x = \cos x$

$2\sin x\cos x = \cos x$

$2\sin x\cos x - \cos x = 0$

$\cos x(2\sin x - 1) = 0$

$\cos x = 0$ or $2\sin x - 1 = 0$

$x = \frac{\pi}{2},\ \frac{3\pi}{2}$ $\qquad$ $\sin x = \frac{1}{2}$

$$x = \frac{\pi}{6},\ \frac{5\pi}{6}$$

41) $2\cos^2 x + 9\cos x - 5 = 0$

$(2\cos x - 1)(\cos x + 5) = 0$

$2\cos x - 1 = 0$ or $\cos x + 5 = 0$

$\cos x = \frac{1}{2}$ $\qquad$ $\cos x = -5$

$x = \frac{\pi}{3}, \frac{5\pi}{3}$ $\qquad$ $-1 \le \sin\theta \le 1, \quad -5 < -1$

No solutions.

45) $2\sin^4 x - 3\sin^2 x + 1 = 0$

$(\sin^2 x - 1)(2\sin^2 x - 1) = 0$

$\sin^2 x - 1 = 0$ or $2\sin^2 x - 1 = 0$

$\sin^2 x = 1$ $\qquad$ $\sin^2 x = \frac{1}{2}$

$\sin x = \pm 1$ $\qquad$ $\sin x = \pm\frac{\sqrt{2}}{2}$

$x = \frac{\pi}{2}, \frac{3\pi}{2}$ $\qquad$ $x = \frac{\pi}{4}, \frac{3\pi}{4}, \frac{5\pi}{4}, \frac{7\pi}{4}$

49) $3\tan^2 x = 7\sec x - 5$

$3(\sec^2 x - 1) - 7\sec x + 5 = 0$

$3\sec^2 x - 7\sec x + 2 = 0$

$(3\sec x - 1)(\sec x - 2) = 0$

$3\sec x - 1 = 0$ or $\sec x - 2 = 0$

$\sec x = \frac{1}{3}$ $\qquad$ $\sec x = 2$

$\cos x = 3$ $\qquad$ $\cos x = \frac{1}{2}$

No solutions since $|\cos x| \le 1$ $\qquad$ $x = \frac{\pi}{3}, \frac{5\pi}{3}$

53) $3\cos 2x + 2\sin^2 x = 2$

$3(1 - 2\sin^2 x) + 2\sin^2 x = 2$

$3 - 6\sin^2 x + 2\sin^2 x = 2$

$-4\sin^2 x = -1$

$\sin^2 x = \frac{1}{4}$

$\sin x = \pm\frac{1}{2}$

$x = \frac{\pi}{6}, \frac{5\pi}{6}, \frac{7\pi}{6}, \frac{11\pi}{6}$

57) $\sin x + \cos x = 1$

$\sin^2 x + 2 \sin x \cos x + \cos^2 x = 1$ (squaring both sides)

$\sin^2 x + \sin 2x + (1 - \sin^2 x) = 1$

$\sin 2x = 0$

If $0 \leq x < 2\pi$, then $0 \leq 2x < 4\pi$, and $2x = 0, \pi, 2\pi, 3\pi$

$x = 0, \frac{\pi}{2}, \pi, \frac{3\pi}{2}$

Extraneous solutions have been introduced by squaring. π and $\frac{3\pi}{2}$ do not check.

Solutions: $x = 0, \frac{\pi}{2}$

61) $\sin \frac{x}{2} = .8258$

$\frac{x}{2} = 55^\circ 40'$ or $124^\circ 20'$

$x = 111^\circ 20'$ or $248^\circ 40'$

65) $\cos^2 x - \sin^2 x = -\frac{3}{4}$

$(1 - \sin^2 x) - \sin^2 x = -\frac{3}{4}$

$1 - 2 \sin^2 x = -\frac{3}{4}$

$2 \sin^2 x = \frac{7}{4}$

$\sin^2 x = \frac{7}{8}$

$\sin x = \pm 0.9354$

$x = 69^\circ 20', 110^\circ 40',$

$249^\circ 20', 290^\circ 40'$

69) $y = \sin x$

$y = -\cos x$

$\sin x = -\cos x$

$\frac{\sin x}{\cos x} = -1$

$\tan x = -1$

$x = \frac{3\pi}{4}, y = \frac{\sqrt{2}}{2}$ or $x = \frac{7\pi}{4}, y = -\frac{\sqrt{2}}{2}$

Section 9.6 Inverse Circular Functions

1) $\text{arc} \sin 0 = 0$

5) $\text{arc} \tan (-1) = -\frac{\pi}{4}$

9) arc tan 115

The nearest entry is 114.6.

$\text{arc} \tan 115 = 89^\circ 30'$

or 1.5621 rad.

13) arc cos (0.1475)

The nearest entry is 0.1478.

$\text{arc} \cos (0.1475) = 81^\circ 30'$

or 1.4224 rad.

17) $\cos[\text{arc sin } (-1)] = \cos\left(-\frac{\pi}{2}\right) = 0$

21) $\sin\left(\text{arc cos } \frac{12}{13}\right)$

$\cos\theta = \frac{\text{adjacent}}{\text{hypotenuse}}$ $\qquad 13^2 = 12^2 + x^2$

$25 = x^2$

$5 = x$ is the opposite side

The angle having $\frac{12}{13}$ as its cosine has $\frac{5}{13}$ as its sine.

25) Sin and arcsin are inverse functions.

29) $y = 2 + \arctan x$

Increase the y-values of $y = \arctan x$ by 2 units.

Domain: all real numbers

Range: $\left(-\frac{\pi}{2} + 2, \frac{\pi}{2} + 2\right)$

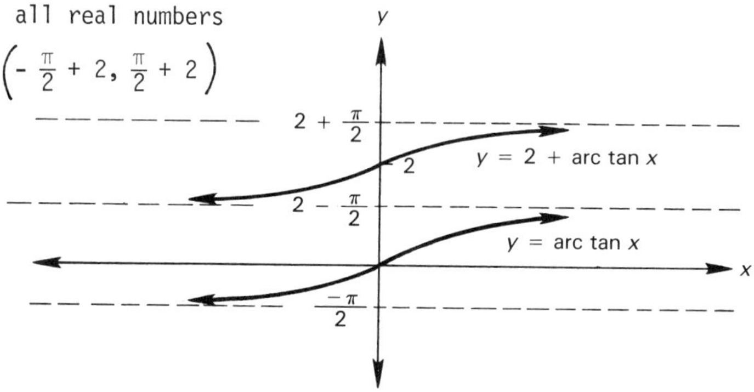

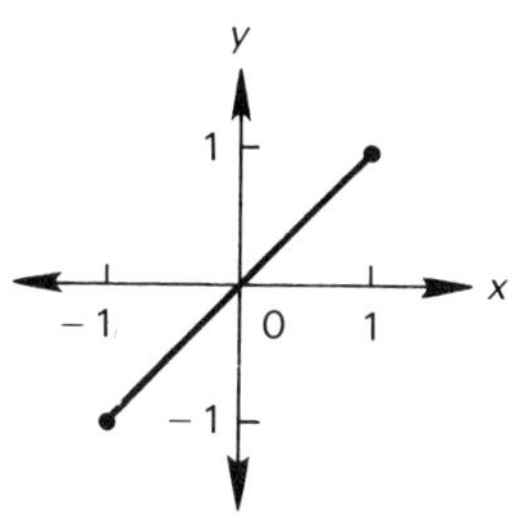

33) $y = \sin(\text{arc sin } x)$

$y = x$

Domain: $-1 \le x \le 1$

$-1 \le x \le 1$

37) $\cos\left(\text{arc sin } \frac{8}{17} - \text{arc cos } \frac{12}{13}\right) = \cos\left(\text{arc sin } \frac{8}{17}\right)\cos\left(\text{arc cos } \frac{12}{13}\right)$

$+ \sin\left(\text{arc sin } \frac{8}{17}\right)\sin\left(\text{arc cos } \frac{12}{13}\right)$

$= \cos\left(\underbrace{\text{arc sin } \frac{8}{17}}_{\alpha}\right)\left(\frac{12}{13}\right) + \left(\frac{8}{17}\right)\sin\left(\underbrace{\text{arc cos } \frac{12}{13}}_{\beta}\right)$

37) continued

$\sin \alpha = \frac{\text{opposite}}{\text{hypotenuse}} = \frac{8}{17}$ By the Pythagorean theorem, the adjacent side is 15. $\cos\left(\text{arc sin } \frac{8}{17}\right) = \frac{15}{17}$

$\cos \beta = \frac{\text{adjacent}}{\text{hypotenuse}} = \frac{12}{13}$ By the Pythagorean theorem, the opposite side is 5. $\sin\left(\text{arc cos } \frac{12}{13}\right) = \frac{5}{13}$

Continuing from above:

$$\cos\left(\text{arc sin } \frac{8}{17}\right)\left(\frac{12}{13}\right) + \left(\frac{8}{17}\right) \sin\left(\text{arc cos } \frac{12}{13}\right) = \left(\frac{15}{17}\right)\left(\frac{12}{13}\right) + \left(\frac{8}{17}\right)\left(\frac{5}{13}\right) = \frac{180 + 40}{221} = \frac{220}{221}$$

41) $\text{arc sin } (x + 2) = \frac{\pi}{6}$

$$x + 2 = \sin \frac{\pi}{6}$$

$$x + 2 = \frac{1}{2}$$

$$x = -\frac{3}{2}$$

Check:

$$\text{arc sin}\left(-\frac{3}{2} + 2\right) = \frac{\pi}{6}$$

$$\text{arc sin}\left(\frac{1}{2}\right) = \frac{\pi}{6}$$

$$\frac{\pi}{6} = \frac{\pi}{6}$$

45) $\tan(\pi + \text{arc tan } x) = \frac{\tan \pi + \tan(\text{arc tan } x)}{1 - \tan \pi \tan(\text{arc tan } x)} = \frac{0 + x}{1 - 0 \cdot x} = x$

49) $\tan^2 x = 4 \tan x - 4$

$\tan^2 x - 4 \tan x + 4 = 0$

$(\tan x - 2)^2 = 0$

$\tan x - 2 = 0$

$\tan x = 2$

$x = \text{arc tan } 2$

$x = 63^\circ\ 30'$ or 1.1083 rad.

53) $f(x) = \text{arc sin } x$ $g(x) = 3x + 2$

$(f \circ g)(x) = f(g(x)) = f(3x + 2) = \text{arc sin } (3x + 2)$

$(g \circ f)(x) = g(f(x)) = g(\text{arc sin } x) = 3 \text{ arc sin } x + 2$

Sample Test Questions for Chapter 9

Part I

1. $y = 2 \sin 2x$

   Amplitude: $|2| = 2$

   Period: $\frac{2\pi}{|2|} = \pi$

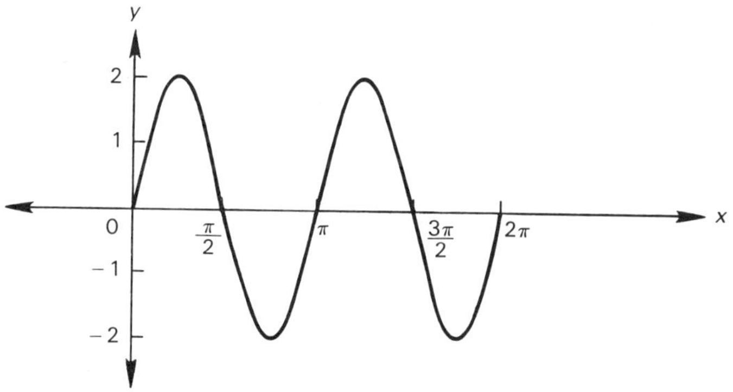

2. $y = \sec x$

   Domain: all $x \neq \frac{\pi}{2} + k\pi$

   Range: $y \geq 1$ or $y \leq -1$

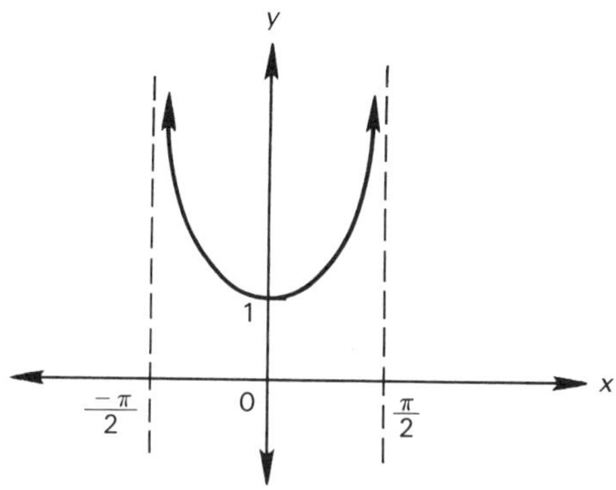

3. $\cos 165^\circ = \cos(135^\circ + 30^\circ) = \cos 135^\circ \cos 30^\circ - \sin 135^\circ \sin 30^\circ$

$$= -\cos 45^\circ \cos 30^\circ - \sin 45^\circ \sin 30^\circ$$

$$= \left(-\frac{\sqrt{2}}{2}\right)\left(\frac{\sqrt{3}}{2}\right) - \left(\frac{\sqrt{2}}{2}\right)\left(\frac{1}{2}\right)$$

$$= -\frac{\sqrt{6}}{4} - \frac{\sqrt{2}}{4} = -\frac{1}{4}(\sqrt{6} + \sqrt{2})$$

4. $\sin\left(-\frac{\pi}{12}\right) = -\sin\frac{\pi}{12} = -\sin\frac{\frac{\pi}{6}}{2} = -\sqrt{\frac{1-\cos\left(\frac{\pi}{6}\right)}{2}}$

$$= -\sqrt{\frac{1-\frac{\sqrt{3}}{2}}{2}} = -\sqrt{\frac{2-\sqrt{3}}{4}} = -\frac{1}{2}\sqrt{2-\sqrt{3}}$$

5. $\sin\theta = \frac{1}{3}$ and θ in I.

$\sqrt{3^2 - 1^2} = \sqrt{8} = 2\sqrt{2}$ is the length of the adjacent side.

$\sin 2\theta = 2\sin\theta\cos\theta = 2\left(\frac{1}{3}\right)\left(\frac{2\sqrt{2}}{3}\right) = \frac{4\sqrt{2}}{9}$

6. $\dfrac{\tan 115^\circ - \tan 55^\circ}{1 + \tan 115^\circ \tan 55^\circ} = \tan(115^\circ - 55^\circ) = \tan 60^\circ = \sqrt{3}$

7. $\sin\left(\frac{\pi}{4} - \theta\right) = \sin\frac{\pi}{4}\cos\theta - \cos\frac{\pi}{4}\sin\theta = \frac{\sqrt{2}}{2}\cos\theta - \frac{\sqrt{2}}{2}\sin\theta$

$$= \frac{\sqrt{2}}{2}(\cos\theta - \sin\theta)$$

8. $\csc 4x = \dfrac{1}{\sin 4x} = \dfrac{1}{2\sin 2x\cos 2x} = \dfrac{1}{2(2\sin x\cos x)(\cos 2x)}$

$$= \frac{1}{4\sin x\cos x\cos 2x} = \frac{1}{4}\left(\frac{1}{\sin x}\right)\left(\frac{1}{\cos x}\right)\left(\frac{1}{\cos 2x}\right)$$

$$= \frac{1}{4}\csc x\sec x\sec 2x$$

9. $\dfrac{\tan x - \sin x}{2\tan x} = \dfrac{\frac{\sin x}{\cos x} - \sin x}{2\left(\frac{\sin x}{\cos x}\right)} = \dfrac{\frac{\sin x - \sin x\cos x}{\cos x}}{\frac{2\sin x}{\cos x}}$

$$= \frac{\sin x(1-\cos x)}{2\sin x} = \frac{1-\cos x}{2} = \sin^2\left(\frac{x}{2}\right)$$

10. $\sin^2 x - \cos^2 x + \sin x = 0$

$$\sin^2 x - (1 - \sin^2 x) + \sin x = 0$$
$$2 \sin^2 x + \sin x - 1 = 0$$
$$(2 \sin x - 1)(\sin x + 1) = 0$$
$$2 \sin x - 1 = 0 \quad \text{or} \quad \sin x + 1 = 0$$
$$\sin x = \frac{1}{2} \qquad\qquad \sin x = -1$$
$$x = \frac{\pi}{6}, \ \frac{5\pi}{6} \qquad\qquad x = \frac{3\pi}{2}$$

11. $\cos 2x = 2 \sin^2 x - 2$

$$0 = 2 \sin^2 x - \cos 2x - 2$$
$$0 = 2 \sin^2 x - (1 - 2 \sin^2 x) - 2$$
$$0 = 4 \sin^2 x - 3$$
$$\frac{3}{4} = \sin^2 x$$
$$\pm \frac{\sqrt{3}}{2} = \sin x$$
$$x = \begin{cases} \frac{\pi}{3} + k\pi \\ -\frac{\pi}{3} + k\pi \end{cases} \qquad k \text{ any integer}$$

12. $y = \arctan x$

Domain: all real numbers

Range: $\left(-\frac{\pi}{2}, \frac{\pi}{2}\right)$

b) $\arctan\left(-\frac{\sqrt{3}}{3}\right) = \arctan\left(-\frac{1}{\sqrt{3}}\right) = -\frac{\pi}{6}$

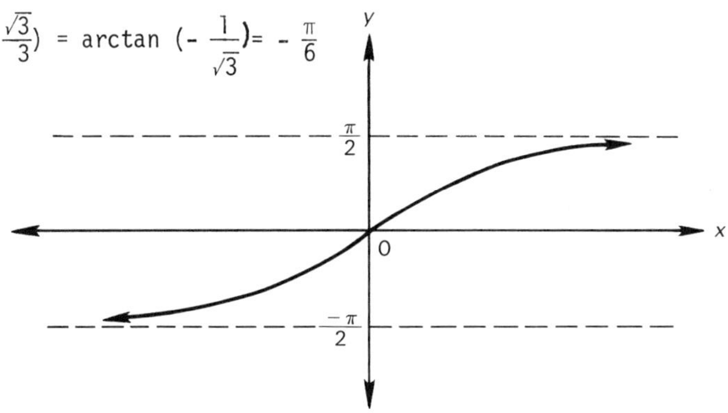

13. a) $\arcsin \frac{\sqrt{2}}{2} = \frac{\pi}{4}$

b) $\sin\left(\arccos \frac{3}{7}\right) = \sin \theta$

$\cos \theta = \frac{\text{adjacent}}{\text{hypotenuse}} = \frac{3}{7}$, $\sqrt{7^2 - 3^2} = 2\sqrt{10}$ is the opposite side.

$\sin\left(\arccos \frac{3}{7}\right) = \frac{2\sqrt{10}}{7}$

c) $\cos\left(2 \arcsin \frac{1}{2}\right) = \cos\left(2 \cdot \frac{\pi}{6}\right) = \cos \frac{\pi}{3} = \frac{1}{2}$

14. $a = \frac{1}{2}$ $\qquad p = \frac{2\pi}{|b|}$

$\frac{2\pi}{3} = \frac{2\pi}{|b|}$

$3 = b$

$y = \frac{1}{2} \cos 3x$

15. $y = \tan 2x$

$p = \frac{\pi}{|2|} = \frac{\pi}{2}$

$x = \frac{\pi}{4}$, $x = \frac{3\pi}{4}$

16. $\sin \alpha = \frac{3}{5}$, $\quad \alpha$ in II.

$\sin \alpha = \frac{\text{opposite}}{\text{hypotenuse}} = \frac{3}{5}$; $\quad -\sqrt{5^2 - 3^2} = -4$ is the adjacent side in quadrant II.

$\tan \beta = \frac{12}{5}$, β in III

$\tan \beta = \frac{\text{opposite}}{\text{adjacent}} = \frac{-12}{-5}$; $\sqrt{12^2 + 5^2} = 13$ is the hypotenuse.

$\cos(\alpha + \beta) = \cos \alpha \cos \beta - \sin \alpha \sin \beta$

$$= \left(\frac{-4}{5}\right)\left(\frac{-5}{13}\right) - \left(\frac{3}{5}\right)\left(\frac{-12}{13}\right) = \frac{20 + 36}{65} = \frac{56}{65}$$

17. $\tan A = \frac{a}{b}$

$\tan 15^{\circ} = \frac{8}{b}$

$\frac{\sin 30^{\circ}}{1 + \cos 30^{\circ}} = \frac{8}{b}$

$$\frac{\frac{1}{2}}{1 + \frac{\sqrt{3}}{2}} = \frac{8}{b}$$

Multiply the left side by $\frac{2}{2}$ to obtain $\frac{1}{2 + \sqrt{3}} = \frac{8}{b}$

$$b = 8(2 + \sqrt{3})$$

18. $y = -\frac{1}{2} \sin \left(3x - \frac{\pi}{2}\right) = -\frac{1}{2} \sin 3\left(x - \frac{\pi}{6}\right)$

Period: $p = \frac{2\pi}{|3|} = \frac{2\pi}{3}$

Amplitude: $\left|-\frac{1}{2}\right| = \frac{1}{2}$

Phase shift: $\frac{\pi}{6}$

19. $y = 2 \cos (2x + \pi) = 2 \cos 2\left(x + \frac{\pi}{2}\right)$

Period: $p = \frac{2\pi}{|b|} = \frac{2\pi}{|2|} = \pi$

Amplitude: $|2| = 2$

Phase shift: $-\frac{\pi}{2}$

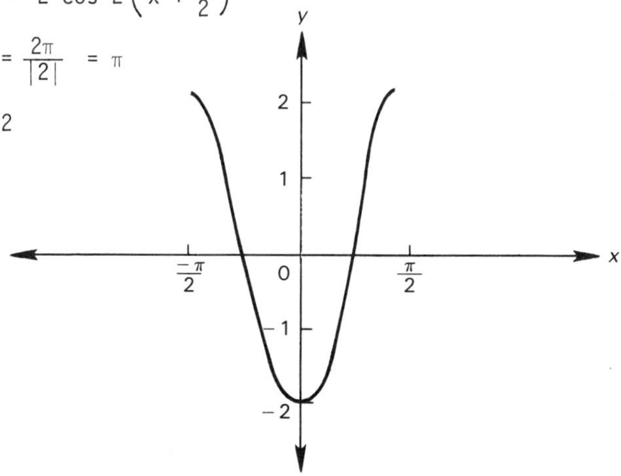

20. $\tan \frac{\pi}{8} = \frac{h}{4}$

$$\tan \frac{\frac{\pi}{4}}{2} = \frac{h}{4}$$

$$\frac{1 - \cos \frac{\pi}{4}}{\sin \frac{\pi}{4}} = \frac{h}{4}$$

$$\frac{1 - \frac{1}{\sqrt{2}}}{\frac{1}{\sqrt{2}}} = \frac{h}{4}$$

Multiply the left side by $\frac{\sqrt{2}}{\sqrt{2}}$ to obtain $\frac{\sqrt{2} - 1}{1} = \frac{h}{4}$

$$4(\sqrt{2} - 1) = h$$

Sample Test Questions for Chapter 9

Part II

1. (d) (2) (a)

3. (d) $p = \frac{2\pi}{|b|} = \frac{2\pi}{|4|} = \frac{\pi}{2}$ (4) (b)

5. (e)

6. (a) $\cos\frac{5\pi}{4}\cos\frac{\pi}{2} - \sin\frac{5\pi}{4}\sin\frac{\pi}{2} = \cos\left(\frac{5\pi}{4} + \frac{\pi}{2}\right) = \cos\frac{7\pi}{4}$

7. (a)

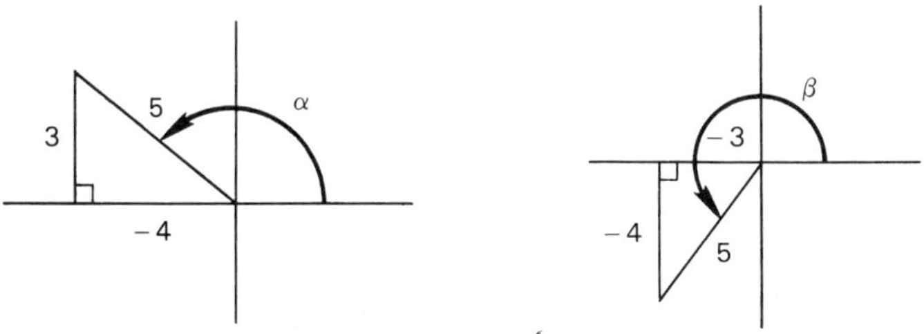

The pythagorean theorem was used to obtain the side -4 in the reference triangle for α, and the side -3 in the reference triangle for β. Then

$$\sin(\alpha - \beta) = \sin\alpha\cos\beta - \cos\alpha\sin\beta$$
$$= \left(\frac{3}{5}\right)\left(-\frac{3}{5}\right) - \left(-\frac{4}{5}\right)\left(-\frac{4}{5}\right)$$
$$= -\frac{9}{25} - \frac{16}{25} = -\frac{25}{25} = -1$$

8. (c)

9. (c) $\cos 2\theta = \cos^2\theta - \sin^2\theta = 2\cos^2\theta - 1 = 1 - 2\sin^2\theta$

10. (d)

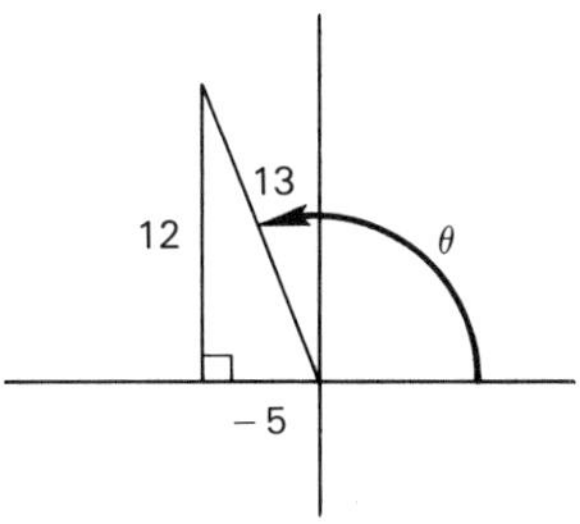

The Pythagorean theorem was used to obtain side 12 in the reference triangle for θ. Then

$$\sin 2\theta = 2\sin\theta\cos\theta$$
$$= 2\left(\frac{12}{13}\right)\left(-\frac{5}{13}\right) = -\frac{120}{169}$$

11. (c) $\sin \frac{\pi}{8} = \sin \frac{\frac{\pi}{4}}{2} = \sqrt{\frac{1 - \cos \frac{\pi}{4}}{2}} = \sqrt{\frac{1 - \frac{\sqrt{2}}{2}}{2}}$

$$= \sqrt{\frac{2 - \sqrt{2}}{4}} = \frac{1}{2}\sqrt{2 - \sqrt{2}}$$

12. (b) $\frac{\sin 2x}{1 - \cos^2 x} = \frac{2 \sin x \cos x}{1 - \cos^2 x} = \frac{2 \sin x \cos x}{\sin^2 x} = \frac{2 \cos x}{\sin x} = 2 \cot x$

13. (a) $\sin 2x = 1$

$2x = \frac{\pi}{2} + 2k\pi$, k any integer

$x = \frac{\pi}{4} + k\pi$

14. (e) $\sin^2 x = \sin x$

$\sin^2 x - \sin x = 0$

$\sin x (\sin x - 1) = 0$

$\sin x = 0$ or $\sin x - 1 = 0$

$x = 0$ $\qquad \sin x = 1$

$x = \frac{\pi}{2}$

15. (b) Let $\theta = \text{arc sin } \frac{5}{7}$ and observe that θ is an acute angle, since for $0 < x < 1$, $0 < \text{arc sin } x < \frac{\pi}{2}$. Now use a right triangle as follows:

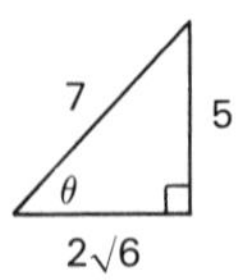

$$\sqrt{7^2 - 5^2} = \sqrt{24} = 2\sqrt{6}$$

Then

$$\tan\left(\text{arc sin } \frac{5}{7}\right) = \tan \theta = \frac{5}{2\sqrt{6}} = \frac{5\sqrt{6}}{12}$$

CHAPTER 10

ADDITIONAL USES OF TRIGONOMETRY

Section 10.1 The Law of Cosines

1) $a^2 = 4^2 + 11^2 - 2 \cdot 4 \cdot 11 \cos 60^{\circ}$

$a^2 = 16 + 121 - 88\left(\frac{1}{2}\right)$

$a^2 = 93$

$a = \sqrt{93}$

5) $a^2 = 8^2 + 13^2 - 2 \cdot 8 \cdot 13 \cos 20^{\circ}$

$a^2 = 64 + 169 - 208(0.9397)$

$a^2 = 37.54$

$a = 6.1$

$\cos \beta = \dfrac{(6.1)^2 + 13^2 - 8^2}{2(6.1)(13)}$

$\cos \beta = \dfrac{37.2 + 169 - 64}{158.6}$

$\cos \beta = 0.8966$

$\beta = 26^{\circ}20'$

9) $\cos \beta = \dfrac{18^2 + 4^2 - 15^2}{2 \cdot (18)(4)}$

$\cos \beta = \dfrac{115}{144}$

$\cos \beta = 0.7986$

$\beta = 37^{\circ}$

$\cos \gamma = \dfrac{18^2 + 15^2 - 4^2}{2(18)(15)}$

$\cos \gamma = \dfrac{533}{540}$

$\cos \gamma = 0.9870$

$\gamma = 9^{\circ}10'$

13) $a^2 = (2.2)^2 + (6.4)^2 - 2(2.2)(6.4)\cos 42^0$

$a^2 = 4.84 + 40.96 - 28.16(0.7431)$

$a^2 = 24.87$

$a = 5.0$

$\cos\beta = \dfrac{5^2 + (6.4)^2 - (2.2)^2}{2(5)(6.4)}$

$\cos\beta = \dfrac{61.12}{64}$

$\cos\beta = 0.955$

$\beta = 17^0 20'$

17) $(AB)^2 = (2.2)^2 + (3.5)^2 - 2(2.2)(3.5)\cos 112^0$

$(AB)^2 = 4.84 + 12.25 - 15.4(-0.3746)$

$(AB)^2 = 22.85$

AB = 4.8 miles

21) Label the vertex of the 26^0 angle D and the vertex of the right angle E.

$\angle DCE = 180^0 - (90^0 + 26^0) = 64^0$

$\angle DCB$ is the supplement of $\angle DCE$, so $\angle DCB = 116^0$

Now apply the law of cosines.

$(AB)^2 = 20^2 + 30^2 - 2 \cdot 20 \cdot 30 \cos 116^0$

$(AB)^2 = 400 + 900 - 1200(-0.4384)$

$(AB)^2 = 1826$

$AB = 42.7$

25) Two sides of this triangle can be found by computing the distances the trains travel.

They run for 20 minutes of 1/3 hour.

Distance of Train A: 1/3(54) = 18 miles

Distance of Train B: 1/3(60) = 20 miles

Let d represent the distance between the trains at 11:20 and apply the law of cosines.

$d^2 = 20^2 + 18^2 - 2 \cdot 20 \cdot 18 \cos 124^0$

$= 400 + 324 - 720(-0.5594) = 1126.62$

$d = 33.6$

At 11:20 the trains will be 33.6 miles apart.

29) $K = \frac{1}{2}(8.4)(12.6)\sin 40^{\circ}30'$

$= 52.92(0.6494) = 34.4$

33) To use $A = \frac{1}{2}bh$, first find h.

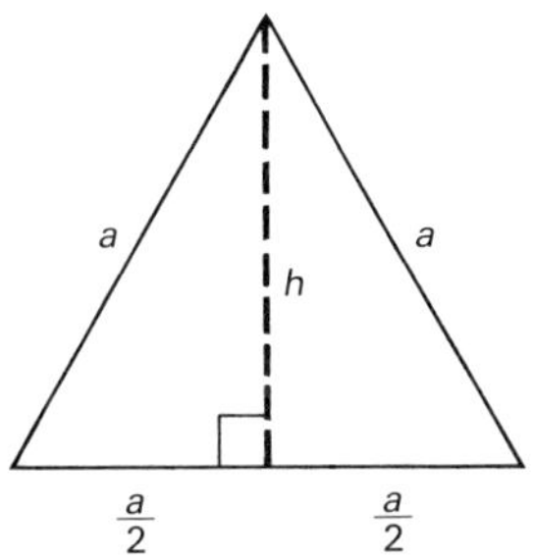

$$a^2 = h^2 + \left(\frac{a}{2}\right)^2$$

$$a^2 = h^2 + \frac{a^2}{4}$$

$$h^2 = a^2 - \frac{1}{4}a^2$$

$$h^2 = \frac{3}{4}a^2$$

$$h = \frac{\sqrt{3}}{2}a$$

$$A = \frac{1}{2}(a)\left(\frac{\sqrt{3}}{2}a\right)$$

$$A = \frac{\sqrt{3}}{4}a^2$$

Using $K = \sqrt{s(s-a)(s-b)(s-c)}$

$$s = \frac{1}{2}(a + a + a)$$

$$= \frac{3}{2}a$$

$$K = \sqrt{\frac{3}{2}a\left(\frac{3}{2}a - a\right)^3}$$

$$K = \sqrt{\frac{3}{2}a\left(\frac{a}{2}\right)^3}$$

$$K = \sqrt{\frac{3}{2}a\left(\frac{a^3}{8}\right)}$$

$$K = \sqrt{\frac{3}{16}a^4}$$

$$K = \frac{\sqrt{3}}{4}a^2$$

37) $(AD)^2 = 20^2 + 30^2 - 2 \cdot 20 \cdot 30 \cos 75^\circ = 400 + 900 - 1200\ (0.2588)$
$= 989.44$

$AD = 31.46$

$S_1 = \frac{1}{2}(31.46 + 30 + 20) = 40.73$

$K_1 = \sqrt{40.73(9.27)(10.73)(20.73)} = \sqrt{83983.35} = 289.8$

$(DC)^2 = 30^2 + 15^2 - 2 \cdot 30 \cdot 15 \cos 40^\circ$
$= 900 + 225 - 900(0.7660) = 435.6$

$DC = 20.87$

$S_2 = \frac{1}{2}(20.87 + 15 + 30) = 32.94$

$K_2 = \sqrt{32.94(2.94)(17.94)(12.07)} = \sqrt{20970.11} = 144.8$

$K_1 + K_2 = 289.8 + 144.8 = 434.6 \approx 435$

Section 10.2 The Law of Sines

1) If $\alpha = 30^\circ$ and $\beta = 120^\circ$, then $\alpha = 180^\circ - 150^\circ = 30^\circ$
Therefore $a = c$.

$$a = c = \frac{54 \sin 30^\circ}{\sin 120^\circ} = \frac{54\left(\frac{1}{2}\right)}{\frac{\sqrt{3}}{2}} = \frac{54\sqrt{3}}{3} = 18\sqrt{3}$$

5) If $\alpha = 25^\circ$ and $\gamma = 55^\circ$, then $\beta = 180^\circ - 80^\circ = 100^\circ$

$$a = \frac{12 \sin 25^\circ}{\sin 100^\circ} = \frac{12(0.4226)}{0.9848} = 5.1$$

$$c = \frac{12 \sin 55^\circ}{\sin 100^\circ} = \frac{12(0.8192)}{0.9848} = 10.0$$

9) $b \sin \alpha = 10 \sin 32^\circ = 10(0.5299) = 5.299 > 5.1 = a$
No solution.

13) $b \sin \alpha = 25 \sin 126^{\circ} = 25(0.8090) = 20.225 > 20 = a$

No solution.

17) $b \sin \alpha = 25 \sin 75^{\circ} = 25(0.9659) = 24.1475 > 7 = a$.

No solution.

21) By alternate interior angles of a transversal, the angle opposite the 40 unit side is 15°.

Solve first for the angle opposite the 80 unit side.

$$\frac{80}{\sin x} = \frac{40}{\sin 15^{\circ}}$$

$$\sin x = \frac{80 \sin 15^{\circ}}{40} = 2 \sin 15^{\circ} = 2(0.2588) = 0.5176$$

$$x = 31^{\circ}\ 10'$$

The obtuse angle has measure $180^{\circ} - 31^{\circ}\ 10' - 15^{\circ} = 133'\ 50^{\circ}$.

25)

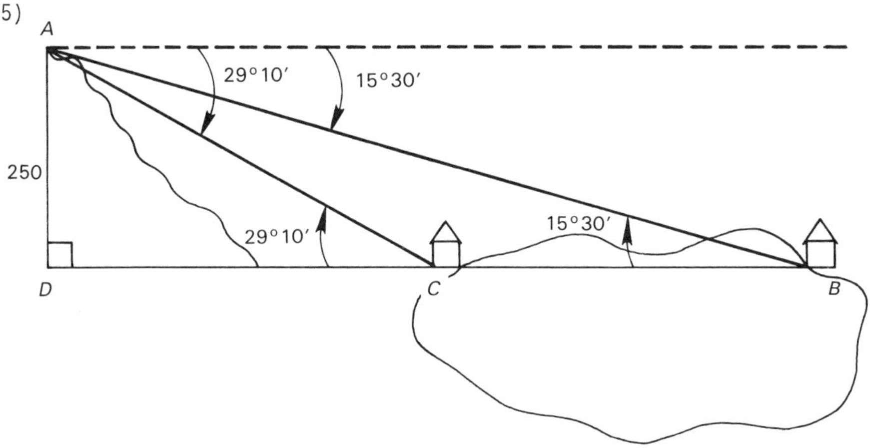

Draw a sketch to record the information given.

By alternate interior angles on transversal AC, the measure of $\angle$ACD is $29^{\circ}\ 10'$. The measure of $\angle$DAC is $180^{\circ} - 90^{\circ} - 29^{\circ}\ 10' = 60^{\circ}\ 50'$.

Using the law of sines in $\triangle$ACD, solve for DC.

25) (continued)

$$DC = \frac{250 \sin 60^\circ 50'}{\sin 29^\circ 10'}$$

$$DC = \frac{250(0.8732)}{0.4874} = 448$$

Similarly, by alternate interior angles on transversal AB, the measure of $\angle$ABD is $15^\circ 30'$. The measure of $\angle$DAB is $180^\circ - 90^\circ - 15^\circ 30' = 74^\circ 30'$. Use the law of sines in $\triangle$ABD to solve for DB.

$$DB = \frac{250 \sin 74^\circ 30'}{\sin 15^\circ 30'}$$

$$DB = \frac{250(0.9636)}{0.2672} = 902$$

The distance between the cottages is the difference DB - DC or 902 - 448 = 454 feet.

29) Let x represent the measure of DB.

$(AD)^2 = (AB)^2 + (DB)^2 - 2(AB)(DB)\cos B$

$(10)^2 = (12)^2 + x^2 - 2(12)(x)\cos 20^\circ$

$100 = 144 + x^2 - 24x(0.9397)$

$0 = x^2 - 22.55x + 44$

$$x = \frac{-(-22.55) \pm \sqrt{(-22.55)^2 - 4(1)(44)}}{2(1)}$$

$$x = \frac{22.55 \pm \sqrt{332.33}}{2}$$

$$x = \frac{22.55 \pm 18.23}{2}$$

$x = 20.39$ or 2.16

($2.16 < 4$ so this answer is rejected.)

If DB $\approx$ 20 and CB = 4, then DC = 16.

(Note: Another solution begins by finding AC using the Law of Cosines. Next find $\angle$BCA using either the Law of Sines or Cosines, and subtract from 180° to find $\angle$ACD. Now find $\angle$D using the Law of Sines, and finally the Law of Cosines gives CD.)

33) 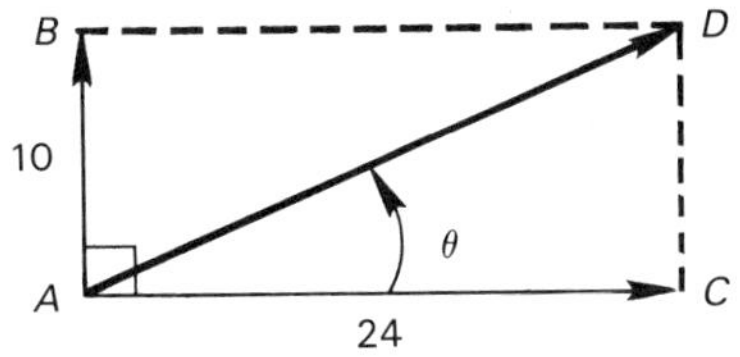

The magnitude of the resultant force can be found using the Pythagorean theorem.

$AD^2 = 10^2 + 24^2 = 676$

AD = 26 pounds

$\sin \theta = \frac{10}{26} = 0.3846$

$\theta \approx 23^o$

37) 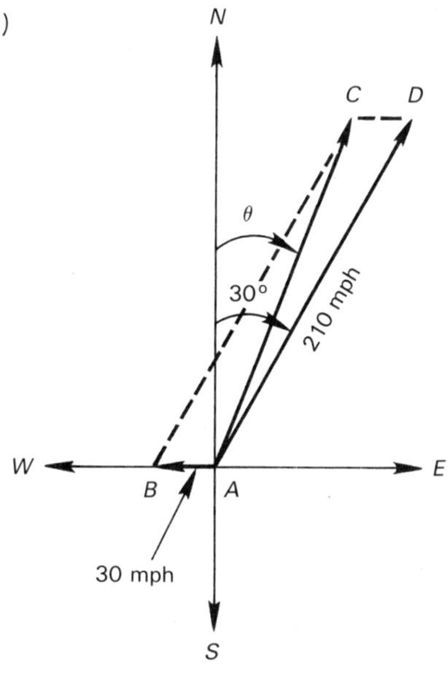

In the diagram,

$\overrightarrow{AD}$ is the air speed.

$\overrightarrow{AB}$ is the wind velocity.

$\overrightarrow{AC}$, the resultant vector will be the ground speed.

θ will be the final bearing.

The angle formed by DA and the WE axis is $90^o - 30^o = 60^o$.

CD||BA, so $\angle CDA = 60^o$.

Use the law of cosines to solve for AC in △ACD.

$AC^2 = 30^2 + 210^2 - 2(30)(210)(\cos 60^o)$

$= 900 + 44100 - 12600(0.5)$

$= 38700$

AC = 197 mph

Now use the law of sines to solve for $\angle CAD$.

$\sin \angle CAD = \frac{30 \sin 60^o}{197} = 0.1319$

$\angle CAD = 8^o$

$\theta = \angle QAD - \angle CAD = 30^o - 8^o = 22^o$

Therefore, the plane's ground speed is 197 mph and its heading is $N22^oE$.

41)

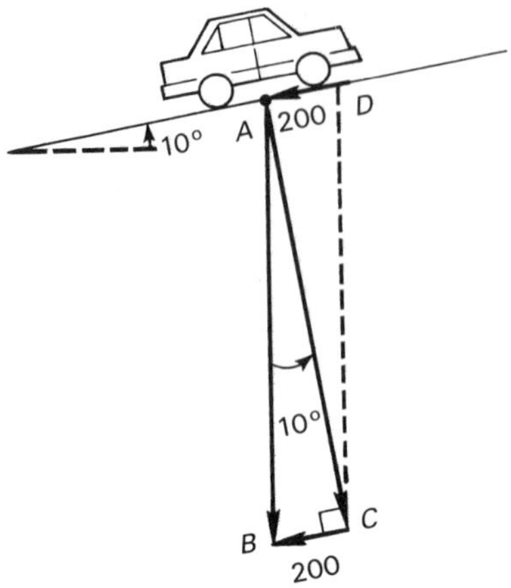

In the sketch $\overrightarrow{AB}$ is the weight or gravity vector, to be found. This vector is the resultant of $\overrightarrow{CB} + \overrightarrow{AC}$ where $\overleftrightarrow{AC}$ is the force perpendicular to the driveway and $\overrightarrow{CB}$ is the force parallel to the driveway, given as 200 lbs. Use the sine relationship to solve for AB.

$$AB = \frac{200}{\sin 10^{0}} = \frac{200}{0.1736} = 1152$$

The vehicle weighs about 1150 pounds.

Section 10.3	Trigonometric Form of Complex Numbers

1)

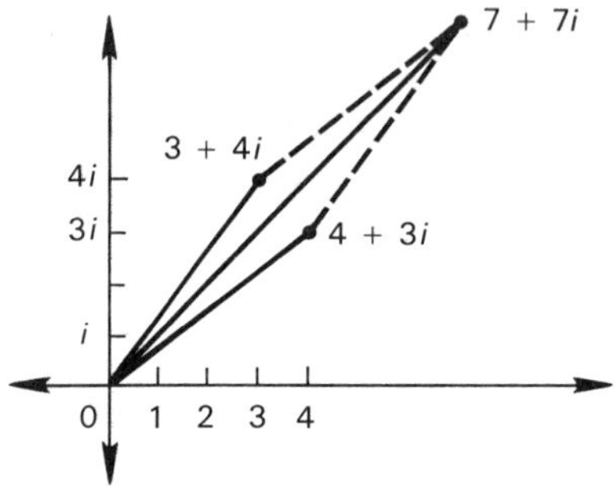

Graph 3 + 4i and 4 + 3i.
Apply the parallelogram rule:
Begin at 3 + 4i and add 4 + 3i by counting right 4 and up 3.
This gives the point 7 + 7i.

$$\begin{aligned}(3 + 4i) + (4 + 3i) &= (3 + 4) + (4 + 3)i \\ &= 7 + 7i\end{aligned}$$

5)

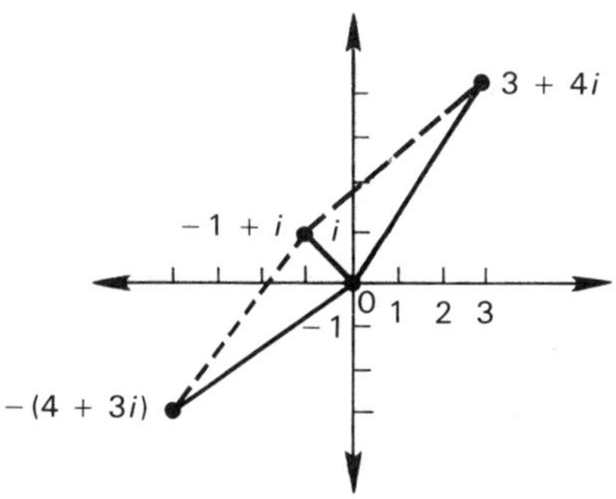

Graph 3 + 4i and -(4 + 3i) = -4 - 3i.
Apply the parallelogram rule:
Begin at -4 - 3i and add 3 + 4i by counting right 3 and up 4.
This gives the point -1 + i.

$$\begin{aligned}(3 + 4i) - (4 + 3i) &= (3 - 4) + (4 - 3)i \\ &= -1 + i\end{aligned}$$

9) 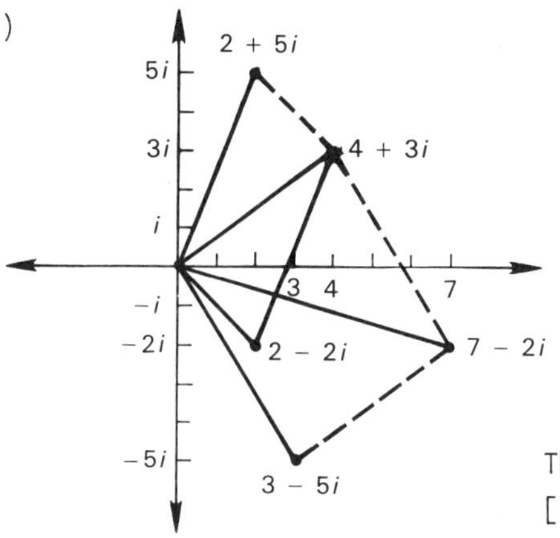

Graph 2 + 5i and 2 - 2i. Apply the parallelogram rule:
Begin at 2 + 5i and add 2 - 2i by counting right 2 and down 2.
This gives the point 4 + 3i.
Graph 3 - 5i.
Begin at 4 + 3i and add 3 - 5i by counting right 3 and down 5.
This gives the point 7 - 2i.

Therefore:

$[(2+5i)+(2-2i)]+(3-5i)=7-2i$

13) 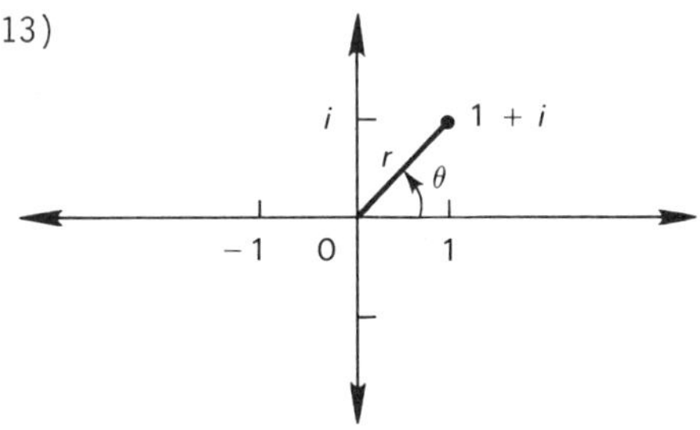

First graph 1 + i and observe that it forms a $45^0 - 45^0 - 90^0$ triangle with $\theta = \frac{\pi}{4}$.

r is the hypotenuse; $\sqrt{2}$.

$1 + i = \sqrt{2}\left(\cos\frac{\pi}{4} + i\sin\frac{\pi}{4}\right)$

17) 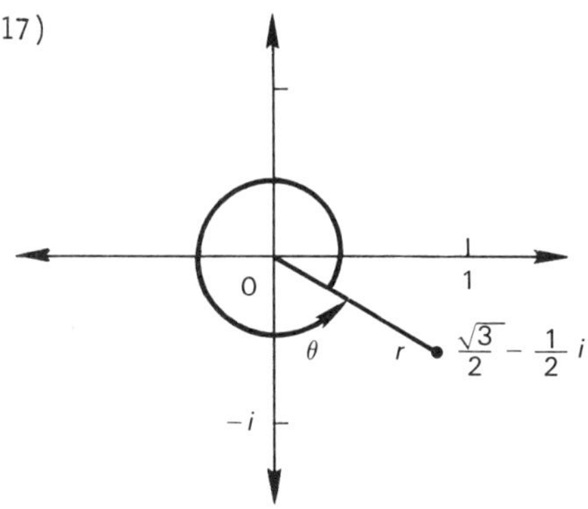

Graph $\frac{\sqrt{3}}{2} - \frac{1}{2}i$ and observe that it forms a $30^0 - 60^0 - 90^0$ triangle.

$\theta = \frac{11\pi}{6}$

r is the hypotenuse; 1.

$\frac{\sqrt{3}}{2} - \frac{1}{2}i = \cos\frac{11\pi}{6} + i\sin\frac{11\pi}{6}$

21)

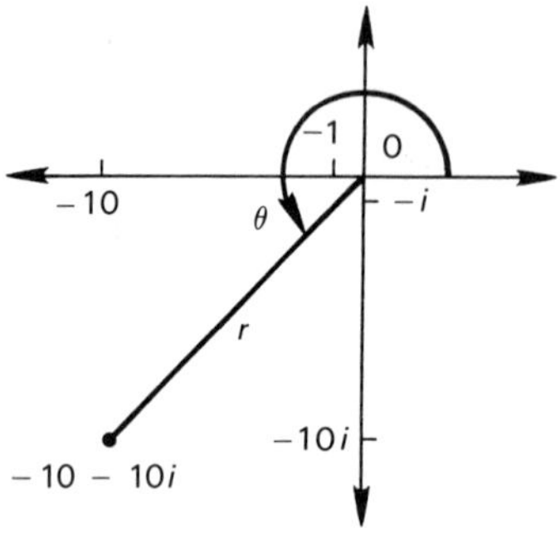

Graph -10 - 10i and observe that it forms a 45^0 - 45^0 - 90^0 triangle.

$\theta = \frac{5\pi}{4}$

r is the hypotenuse; $10\sqrt{2}$.

$$-10 - 10i = 10\sqrt{2}\left(\cos\frac{5\pi}{4} + i\sin\frac{5\pi}{4}\right)$$

25) $5\left(\cos\frac{3\pi}{2} + i\sin\frac{3\pi}{2}\right) = 5\left(\cos\frac{3\pi}{2}\right) + 5\left(\sin\frac{3\pi}{2}\right)i$

$= 5(0) + 5(-1)i = -5i$

29) $4(\cos 315^0 + i\sin 315^0) = 4(\cos 315^0) + 4(\sin 315^0)i$

$= 4(\cos 45^0) + 4(-\sin 45^0)i$

$= 4\left(\frac{\sqrt{2}}{2}\right) + 4\left(-\frac{\sqrt{2}}{2}\right)i$

$= 2\sqrt{2} - 2\sqrt{2}\ i$

33) $zw = \frac{1}{2}(\cos 100^0 + i\sin 100^0) \cdot 10(\cos 50^0 + i\sin 50^0)$

$= \frac{1}{2} \cdot 10[\cos(100^0 + 50^0) + i\sin(100^0 + 50^0)]$

$= 5(\cos 150^0 + i\sin 150^0) = 5(\cos 150^0) + 5(\sin 150^0)i$

$= 5\left(-\frac{\sqrt{3}}{2}\right) + 5\left(\frac{1}{2}\right)i = -\frac{5\sqrt{3}}{2} + \frac{5}{2}i$

$\frac{z}{w} = \frac{1}{2}(\cos 100^0 + i\sin 100^0) \div 10(\cos 50^0 + i\sin 50^0)$

$= \frac{\frac{1}{2}}{10}[\cos(100^0 - 50^0) + i\sin(100^0 - 50^0)]$

$= \frac{1}{20}(\cos 50^0 + i\sin 50^0) = \frac{1}{20}(\cos 50^0) + \frac{1}{20}(\sin 50^0)i$

$= \frac{1}{20}(0.6428) + \frac{1}{20}(0.7660)i = 0.0321 + 0.0383i$

33) continued

$$\frac{w}{z} = 10\ (\cos 50^{\circ} + i \sin 50^{\circ}) \div \frac{1}{2}\ (\cos 100^{\circ} + i \sin 100^{\circ})$$

$$= \frac{10}{\frac{1}{2}}\ [\cos(50^{\circ} - 100^{\circ}) + i \sin (50^{\circ} - 100^{\circ})]$$

$$= 20[\cos(-50^{\circ}) + i \sin (-50^{\circ})] = 20[\cos 310^{\circ} + i \sin 310^{\circ}],$$

(310° is the principal argument)

$$= 20(\cos 310^{\circ}) + 20(\sin 310^{\circ})i = 20(0.6428) + 20(-0.7660)i$$

$$= 12.856 - 15.32i$$

37) $z = 7i = 7\left(\cos \frac{\pi}{2} + i \sin \frac{\pi}{2}\right)$

$$w = \sqrt{3} + i = 2\left(\cos \frac{\pi}{6} + i \sin \frac{\pi}{6}\right)$$

$$zw = 7\left(\cos \frac{\pi}{2} + i \sin \frac{\pi}{2}\right) \bullet 2\left(\cos \frac{\pi}{6} + i \sin \frac{\pi}{6}\right)$$

$$= 7 \cdot 2\left[\cos\left(\frac{\pi}{2} + \frac{\pi}{6}\right) + i \sin\left(\frac{\pi}{2} + \frac{\pi}{6}\right)\right]$$

$$= 14\left(\cos \frac{2\pi}{3} + i \sin \frac{2\pi}{3}\right) = 14\left(\cos \frac{2\pi}{3}\right) + 14\left(\sin \frac{2\pi}{3}\right)i$$

$$= 14\left(-\frac{1}{2}\right) + 14\left(\frac{\sqrt{3}}{2}\right)i = -7 + 7\sqrt{3}\ i$$

$$\frac{z}{w} = 7\left(\cos \frac{\pi}{2} + i \sin \frac{\pi}{2}\right) \div 2\left(\cos \frac{\pi}{6} + i \sin \frac{\pi}{6}\right)$$

$$= \frac{7}{2}\left[\cos\left(\frac{\pi}{2} - \frac{\pi}{6}\right) + i \sin\left(\frac{\pi}{2} - \frac{\pi}{6}\right)\right]$$

$$= \frac{7}{2}\left(\cos \frac{\pi}{3} + i \sin \frac{\pi}{3}\right) = \frac{7}{2}\left(\cos \frac{\pi}{3}\right) + \frac{7}{2}\left(\sin \frac{\pi}{3}\right)i$$

$$= \frac{7}{2}\left(\frac{1}{2}\right) + \frac{7}{2}\left(\frac{\sqrt{3}}{2}\right)i = \frac{7}{4} + \frac{7\sqrt{3}}{4}\ i$$

41) $z = -1 + i = \sqrt{2}\left(\cos \frac{3\pi}{4} + i \sin \frac{3\pi}{4}\right)$

$$w = 1 - i = \sqrt{2}\left(\cos \frac{7\pi}{4} + i \sin \frac{7\pi}{4}\right)$$

$$zw = \sqrt{2}\left(\cos \frac{3\pi}{4} + i \sin \frac{3\pi}{4}\right) \cdot \sqrt{2}\left(\cos \frac{7\pi}{4} + i \sin \frac{7\pi}{4}\right)$$

$$= \sqrt{2} \cdot \sqrt{2}\left[\cos\left(\frac{3\pi}{4} + \frac{7\pi}{4}\right) + i \sin\left(\frac{3\pi}{4} + \frac{7\pi}{4}\right)\right]$$

41) (continued)

$$= 2\left(\cos\frac{\pi}{2} + i\sin\frac{\pi}{2}\right) = 2\left(\cos\frac{\pi}{2}\right) + 2\left(\sin\frac{\pi}{2}\right)i$$

$$= 2(0) + 2(1)i = 2i$$

$$\frac{z}{w} = \sqrt{2}\left(\cos\frac{3\pi}{4} + i\sin\frac{3\pi}{4}\right) \div \sqrt{2}\left(\cos\frac{7\pi}{4} + i\sin\frac{7\pi}{4}\right)$$

$$= \frac{\sqrt{2}}{\sqrt{2}}\left[\cos\left(\frac{3\pi}{4} - \frac{7\pi}{4}\right) + i\sin\left(\frac{3\pi}{4} - \frac{7\pi}{4}\right)\right]$$

$$= \cos\pi + i\sin\pi = -1$$

45) $\left[\frac{2}{3}(\cos 122^\circ + i\sin 122^\circ)\right]\left[\frac{9}{4}(\cos 77^\circ + i\sin 77^\circ)\right]$

$$= \frac{2}{3}\cdot\frac{9}{4}[\cos(122^\circ + 77^\circ) + i\sin(122^\circ + 77^\circ)]$$

$$= \frac{3}{2}(\cos 199^\circ + i\sin 199^\circ) = \frac{3}{2}\cos 199^\circ + \frac{3}{2}(\sin 199^\circ)i$$

$$= \frac{3}{2}(-0.9455) + \frac{3}{2}(-0.3256)i = -1.4183 - 0.4884i$$

49) $|zw| = |(a + bi)(c + di)| = |(ac - bd) + (bc + ad)i|$

$$= \sqrt{(ac - bd)^2 + (bc + ad)^2} = \sqrt{a^2c^2 + b^2d^2 + b^2c^2 + a^2d^2}$$

$$|z||w| = |a + bi||c + di| = \sqrt{a^2 + b^2}\cdot\sqrt{c^2 + d^2}$$

$$= \sqrt{(a^2 + b^2)(c^2 + d^2)} = \sqrt{a^2c^2 + b^2d^2 + b^2c^2 + a^2d^2}$$

53) $|z - (2 + 3i)| = 1$

$|(x + yi) - (2 + 3i)| = 1$

$|(x - 2) + (y - 3)i| = 1$

$(x - 2)^2 + (y - 3)^2 = 1$

Circle with center (2, 3) and $r = 1$.

Section 10.4 DeMoivre's Theorem

1) $(\cos 6^\circ + i \sin 6^\circ)^{10} = 1^{10}(\cos 60^\circ + i \sin 60^\circ)$

$$= 1\left(\frac{1}{2} + i\,\frac{\sqrt{3}}{2}\right) = \frac{1}{2} + \frac{\sqrt{3}}{2}\,i$$

5) $\left[\frac{1}{2}\left(\cos \frac{\pi}{8} + i \sin \frac{\pi}{8}\right)\right]^6 = \left(\frac{1}{2}\right)^6\left(\cos \frac{3\pi}{4} + i \sin \frac{3\pi}{4}\right)$

$$= \frac{1}{64}\left(-\frac{\sqrt{2}}{2} + i\frac{\sqrt{2}}{2}\right) = -\frac{\sqrt{2}}{128} + \frac{\sqrt{2}}{128}\,i$$

9) $(-\sqrt{3} + \sqrt{3}\,i)^6 = \left[\sqrt{6}\left(\cos \frac{3\pi}{4} + i \sin \frac{3\pi}{4}\right)\right]^6$

$$= (\sqrt{6})^6\left(\cos \frac{9\pi}{2} + i \sin \frac{9\pi}{2}\right)$$

$$= 216\left(\cos \frac{\pi}{2} + i \sin \frac{\pi}{2}\right) = 216(0 + i) = 216i$$

13) $\left(-\frac{\sqrt{3}}{2} + \frac{1}{2}\,i\right)^{12} = \left(\cos \frac{5\pi}{6} + i \sin \frac{5\pi}{6}\right)^{12} = \cos 10\pi + i \sin 10\pi$

$$= \cos 0 + i \sin 0 = 1$$

17) The 5th roots of $32\left(\cos \frac{\pi}{8} + i \sin \frac{\pi}{8}\right)$ are

$$32^{1/5}\left(\cos \frac{\frac{\pi}{8} + 2k\pi}{5} + i \sin \frac{\frac{\pi}{8} + 2k\pi}{5}\right) \text{ for } k = 0, 1, 2, 3, 4$$

For k = 0:

$$32^{1/5}\left(\cos \frac{\frac{\pi}{8}}{5} + i \sin \frac{\frac{\pi}{8}}{5}\right) = 2\left(\cos \frac{\pi}{40} + i \sin \frac{\pi}{40}\right)$$

For k = 1:

$$32^{1/5}\left(\cos \frac{\frac{\pi}{8} + 2\pi}{5} + i \sin \frac{\frac{\pi}{8} + 2\pi}{5}\right) = 2\left(\cos \frac{17\pi}{40} + i \sin \frac{17\pi}{40}\right)$$

For k = 2:

$$32^{1/5}\left(\cos \frac{\frac{\pi}{8} + 4\pi}{5} + i \sin \frac{\frac{\pi}{8} + 4\pi}{5}\right) = 2\left(\cos \frac{33\pi}{40} + i \sin \frac{33\pi}{40}\right)$$

For k = 3:

$$32^{1/5}\left(\cos \frac{\frac{\pi}{8} + 6\pi}{5} + i \sin \frac{\frac{\pi}{8} + 6\pi}{5}\right) = 2\left(\cos \frac{49\pi}{40} + i \sin \frac{49\pi}{40}\right)$$

17) continued

For k = 4:

$$32^{1/5}\left(\cos\frac{\frac{\pi}{8}+8\pi}{5}+i\sin\frac{\frac{\pi}{8}+8\pi}{5}\right)=2\left(\cos\frac{65\pi}{40}+i\sin\frac{65\pi}{40}\right)$$

21) $-1 = \cos\pi + i\sin\pi$

The cube roots of $\cos\pi + i\sin\pi$ are given by

$$1^{1/3}\left(\cos\frac{\pi+2k\pi}{3}+i\sin\frac{\pi+2k\pi}{3}\right)\quad \text{for } k = 0, 1, 2$$

For k = 0:

$$1^{1/3}\left(\cos\frac{\pi}{3}+i\sin\frac{\pi}{3}\right)=\frac{1}{2}+\frac{\sqrt{3}}{2}i$$

For k = 1:

$$1^{1/3}\left(\cos\frac{3\pi}{3}+i\sin\frac{3\pi}{3}\right)=\cos\pi+i\sin\pi=-1$$

For k = 2:

$$1^{1/3}\left(\cos\frac{5\pi}{3}+i\sin\frac{5\pi}{3}\right)=\frac{1}{2}-\frac{\sqrt{3}}{2}i$$

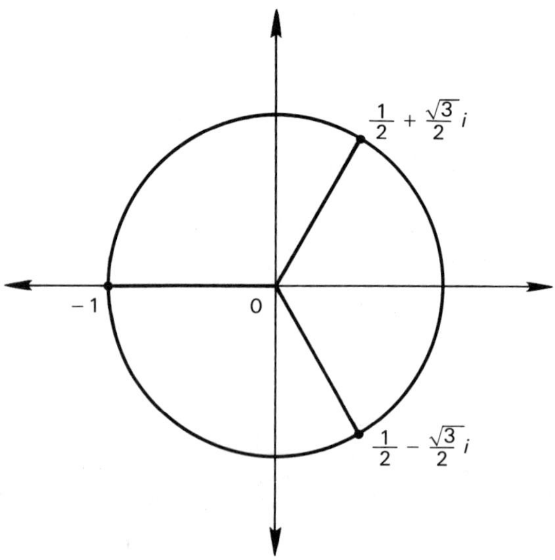

25) $1 = \cos 0 + i \sin 0$

The 5th roots of unity are given by

$1^{1/5}\left(\cos \frac{2k\pi}{5} + i \sin \frac{2k\pi}{5}\right)$ for $k = 0, 1, 2, 3, 4$

For $k = 0$:

$1^{1/5}(\cos 0 + i \sin 0) = 1$

For $k = 1$:

$1^{1/5}\left(\cos \frac{2\pi}{5} + i \sin \frac{2\pi}{5}\right) = \cos 72^0 + i \sin 72^0$

For $k = 2$:

$1^{1/5}\left(\cos \frac{4\pi}{5} + i \sin \frac{4\pi}{5}\right) = \cos 144^0 + i \sin 144^0$

For $k = 3$:

$1^{1/5}\left(\cos \frac{6\pi}{5} + i \sin \frac{6\pi}{5}\right) = \cos 216^0 + i \sin 216^0$

For $k = 4$:

$1^{1/5}\left(\cos \frac{8\pi}{5} + i \sin \frac{8\pi}{5}\right) = \cos 288^0 + i \sin 288^0$

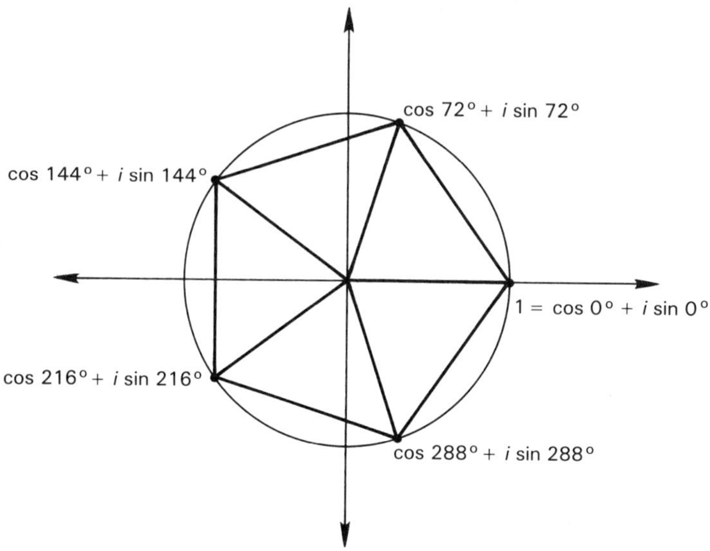

29) $z^3 + 8i = 0$

$z^3 = -8i = -8\left(\cos\frac{3\pi}{2} + i\sin\frac{3\pi}{2}\right)$

$$z = 8^{1/3}\left(\cos\frac{\frac{3\pi}{2} + 2k\pi}{3} + i\sin\frac{\frac{3\pi}{2} + 2k\pi}{3}\right) \quad \text{for } k = 0, 1, 2$$

For $k = 0$:

$$8^{1/3}\left(\cos\frac{3\pi}{6} + i\sin\frac{3\pi}{6}\right) = 2(0 + i) = 2i$$

For $k = 1$:

$$8^{1/3}\left(\cos\frac{7\pi}{6} + i\sin\frac{7\pi}{6}\right) = 2\left(-\frac{\sqrt{3}}{2} - \frac{1}{2}i\right) = -\sqrt{3} - i$$

For $k = 2$:

$$8^{1/3}\left(\cos\frac{11\pi}{6} + i\sin\frac{11\pi}{6}\right) = 2\left(\frac{\sqrt{3}}{2} - \frac{1}{2}i\right) = \sqrt{3} - i$$

33) a) For $k = 0$: $r^{1/n}\left(\cos\frac{\theta + 0}{n} + i\sin\frac{\theta + 0}{n}\right) = r^{1/n}\left(\cos\frac{\theta}{n} + i\sin\frac{\theta}{n}\right)$

For $k = n$: $r^{1/n}\left(\cos\frac{\theta + 2n\pi}{n} + i\sin\frac{\theta + 2n\pi}{n}\right)$

$$= r^{1/n}\left[\cos\left(\frac{\theta}{n} + 2\pi\right) + i\sin\left(\frac{\theta}{n} + 2\pi\right)\right]$$

$$= r^{1/n}\left(\cos\frac{\theta}{n} + i\sin\frac{\theta}{n}\right)$$

b) For $k = 1$: $r^{1/n}\left(\cos\frac{\theta + 2\pi}{n} + i\sin\frac{\theta + 2\pi}{n}\right)$

For $k = n + 1$: $r^{1/n}\left(\cos\frac{\theta + 2(n+1)\pi}{n} + i\sin\frac{\theta + 2(n+1)\pi}{n}\right)$

$$= r^{1/n}\left[\cos\left(\frac{\theta + 2\pi}{n} + 2\pi\right) + i\sin\left(\frac{\theta + 2\pi}{n} + 2\pi\right)\right]$$

$$= r^{1/n}\left(\cos\frac{\theta + 2\pi}{n} + i\sin\frac{\theta + 2\pi}{n}\right)$$

33) continued

c) For k: $r^{1/n}\left(\cos\frac{\theta + 2k\pi}{n} + i\sin\frac{\theta + 2k\pi}{n}\right)$

For k + n: $r^{1/n}\left(\cos\frac{\theta + 2(k + n)\pi}{n} + i\sin\frac{\theta + 2(k + n)\pi}{n}\right)$

$$= r^{1/n}\left[\cos\left(\frac{\theta + 2k\pi}{n} + 2\pi\right) + i\ \sin\left(\frac{\theta + 2k\pi}{n} + 2\pi\right)\right]$$

$$r^{1/n}\left(\cos\frac{\theta + 2k\pi}{n} + i\sin\frac{\theta + 2k\pi}{n}\right)$$

Section 10.5 Polar Coordinates

1)

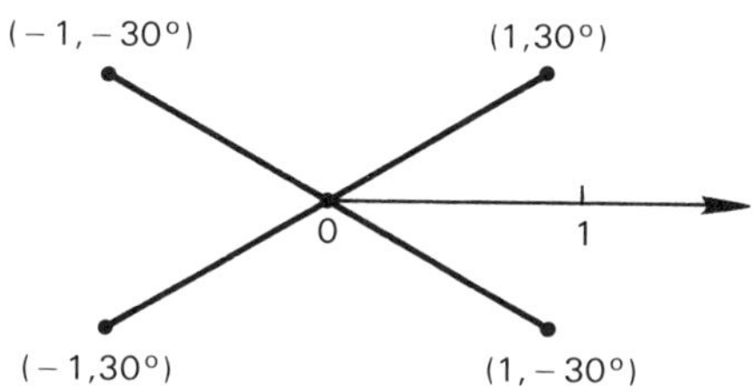

5)

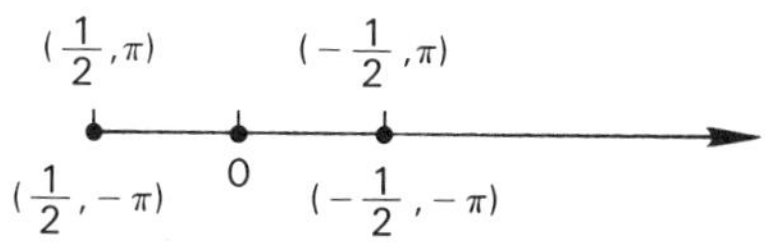

9) a) $\left(-3, \frac{\pi}{6}\right) = \left(3, \frac{\pi}{6} + \pi\right) = \left(3, \frac{7\pi}{6}\right) = \left(3, \frac{7\pi}{6} + 2\pi\right) = \left(3, \frac{19\pi}{6}\right)$

b) $\left(-3, \frac{\pi}{6}\right) = \left(3, -\frac{5\pi}{6}\right)$

c) $\left(-3, \frac{\pi}{6}\right)$

d) $\left(-3, \frac{\pi}{6}\right) = \left(-3, -\frac{11\pi}{6}\right)$

13) $\left(-\frac{3}{2}, \frac{5\pi}{2}\right)$

$x = r \cos \theta = \left(-\frac{3}{2}\right) \cos \frac{5\pi}{2} = \left(-\frac{3}{2}\right)(0) = 0$

$y = r \sin \theta = \left(-\frac{3}{2}\right) \sin \frac{5\pi}{2} = \left(-\frac{3}{2}\right)(1) = -\frac{3}{2}$

$(x, y) = \left(0, -\frac{3}{2}\right)$

17) $(2, 0)$
$r = \sqrt{x^2 + y^2} = \sqrt{2^2 + 0^2} = \sqrt{4} = 2$

$\tan \theta = \frac{y}{x} = \frac{0}{2} = 0$

$\theta = 0$ or $\theta = \pi$ since $0 \le \theta < 2\pi$. We reject $\theta = \pi$ since for $\theta = \pi$, $r = -2$ contrary to the requirement that $r \ge 0$. Thus $(r, \theta) = (2, 0)$.

21) $(-6, -6)$

$r = \sqrt{x^2 + y^2} = \sqrt{6^2 + 6^2} = \sqrt{72} = 6\sqrt{2}$

$\tan \theta = \frac{y}{x} = \frac{-6}{-6} = 1$

$\theta = \frac{\pi}{4}$ or $\frac{5\pi}{4}$. Since P is in quadrant III, $\theta = \frac{5\pi}{4}$.

$(r, \theta) = \left(6\sqrt{2}, \frac{5\pi}{4}\right)$

25) $(\sqrt{3}, -1)$

$r = \sqrt{x^2 + y^2} = \sqrt{(\sqrt{3})^2 + (-1)^2} = \sqrt{4} = 2$

$\tan \theta = \frac{y}{x} = \frac{-1}{\sqrt{3}} = -\frac{\sqrt{3}}{3}$

$\theta = \frac{5\pi}{6}$ or $\frac{11\pi}{6}$; $(\sqrt{3}, -1)$ is in quadrant IV, so $\theta = \frac{11\pi}{6}$

$(r, \theta) = \left(2, \frac{11\pi}{6}\right)$

29) $(-\sqrt{3}, 1)$

$$-r = \sqrt{x^2 + y^2} = \sqrt{(-\sqrt{3})^2 + (-1)^2} = 2$$

$$r = -2$$

$$\tan \theta = \frac{y}{x} = -\frac{1}{\sqrt{3}} = -\frac{\sqrt{3}}{3}$$

$\theta = \frac{5\pi}{6}$ or $\frac{11\pi}{6}$ $(-\sqrt{3}, 1)$ is in quadrant II on the negative extension of ray θ, so the positive point of ray θ is in quadrant IV. Then $\theta = \frac{11\pi}{6}$ and $(r, \theta) = \left(-2, \frac{11\pi}{6}\right)$.

33) $(0, -10)$

$$r = \sqrt{x^2 + y^2} = \sqrt{0^2 + (-10)^2} = \sqrt{100} = 10$$

$\tan \theta = \frac{y}{x} = \frac{-10}{0}$ undefined

$\theta = 90^o$ or -90° $(0, -10)$ is on the negative part of the y-axis, so $\theta = -90^o$.

$(r, \theta) = (10, -90^o)$

37) For $(r, \theta) = (-r, \theta + \pi)$ when $r < 0$:

$$\begin{aligned} x = -r \cos(\theta + \pi) &= -r(\cos\theta \cos\pi - \sin\theta \sin\pi) \\ &= -r(\cos\theta \cdot (-1) - \sin\theta \cdot (0)) \\ &= -r(-\cos\theta) = r\cos\theta \end{aligned}$$

$$\begin{aligned} y = -r \sin(\theta + \pi) &= -r(\sin\theta \cos\pi + \cos\theta \sin\pi) \\ &= -r(\sin\theta \cdot (-1) + \cos\theta \cdot (0)) \\ &= -r(-\sin\theta) = r(\sin\theta) \end{aligned}$$

Section 10.6 Graphing Polar Equations

1) $\theta = 30^o$, $\theta = 210^o$, $\theta = -150^o$, $\theta = -30^o$

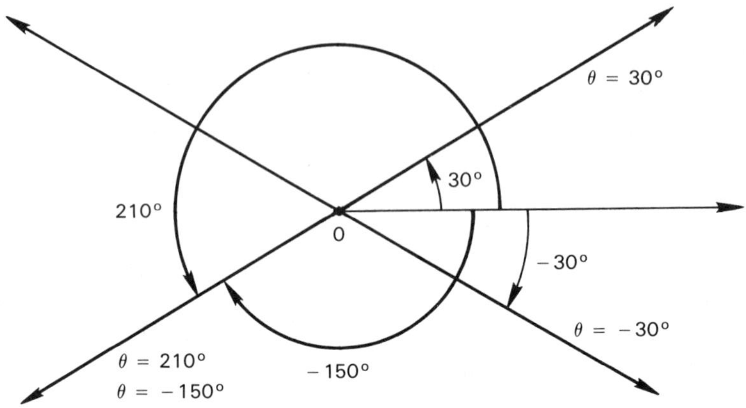

5) $r = -2$, $r = 4$, $r = -4$

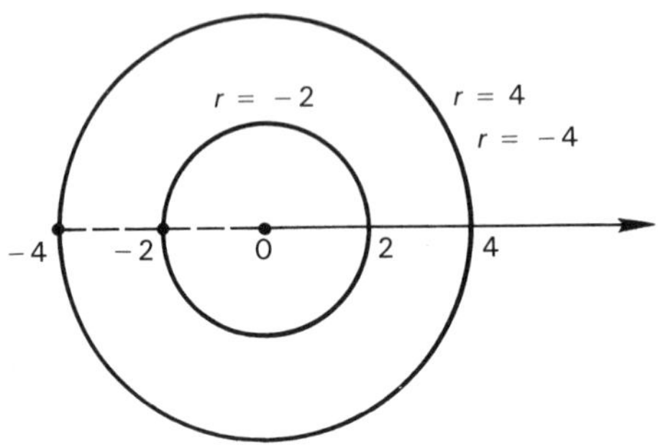

9)

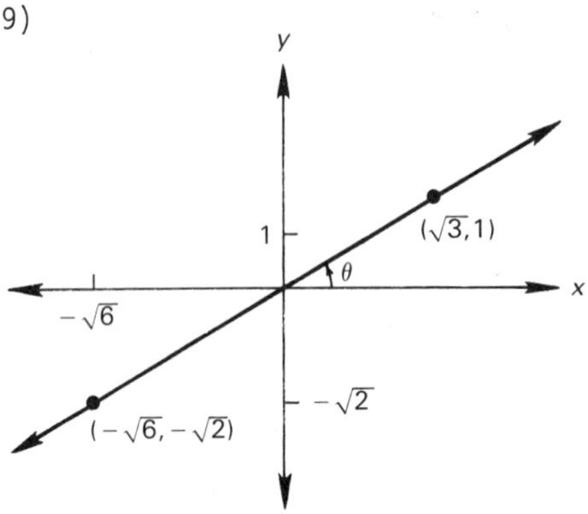

$\tan\theta = \frac{y}{x} = \frac{1}{\sqrt{3}} = \frac{\sqrt{3}}{3}$

Therefore the line makes an angle of $\frac{\pi}{6}$ with the polar axis, and a polar equation of the line is $\theta = \frac{\pi}{6}$.

13) $r = 5 \sin \theta = 2\left(\frac{5}{2}\right)\sin \theta$

$a = \frac{5}{2}$ Because $\sin \theta$ is involved, the center of the circle is on the ray $\theta = \frac{\pi}{2}$.

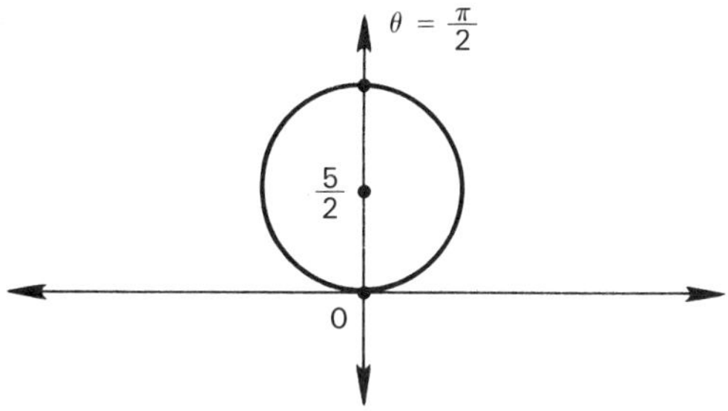

Change to rectangular coordinates:

$$r = 5 \sin \theta$$
$$r^2 = 5r \sin \theta$$
$$x^2 + y^2 = 5y$$
$$x^2 + y^2 - 5y = 0$$
$$x^2 + y^2 - 5y + \frac{25}{4} = \frac{25}{4}$$
$$x^2 + \left(y - \frac{5}{2}\right)^2 = \frac{25}{4}$$

17) a) $r = 2$, $r^2 = 4$, $x^2 + y^2 = 4$

b) $r = -2$, $r^2 = 4$, $x^2 + y^2 = 4$

c) $r = \sqrt{5}$, $r^2 = 5$, $x^2 + y^2 = 5$

21) $r = 2(1 - \cos \theta)$

Because $\cos \theta$ is involved, the polar axis is the axis of symmetry.

θ	0°	30°	60°	90°	120°	150°	180°	210°	240°	270°	300°	330°	360°
$\cos \theta$	1	.87	.5	0	-.5	-.87	-1	-.87	-.5	0	.5	.87	1
$2(1 - \cos \theta)$	0	.3	1	2	3	3.7	4	3.7	3	2	1	.3	0

21) (continued)

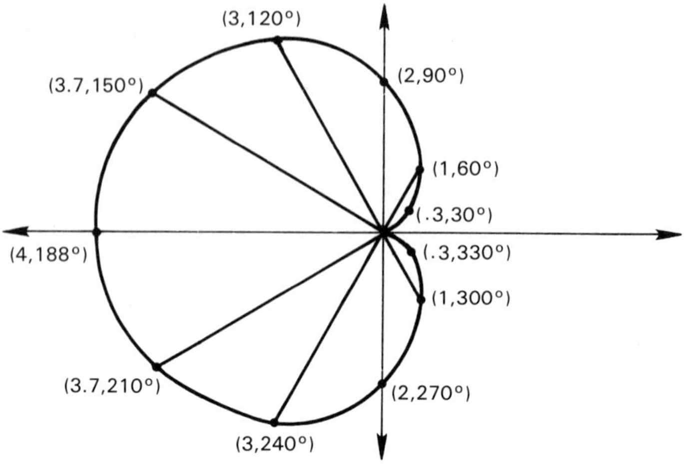

25) $\cot\theta = -3$

$\theta = -18^{0}\ 26'$ (using a calculator)

Or switch to rectangular coordinates:

$$\cot\theta = \frac{x}{y} = -3 \qquad \text{or} \qquad y = -\frac{1}{3}x$$

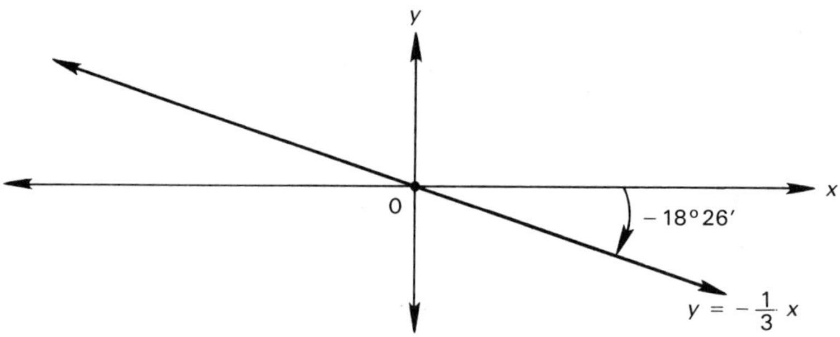

29) $r = \cos 2\theta$

Recall the graph of $y = \cos 2x$ as it appears in rectangular coordinates.

On $0 < x < \frac{\pi}{4}$, cos 2x decreases from 1 to 0.

On $\frac{\pi}{4} < x < \frac{\pi}{2}$, cos 2x decreases from 0 to -1.

On $\frac{\pi}{2} < x < \frac{3\pi}{4}$, cos 2x increases from -1 to 0.

29) (continued)

On $\frac{3\pi}{4} < x < \pi$, cos 2x increases from 0 to 1.

This information can be used to draw the first half of the graph in polar coordinates, as θ goes from 0 to π. A table of values can be used to improve accuracy.

0	0	$\frac{\pi}{12}$	$\frac{\pi}{8}$	$\frac{\pi}{6}$	$\frac{\pi}{4}$	$\frac{\pi}{3}$	$\frac{3\pi}{8}$	$\frac{5\pi}{12}$	$\frac{\pi}{2}$	$\frac{7\pi}{12}$	$\frac{5\pi}{8}$	$\frac{2\pi}{3}$	$\frac{3\pi}{4}$	$\frac{5\pi}{6}$	$\frac{7\pi}{8}$	$\frac{11\pi}{12}$	π
2θ	0	$\frac{\pi}{6}$	$\frac{\pi}{4}$	$\frac{\pi}{3}$	$\frac{\pi}{2}$	$\frac{2\pi}{3}$	$\frac{3\pi}{4}$	$\frac{5\pi}{6}$	π	$\frac{7\pi}{6}$	$\frac{5\pi}{4}$	$\frac{4\pi}{3}$	$\frac{3\pi}{2}$	$\frac{5\pi}{3}$	$\frac{7\pi}{4}$	$\frac{11\pi}{6}$	2π
cos θ	1	.9	.7	.5	0	-.5	-.7	-.9	-1	-.9	-.7	.-5	0	.5	.7	.9	1

The remaining portion of the graph, as θ goes from π to 2π, can be drawn using symmetry. (This portion is shown with a dashed curve.)

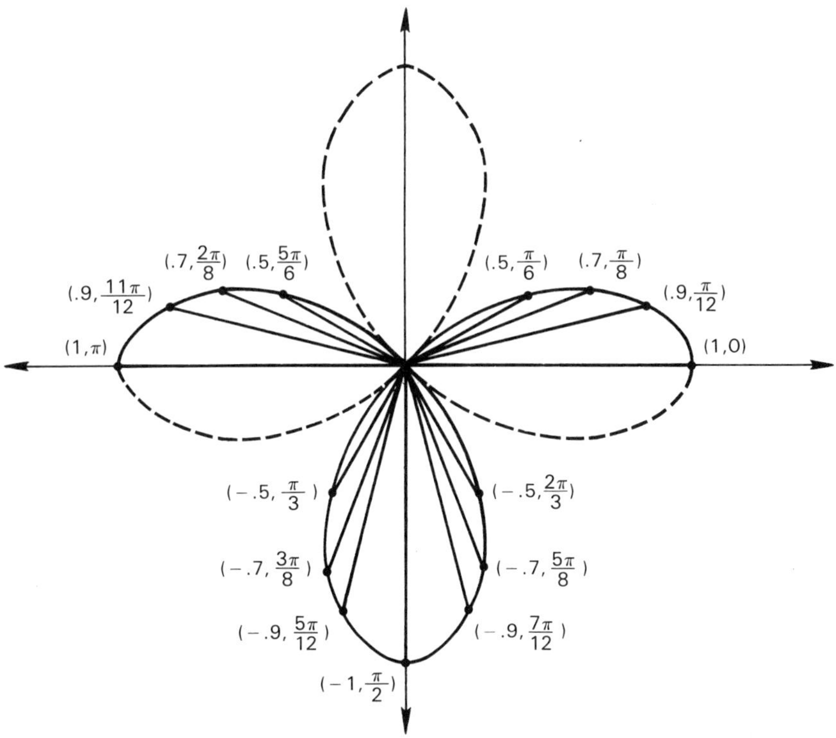

33) $r^2 = 4 \cos 2\theta$

$r = \pm 2\sqrt{\cos 2\theta}$

On $\left(\frac{\pi}{4}, \frac{3\pi}{4}\right)$ and $\left(\frac{5\pi}{4}, \frac{7\pi}{4}\right)$ $\cos 2\theta < 0$, so there will be no real values for r on these intervals. In the remaining intervals, each value of θ will have two associated r-values.

θ	0	$\frac{\pi}{12}$	$\frac{\pi}{8}$	$\frac{\pi}{6}$	$\frac{\pi}{4}$	$\frac{5\pi}{6}$	$\frac{7\pi}{8}$	$\frac{11\pi}{12}$	π
2θ	0	$\frac{\pi}{6}$	$\frac{\pi}{4}$	$\frac{\pi}{3}$	$\frac{\pi}{2}$	$\frac{5\pi}{3}$	$\frac{7\pi}{4}$	$\frac{11\pi}{6}$	2π
$\pm 2\sqrt{\cos 2\theta}$	± 2	± 1.9	± 1.7	± 1.4	0	± 1.4	± 1.7	± 1.9	± 2

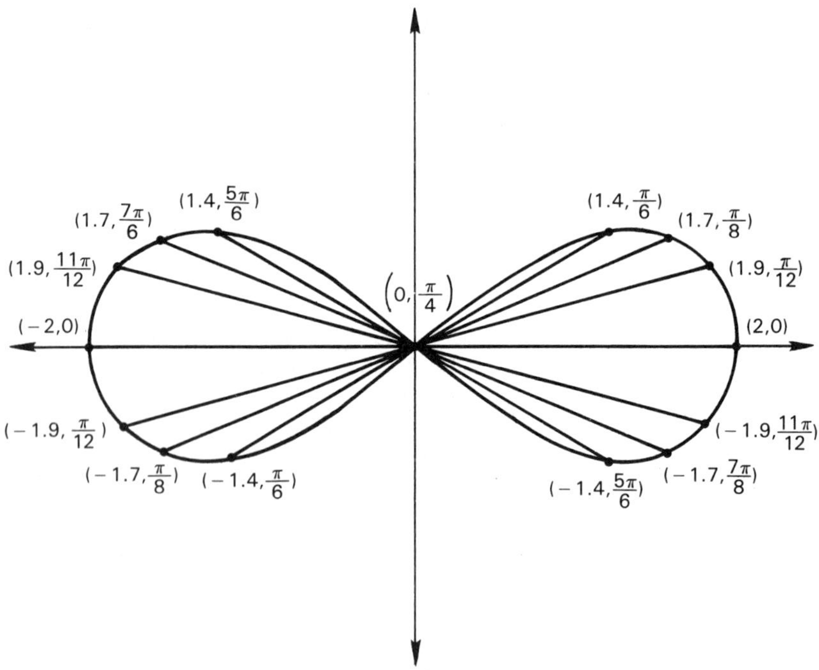

37) $r = \dfrac{4}{1 + \cos\theta}$

$r + r\cos\theta = 4$

$r = 4 - r\cos\theta$

$r = 4 - x$

$r^2 = 16 - 8x + x^2$

$x^2 + y^2 = 16 - 8x + x^2$

$y^2 = 16 - 8x$ (parabola)

41) $r = 6\cos\theta + 8\sin\theta$

$r^2 = 6r\cos\theta + 8r\sin\theta$

$x^2 + y^2 = 6x + 8y$

$x^2 - 6x + y^2 - 8y = 0$

$x^2 - 6x + 9 + y^2 - 8y + 16 = 25$

$(x-3)^2 + (y-4)^2 = 25$ (circle)

45) $r\cos\left(\theta - \frac{\pi}{3}\right) = \sqrt{3}$

$r\left(\cos\theta\cos\frac{\pi}{3} + \sin\theta\sin\frac{\pi}{3}\right) = \sqrt{3}$

$(r\cos\theta)\left(\frac{1}{2}\right) + (r\sin\theta)\left(\frac{\sqrt{3}}{2}\right) = \sqrt{3}$

$\frac{1}{2}x + \frac{\sqrt{3}}{2}y = \sqrt{3}$

$x + \sqrt{3}\,y = 2\sqrt{3}$ (line)

49) $r = 3\cos\theta$ $\qquad$ $r = 1 + \cos\theta$

$3\cos\theta = 1 + \cos\theta$

$2\cos\theta = 1$

$\cos\theta = \frac{1}{2}$

$\theta = \frac{\pi}{3}$ or $\frac{5\pi}{3}$

$r = 3\left(\cos\frac{\pi}{3}\right)$ $\qquad$ $r = 3\left(\cos\frac{5\pi}{3}\right)$

$r = \frac{3}{2}$ $\qquad$ $r = \frac{3}{2}$

Solution: $\left(\frac{3}{2}, \frac{\pi}{3}\right)$, $\left(\frac{3}{2}, \frac{5\pi}{3}\right)$

53) $r = \cos\theta = \frac{1}{2}$ $\qquad\qquad$ $2r\sin\theta = -\sqrt{3}$

Multiply by $2\sin\theta$: $\qquad\qquad$ Multiply by $\cos\theta$:

$2r\sin\theta\cos\theta = \sin\theta$ $\qquad\qquad$ $2r\sin\theta\cos\theta = -\sqrt{3}\cos\theta$

Subtract:

$$\begin{aligned}
2r\sin\theta\cos\theta &= -\sqrt{3}\cos\theta \\
\underline{2r\sin\theta\cos\theta} &= \underline{\sin\theta} \\
0 &= -\sqrt{3}\cos\theta - \sin\theta \\
\sin\theta &= -\sqrt{3}\cos\theta \\
\frac{\sin\theta}{\cos\theta} &= -\sqrt{3} \\
\tan\theta &= -\sqrt{3} \\
\theta &= -\frac{\pi}{3}
\end{aligned}$$

$$r\cos\left(-\frac{\pi}{3}\right) = \frac{1}{2}$$

$$r\left(\frac{1}{2}\right) = \frac{1}{2}$$

$$r = 1$$

Solution: $\left(1, -\frac{\pi}{3}\right)$

(Note: Since $\cos\theta \neq 0$ in the equation $r\cos\theta = \frac{1}{2}$, we may also proceed by dividing by $\cos\theta$ to obtain $r = \frac{1}{2\cos\theta}$. Substitute into the second equation to get $\tan\theta = -\sqrt{3}$).

Sample Test Questions for Chapter 10

Part I

1. a) $a^2 = 8^2 + 12^2 - 2 \cdot 8 \cdot 12 \cos 60^0$

$a^2 = 64 + 144 - 192(0.5)$

$a^2 = 112$

$a = 4\sqrt{7}$

$\angle A = 60^0 = \frac{\pi}{3}$

b) $\frac{8}{\sin B} = \frac{4\sqrt{7}}{\sin 60^0}$

$\sin B = \frac{8 \sin 60^0}{4\sqrt{7}} = \frac{\sqrt{21}}{7}$

2. a) $b \sin \alpha = 12 \sin 30^0 = 12(0.5) = 6 < 5.8 = a$

No solution.

b) $b \sin \alpha = 8 \sin 30^0 = 8(0.5) = 4$

$4 < 4.1 < 8$

$b \sin \alpha < a < b$

2 solutions

3. $\angle C = 180^0 - 75^0 - 45^0 = 60^0$

$\frac{a}{\sin 45^0} = \frac{20}{\sin 60^0}$

$a = \frac{20 \sin 45^0}{\sin 60^0} = \frac{20\left(\frac{\sqrt{2}}{2}\right)}{\frac{\sqrt{3}}{2}} = 10\sqrt{2} \cdot \frac{2}{\sqrt{3}} = \frac{20\sqrt{6}}{3}$

4. $K = \frac{1}{2}(20)(34)\sin 18^0 = 340(0.3090) = 105 \text{ cm}^2$

5. $s = \frac{1}{2}(7 + 9 + 12) = \frac{1}{2}(28) = 14$

$K = \sqrt{14(14 - 7)(14 - 9)(14 - 12)}$

$K = \sqrt{980}$

$K = 31 \text{ cm}^2$

6. 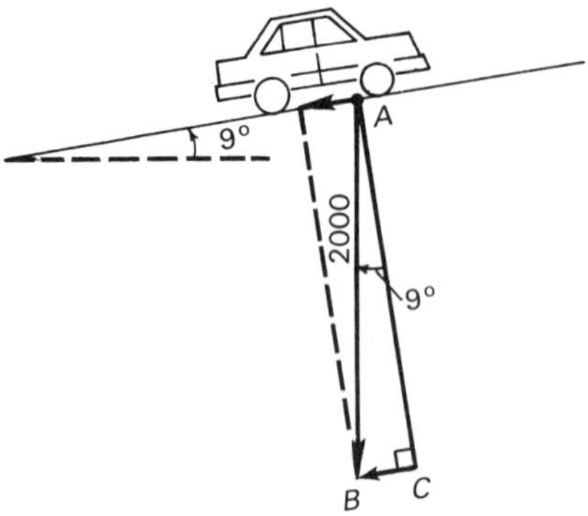

In the sketch $\overrightarrow{AB}$ is the weight or gravity vector. $\overrightarrow{CB}$ is the force parallel to the driveway. Use the sine relationship to solve for CB.

CB $= 2000(\sin 9^{\circ}) = 2000(0.1564)$

$= 313$ pounds.

A 313 pound force will hold the car from rolling.

7. 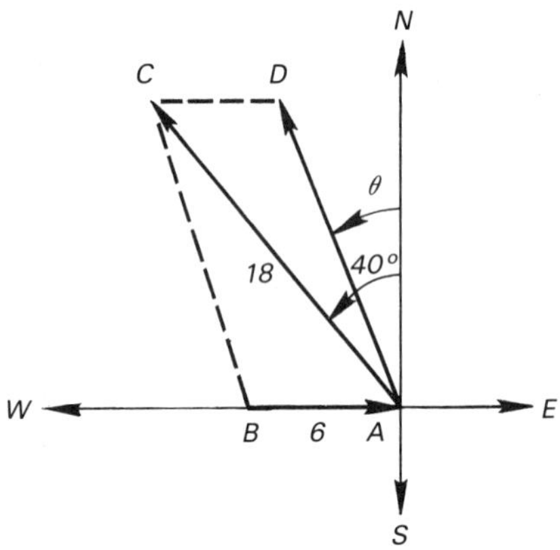

In the diagram,

$\overrightarrow{AC}$ is the boat speed vector.

$\overrightarrow{AB}$ is the water current vector.

$\overrightarrow{AD}$, the resultant will be the boat's actual speed.

θ is the final heading.

$\angle BAC = 90^{\circ} - 40^{\circ} = 50^{\circ}$ CD||BA, so $\angle ACD = 50^{\circ}$.

Use the law of cosines in $\triangle ACD$ to solve for AD.

$$(AD)^2 = 18^2 + 6^2 - 2 \cdot 6 \cdot 18 (\cos 50^{\circ})$$
$$= 324 + 36 - 216(0.6428) = 221$$

AD = 15 mph.

8. Use the information from exercise 7 and the law of sines to solve for $\angle DAC$.

$$\frac{15}{\sin 50^{\circ}} = \frac{6}{\sin \angle DAC}$$
$$\sin \angle DAC = \frac{6 \sin 50^{\circ}}{15}$$
$$\sin \angle DAC = \frac{6(0.7660)}{15}$$

8. (continued)

$$\sin \angle DAC = 0.3064$$
$$\angle DAC = 18^{\circ}$$

$\theta = 40^{\circ} - 18^{\circ} = 22^{\circ}$

The final bearing is N 22° W.

9.

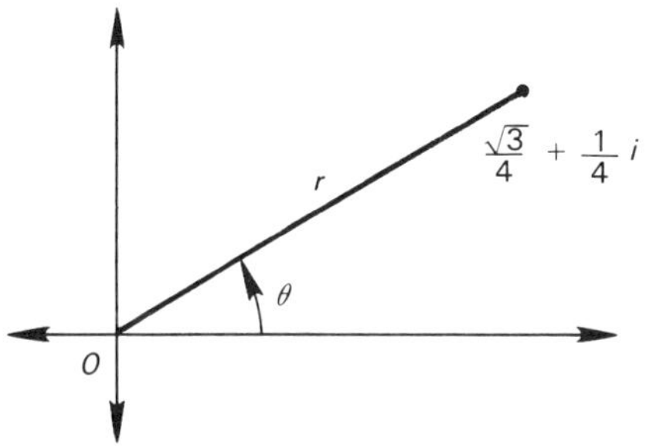

$$\theta = \frac{\pi}{6}$$

$$r = \sqrt{\left(\frac{\sqrt{3}}{4}\right)^2 + \left(\frac{1}{4}\right)^2}$$

$$r = \sqrt{\frac{1}{4}}$$

$$r = \frac{1}{2}$$

$$z = \frac{\sqrt{3}}{4} + \frac{1}{4}i = \frac{1}{2}\left(\cos\frac{\pi}{6} + i\sin\frac{\pi}{6}\right)$$

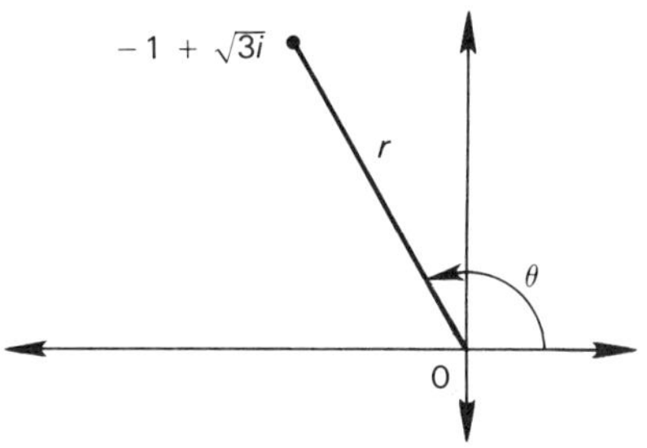

$$\theta = \frac{2\pi}{3}$$

$$r = \sqrt{(-1)^2 + (\sqrt{3})^2}$$

$$r = \sqrt{4}$$

$$r = 2$$

$$w = -1 + \sqrt{3}\,i = 2\left(\cos\frac{2\pi}{3} + i\sin\frac{2\pi}{3}\right)$$

$$\frac{z}{w} = \frac{1}{2}\left(\cos\frac{\pi}{6} + i\sin\frac{\pi}{6}\right) \div 2\left(\cos\frac{2\pi}{3} + i\sin\frac{2\pi}{3}\right)$$

$$= \frac{\frac{1}{2}}{2}\left[\cos\left(\frac{\pi}{6} - \frac{2\pi}{3}\right) + i\sin\left(\frac{\pi}{6} - \frac{2\pi}{3}\right)\right]$$

$$= \frac{1}{4}\left[\cos\left(-\frac{\pi}{2}\right) + i\sin\left(-\frac{\pi}{2}\right)\right]$$

$$= \frac{1}{4}\left(\cos\frac{3\pi}{2} + i\sin\frac{3\pi}{2}\right) \qquad \left(\frac{3\pi}{2} \text{ is the principal argument}\right)$$

$$= -\frac{1}{4}i$$

10. $z^5 = \left(\frac{1}{2}\right)^5\left(\cos\frac{\pi}{6} + i\sin\frac{\pi}{6}\right)^5 = \frac{1}{32}\left(\cos\frac{5\pi}{6} + i\sin\frac{5\pi}{6}\right)$

$= \frac{1}{32}\left(\cos\frac{5\pi}{6}\right) + \frac{1}{32}\left(\sin\frac{5\pi}{6}\right)i = \frac{1}{32}\left(-\frac{\sqrt{3}}{2}\right) + \frac{1}{32}\left(\frac{1}{2}\right)i$

$= -\frac{\sqrt{3}}{64} + \frac{1}{64}i$

11. $8^{\frac{1}{3}} = [8(\cos 0\ ) + i(\sin 0\ ]^{1/3}$

$= 8^{\frac{1}{3}}\left[\cos\left(\frac{0 + 2k\pi}{3}\right) + i\sin\left(\frac{0 + 2k\pi}{3}\right)\right]$ for $k = 0, 1, 2$

For $k = 0$:

$8^{1/3}(\cos 0 + i\sin 0) = 2$

For $k = 1$:

$8^{1/3}\left[\cos\left(\frac{0 + 2\pi}{3}\right) + i\sin\left(\frac{0 + 2\pi}{3}\right)\right] = 2\left(\cos\frac{2\pi}{3} + i\sin\frac{2\pi}{3}\right)$

$= 2\left(\cos\frac{2\pi}{3}\right) + 2\left(\sin\frac{2\pi}{3}\right)i$

$= 2\left(-\frac{1}{2}\right) + 2\left(\frac{\sqrt{3}}{2}\right)i$

$= -1 + \sqrt{3}\ i$

For $k = 2$:

$8^{1/3}\left[\cos\left(\frac{0 + 4\pi}{3}\right) + i\sin\left(\frac{0 + 4\pi}{3}\right)\right] = 2\left(\cos\frac{4\pi}{3} + i\sin\frac{4\pi}{3}\right)$

$= 2\left(\cos\frac{4\pi}{3}\right) + 2\left(\sin\frac{4\pi}{3}\right)i$

$= 2\left(-\frac{1}{2}\right) + 2\left(-\frac{\sqrt{3}}{2}\right)i$

$= -1 - \sqrt{3}\ i$

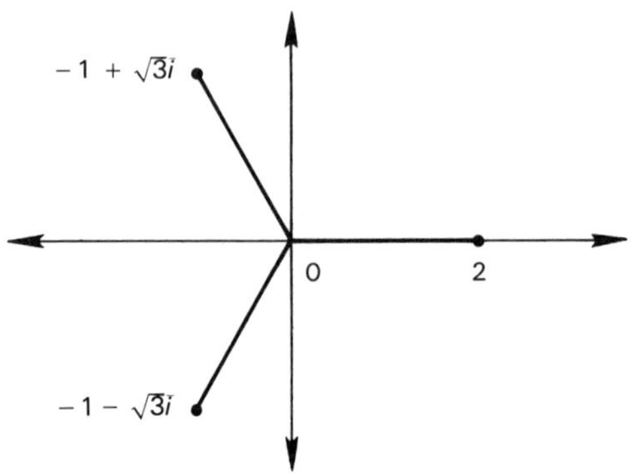

12. $-8\sqrt{2} + 8\sqrt{2}i$

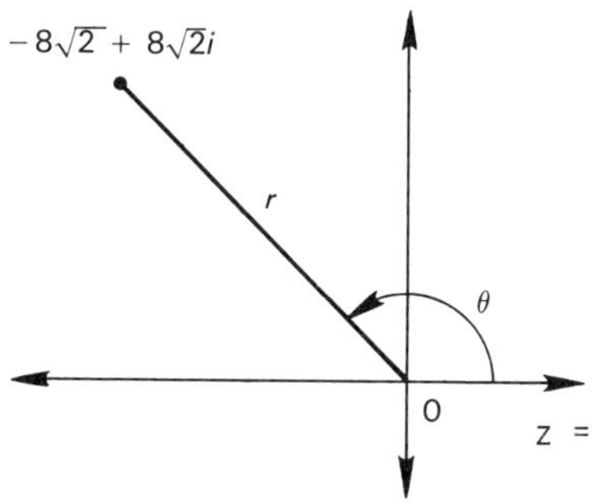

$\theta = \frac{3\pi}{4}$

$r = \sqrt{(-8\sqrt{2})^2 + (8\sqrt{2})^2}$

$r = \sqrt{256}$

$r = 16$

$z = -8\sqrt{2} + 8\sqrt{2}\,i = 16\left(\cos\frac{3\pi}{4} + i\sin\frac{3\pi}{4}\right)$

$$z^{\frac{1}{4}} = \left[16\left(\cos\frac{3\pi}{4} + i\sin\frac{3\pi}{4}\right)\right]^{\frac{1}{4}}$$

$$= 16^{1/4}\ \cos\left(\frac{\frac{3\pi}{4} + 2k\pi}{4} + i\sin\frac{\frac{3\pi}{4} + 2k\pi}{4}\right) \quad \text{for } k = 0, 1, 2, 3$$

For k = 1:

$$16^{1/4}\left(\cos\frac{\frac{3\pi}{4}}{4} + i\sin\frac{\frac{3\pi}{4}}{4}\right) = 2\left(\cos\frac{3\pi}{16} + i\sin\frac{3\pi}{16}\right)$$

For k = 2:

$$16^{1/4}\left(\cos\frac{\frac{3\pi}{4} + 2\pi}{4} + i\sin\frac{\frac{3\pi}{4} + 2\pi}{4}\right) = 2\left(\cos\frac{11\pi}{16} + i\sin\frac{11\pi}{16}\right)$$

For k = 3:

$$16^{1/4}\left(\cos\frac{\frac{3\pi}{4} + 4\pi}{4} + i\sin\frac{\frac{3\pi}{4} + 4\pi}{4}\right) = 2\left(\cos\frac{19\pi}{16} + i\sin\frac{19\pi}{16}\right)$$

For k = 4:

$$16^{1/4}\left(\cos\frac{\frac{3\pi}{4} + 6\pi}{4} + i\sin\frac{\frac{3\pi}{4} + 6\pi}{4}\right) = 2\left(\cos\frac{27\pi}{16} + i\sin\frac{27\pi}{16}\right)$$

13. a) $(-3, 60^{\circ}) = (-3, -300^{\circ})$

b) $(-3, 60^{\circ}) = (3, 240^{\circ})$

14. a) $x = \frac{1}{3}\cos\left(-\frac{\pi}{6}\right) = \frac{1}{3}\left(\frac{\sqrt{3}}{2}\right) = \frac{\sqrt{3}}{6}$

$y = \frac{1}{3}\sin\left(-\frac{\pi}{6}\right) = \frac{1}{3}\left(-\frac{1}{2}\right) = -\frac{1}{6}$

$(x, y) = \left(\frac{\sqrt{3}}{6}, -\frac{1}{6}\right)$

b) $r = \sqrt{(-7\sqrt{2})^2 + (7\sqrt{2})^2} = \sqrt{196} = 14$

$\tan\theta = \frac{y}{x} = \frac{7\sqrt{2}}{-7\sqrt{2}} = -1$

$\theta = \frac{3\pi}{4}$ or $\frac{7\pi}{4}$ $(-7\sqrt{2}, 7\sqrt{2})$ is in quadrant II,

so $\theta = \frac{3\pi}{4}$

$(r, \theta) = \left(14, \frac{3\pi}{4}\right)$

15. $r = 2\sin\theta$

$r^2 = 2r\sin\theta$

$x^2 + y^2 = 2y$

$x^2 + y^2 - 2y = 0$

$x^2 + y^2 - 2y + 1 = 1$

$x^2 + (y - 1)^2 = 1$ (circle)

16. $2r\sin\theta = 6$

$2y = 6$

$y = 3$

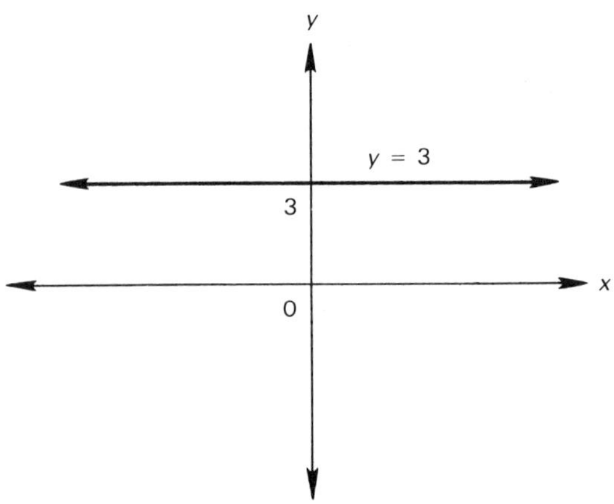

17. $r = 2 + 2 \cos 0$

$r = 2(1 + \cos \theta)$

This is the equation of a cardioid. Because cosine is involved, the symmetry is around the polar axis.

θ	0	$\frac{\pi}{6}$	$\frac{\pi}{4}$	$\frac{\pi}{3}$	$\frac{\pi}{2}$	$\frac{2\pi}{3}$	$\frac{3\pi}{4}$	$\frac{5\pi}{6}$	π
$2(1 + \cos \theta)$	4	3.7	3.4	3	2	1	0.6	0.3	0

Graph the points for $0 \leq \theta \leq \pi$ and complete the graph using the symmetry of the function.

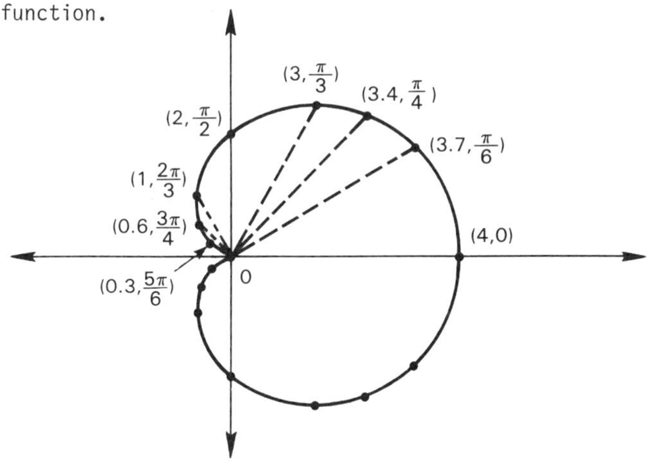

18. $r = 4 \sin 2\theta$

θ	0	$\frac{\pi}{6}$	$\frac{\pi}{4}$	$\frac{\pi}{3}$	$\frac{\pi}{2}$	$\frac{2\pi}{3}$	$\frac{3\pi}{4}$	$\frac{5\pi}{6}$	π
2θ	0	$\frac{\pi}{3}$	$\frac{\pi}{2}$	$\frac{2\pi}{3}$	π	$\frac{4\pi}{3}$	$\frac{3\pi}{2}$	$\frac{5\pi}{3}$	2π
$4 \sin 2\theta$	0	3.5	4	3.5	0	-3.5	-4	-3.5	0

Graph the points for $0 \leq \theta \leq \pi$ and complete the graph using the symmetry of the function.

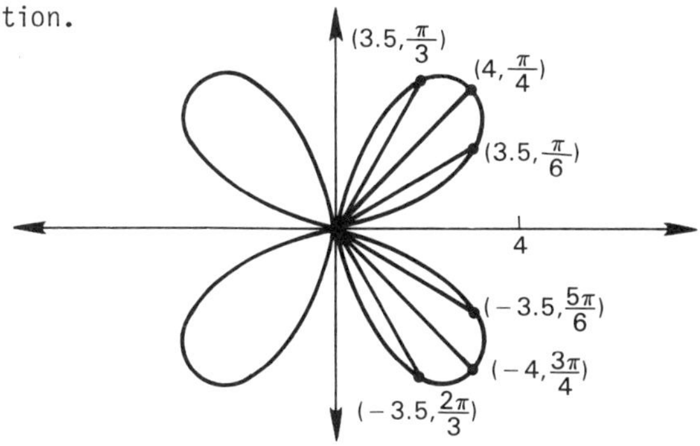

Sample Test Questions for Chapter 10

Part II

1. (e)

2. (d) $\cos B = \dfrac{4^2 + (5\sqrt{2})^2 - (\sqrt{106})^2}{2(4)(5\sqrt{2})} = \dfrac{-40}{2(4)(5\sqrt{2})} = -\dfrac{1}{\sqrt{2}}$

$\angle B = \dfrac{3\pi}{4}$

3. (b) $S = \frac{1}{2}(8 + 8 + 12) = 14$

$K = \sqrt{14(14-8)^2(14-12)} = \sqrt{7 \cdot 2^2 \cdot 6^2} = 12\sqrt{7}$

4. (b) $\angle C = 180^\circ - (75^\circ + 40^\circ) = 65^\circ$

$\dfrac{c}{\sin 65^\circ} = \dfrac{40}{\sin 40^\circ}$

$c = \dfrac{40 \sin 65^\circ}{\sin 40^\circ}$

5. (c) $b \sin A = 7 \sin 30^\circ = 7(.5) = 3.5 < 4.2 = a < 7 = b$

Two triangles are possible.

6. (a) The required force x is the magnitude of a vector that is the adjacent side to a 75° angle in a right triangle with hypotenuse = 300.

$\cos 75^\circ = \dfrac{x}{300}$, $x = 300 \cos 75^\circ$

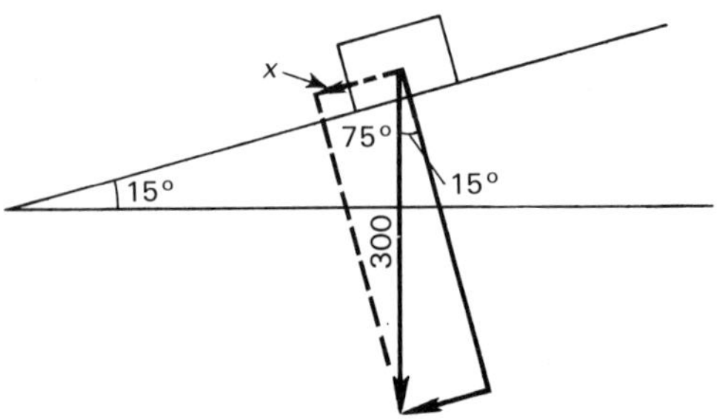

7. (c) $r = \sqrt{1^2 + (-1)^2} = \sqrt{2}$

$\tan \theta = \frac{-1}{1} = -1$. Also, since $1 - i$ is in quadrant IV,

$\theta = \frac{7\pi}{4}$. Then, $1 - i = \sqrt{2}\left(\cos \frac{7\pi}{4} + i \sin \frac{7\pi}{4}\right)$

8. (d) $z^3 = [2(\cos 20^\circ + i \sin 20^\circ)]^3 = 8(\cos 60^\circ + i \sin 60^\circ)$

$= 8\left(\frac{1}{2} + \frac{\sqrt{3}}{2} i\right) = 4 + 4\sqrt{3}\, i$

9. (c) $\frac{z}{w} = \dfrac{8\left(\cos \frac{4\pi}{3} + i \sin \frac{4\pi}{3}\right)}{24\left(\cos \frac{11\pi}{6} + i \sin \frac{11\pi}{6}\right)}$

$= \frac{1}{3}\left[\cos\left(\frac{4\pi}{3} - \frac{11\pi}{6}\right) + i \sin\left(\frac{4\pi}{3} - \frac{11\pi}{6}\right)\right]$

$= \frac{1}{3}\left[\cos\left(-\frac{\pi}{2}\right) + i \sin\left(-\frac{\pi}{2}\right)\right]$

$= \frac{1}{3}\left[\cos \frac{3\pi}{2} + i \sin \frac{3\pi}{2}\right]$

10. (a) $x = r \cos \theta = -\frac{1}{2} \cos \frac{\pi}{3} = -\frac{1}{2}\left(\frac{1}{2}\right) = -\frac{1}{4}$

$y = r \sin \theta = -\frac{1}{2} \cos \frac{\pi}{3} = -\frac{1}{2}\left(\frac{\sqrt{3}}{2}\right) = -\frac{\sqrt{3}}{4}$

$(x, y) = \left(-\frac{1}{4}, -\frac{\sqrt{3}}{4}\right)$

11. (b) $r = \sqrt{(-3)^2 + (3\sqrt{3})^2} = \sqrt{36} = 6$.

$\tan \theta = \frac{3\sqrt{3}}{-3} = -\sqrt{3}$. Also, since $(-3, 3\sqrt{3})$ is in

quadrant II, $\theta = \frac{2\pi}{3}$ and the polar coordinates are $\left(6, \frac{2\pi}{3}\right)$

12. (d) $-64 = 64(\cos \pi + i \sin \pi)$

Using $\theta = \pi$, $r = 64$, and $n = 3$ in the nth-root formula, the cube roots are:

$$64^{\frac{1}{3}}\left(\cos \frac{\pi + 2k\pi}{3} + i \sin \frac{\pi + 2k\pi}{3}\right)$$

For $k = 0, 1, 2$. When $k = 2$, the root is

$$4\left(\cos \frac{5\pi}{3} + i \sin \frac{5\pi}{3}\right) = 4\left(\frac{1}{2} - \frac{\sqrt{3}}{2} i\right)$$

$$= 2 - 2\sqrt{3}\, i$$

The cube roots obtained for $k = 0$ or 1 are not among the choices listed.

13. (c) I: $r = -5$

$r^2 = x^2 + y^2 = 25$ (circle)

II. $r = -\frac{1}{2} \sin \theta = -2\left(\frac{1}{4}\right)\sin \theta = -2a \sin \theta$ (circle)

III. $6 = r \cos \theta = x$ (line)

14. (a) $r = 3 - 3 \cos \theta = 3(1 - \cos \theta) = a(1 - \cos \theta)$

15. (b) $r = -8 \cos \theta$

$r^2 = -8r \cos \theta$ (multiply by r)

$x^2 + y^2 = -8x$ ($r^2 = x^2 + y^2$; $r \cos \theta = x$)

$x^2 + 8x + y^2 = 0$

$x^2 + 8x + 16 + y^2 = 16$

$(x + 4)^2 + y^2 = 16$

CHAPTER 11

LINEAR SYSTEMS, MATRICES, AND DETERMINANTS

Section 11.1 Systems of Linear Equations

1) $2x + y = -10$ $\qquad$ $6x - 3y = 6$

$$y = -2x - 10$$
$$y = -2x - 10$$

Substitute $(-2x - 10)$ for y in equation 2.

$$6x - 3(-2x - 10) = 6$$
$$6x + 6x + 30 = 6$$
$$12x = -24$$
$$x = -2$$

In equation 1, let $x = -2$.

$$2(-2) + y = -10$$
$$-4 + y = -10$$
$$y = -6$$

Solution: $(-2, -6)$

5) $x + 5y = -9$ $\qquad$ $4x - 3y = -13$

$$x = -5y - 9$$

Substitute $(-5y - 9)$ for x in equation 2.

$$4(-5y - 9) - 3y = -13$$
$$-20y - 36 - 3y = -13$$
$$-23y = 23$$
$$y = -1$$

In equation 1, let $y = -1$.

$$x + 5(-1) = -9$$
$$x - 5 = -9$$
$$x = -4$$

Solution: $(-4, -1)$

9) $3y - 9x = 30$ $\quad$ (multiply by $\frac{1}{3}$) $\quad$ $y - 3x = 10$

$-4y + 8x = 24$ $\quad$ (multiply by $\frac{1}{4}$) $\quad$ $-y + 2x = 6$

$$-x = 16$$
$$x = -16$$

In the simplified form of equation 1, let $x = -16$.

9) continued

$$y - 3(-16) = 10$$
$$y + 48 = 10$$
$$y = -38$$

Solution: (-16, -38)

13) $x - 2y = 3$ $\qquad$ $y - 3x = -14$

$$x = 2y + 3$$

Substitute (2y + 3) for x in equation 2.

$$y - 3(2y + 3) = -14$$
$$y - 6y - 9 = -14$$
$$-5y = -5$$
$$y = 1$$

In equation 1, let y = 1.

$$x - 2(1) = 3$$
$$x - 2 = 3$$
$$x = 5$$

Solution: (5, 1)

17) $16x - 5y = 103$ (multiply by 19) $\qquad 304x - 95y = 1957$

$7x + 19y = -188$ (multiply by 5) $\qquad 35x + 95y = -940$

$$339x = 1017$$
$$x = 3$$

In equation 1, let x = 3.

$$16(3) - 5y = 103$$
$$48 - 5y = 103$$
$$-5y = 55$$
$$y = -11$$

Solution: (3, -11)

21) $\frac{1}{4}x + \frac{1}{3}y = \frac{5}{12}$ (multiply by -3) $\qquad -\frac{3}{4}x - y = -\frac{5}{4}$

$\frac{1}{2}x + y = 1$ $\qquad \frac{1}{2}x + y = 1$

$$-\frac{1}{4}x = -\frac{1}{4}$$
$$x = 1$$

In equation 2, let x = 1.

21) continued

$\frac{1}{2}(1) + y = 1$

$\frac{1}{2} + y = 1$

$y = \frac{1}{2}$

Solution: $(1, \frac{1}{2})$

25) $2(x - y - 1) = 1 - 2x$ $\quad$ $6(x - y) = 4 - 3(3y - x)$

$2x - 2y - 2 = 1 - 2x$ $\quad$ $6x - 6y = 4 - 9y + 3x$

$4x - 2y = 3$ $\quad$ $3x + 3y = 4$

$4x - 2y = 3$ (multiply by 3) $\quad 12x - 6y = 9$

$3x + 3y = 4$ (multiply by 2) $\quad 6x + 6y = 8$

$$18x = 17$$

$$x = \frac{17}{18}$$

In the simplified form of equation 2, let $x = \frac{17}{18}$.

$3(\frac{17}{18}) + 3y = 4$

$\frac{17}{6} + 3y = 4$

$3y = \frac{7}{6}$

$y = \frac{7}{18}$

Solution: $(\frac{17}{18}, \frac{7}{18})$

29) $2x - 3y = 8$ $\quad$ $2x - 3y = 8$

$-8x + 12y = 33$ (Divide by 4) $\quad -2x + 3y = \frac{33}{4}$

$$0 = \frac{65}{4}$$

Inconsistent.

33) $4x - 12y = 3$ $\quad$ $4x - 12y = 3$

$x + \frac{1}{3}y = 3$ (multiply by -4) $\quad -4x - \frac{4}{3}y = -12$

$$-\frac{40}{3}y = -9$$

$$y = \frac{27}{40}$$

33) continued

In equation 2, let $y = \frac{27}{40}$.

$$x + \frac{1}{3}\left(\frac{27}{40}\right) = 3$$

$$x + \frac{9}{40} = 3$$

$$x = \frac{111}{40}$$

Consistent.

37) $x - 5y = 15$

$0.01x - 0.05y = 0.5$ (multiply by -100)

$$\begin{aligned} x - 5y &= 15 \\ -x + 5y &= -50 \\ \hline 0 &= -35 \end{aligned}$$

Inconsistent

41) (Eq. 1) $x + 2y + 3z = -4$

(Eq. 2) $4x + 5y + 6z = -4$

$$\begin{aligned} -4x - 8y - 12z &= 16 \\ 4x + 5y + 6z &= -4 \\ \hline -3y - 6z &= 12 \\ \text{(Eq. 4)} \quad y + 2z &= -4 \end{aligned}$$

(Eq. 1) $x + 2y + 3z = -4$

(Eq. 3) $7x - 15y - 9z = 4$

$$\begin{aligned} -7x - 14y - 21z &= 28 \\ 7x - 15y - 9z &= 4 \\ \hline \text{(Eq. 5)} \ -29y - 30z &= 32 \end{aligned}$$

(Eq. 4) $y + 2z = -4$

(Eq. 5) $-29y - 30z = 32$

$$\begin{aligned} 15y + 30z &= -60 \\ -29y - 30z &= 32 \\ \hline -14y \quad &= -28 \\ y &= 2 \end{aligned}$$

In equation 4, let $y = 2$.

$$(2) + 2z = -4$$

$$2z = -6$$

$$z = -3$$

In equation 1, let $y = 2$ and $z = -3$.

$$x + 2(2) + 3(-3) = -4$$

$$x + 4 - 9 = -4$$

$$x = 1$$

Solution: $(1, 2, -3)$

45) (Eq. 1) $4x - 2y - z = 1$

-(Eq. 3) $-(x - 3y - z = \frac{3}{2})$

(Eq. 4) $3x + y = -\frac{1}{2}$

(Eq. 1) $4x - 2y - z = 1$ $\quad 8x - 4y - 2z = 2$

(Eq. 2) $2x + y + 2z = 9$ $\quad 2x + y + 2z = 9$

(Eq. 5) $10x - 3y = 11$

(Eq. 4) $3x + y = -\frac{1}{2}$ $\quad 9x + 3y = -\frac{3}{2}$

(Eq. 5) $10x - 3y = 11$ $\quad 10x - 3y = 11$

$$19x = \frac{19}{2}$$

$$x = \frac{1}{2}$$

In equation 5, let $x = \frac{1}{2}$.

$$10\left(\frac{1}{2}\right) - 3y = 11$$
$$5 - 3y = 11$$
$$-3y = 6$$
$$y = -2$$

In equation 1, let $x = \frac{1}{2}$ and $y = -2$.

$$4\left(\frac{1}{2}\right) - 2(-2) - z = 1$$
$$2 + 4 - z = 1$$
$$5 = z$$

Solution: $\left(\frac{1}{2}, -2, 5\right)$

49) (Eq. 1) $2x + 3y - z = 8$

(Eq. 2) $x + y + z = 5$

(Eq. 4) $3x + 4y = 13$

(Eq. 2) $x + y + z = 5$ $\quad -2x - 2y - 2z = -10$

(Eq. 3) $-4x - 6y + 2z = 3$ $\quad -4x - 6y + 2z = 3$

(Eq. 5) $-6x - 8y = -7$

(Eq. 4) $3x + 4y = 13$ $\quad 6x + 8y = 26$

(Eq. 5) $-6x - 8y = -7$ $\quad -6x - 8y = -7$

$$0 \neq 19$$

Inconsistent. No solutions.

53) $$\begin{aligned} -x + y &= 6c \\ x + y &= 3c \\ \hline 2y &= 9c \\ y &= \frac{9c}{2} \end{aligned}$$

In equation 2, let $y = \frac{9c}{2}$.

$$x + \left(\frac{9c}{2}\right) = 3c$$

$$x = -\frac{3c}{2}$$

Solution: $\left(-\frac{3c}{2}, \frac{9c}{2}\right)$

57) (6, 3) and (-1, -1) both satisfy $ax + by = 3$. Substitute these values for x and y to obtain two equations in a and b.

$$\begin{aligned} 6a + 3b &= 3 \\ -a - b &= 3 \end{aligned} \qquad \begin{aligned} 6a + 3b &= 3 \\ -3a - 3b &= 9 \\ \hline 3a \quad &= 12 \\ a &= 4 \end{aligned}$$

In equation 2, let a = 4.

$$-(4) - b = 3$$

$$b = -7$$

Solution: $a = 4$, $b = -7$

Section 11.2 Applications of Linear Equations

1) Let x represent the number of free throws.
Let y represent the number of field goals.
x is the points scored in free throws.
2y is the points scored in field goals.
The relationship of the number of goals and free throws:

$$y = 2.5x$$

Total points: $x + 2y = 96$

$$\begin{aligned} 2.5x - y &= 0 \\ x + 2y &= 96 \end{aligned} \qquad \begin{aligned} 5x - 2y &= 0 \\ x + 2y &= 96 \\ \hline 6x \quad &= 96 \\ x &= 16 \end{aligned}$$

In equation 2, let x = 16.

1) continued

$$16 + 2y = 96$$
$$2y = 80$$
$$y = 40$$

Answer: The team made 40 field goals and 16 free throws.

5) Let x represent the number of pounds of potatoes.

Let y represent the number of pounds of beans.

35x is the cost of the potatoes, in cents.

68y is the cost of the beans.

Total weight: $x + y = 9.5$

Total cost (change \$3.82 to cents): $35x + 68y = 382$

$$x + y = 9.5 \qquad\qquad -35x - 35y = -332.5$$
$$35x + 68y = 382 \qquad\qquad \underline{35x + 68y = 382.0}$$
$$33y = 49.5$$
$$y = 1.5$$

In equation 1, let $y = 1.5$.

$$x + (1.5) = 9.5$$
$$x = 8$$

Answer: The shopper bought 8 pounds of potatoes and 1 1/2 pounds of beans.

9) Let x represent the number of nickels.

Let y represent the number of quarters.

5x is the value of the nickels, in cents.

25y is the value of the quarters.

Total number: $x + y = 22$

Total value (change \$3.90 to cents): $5x + 25y = 390$

$$x + y = 22 \qquad\qquad 5x + 5y = 110$$
$$5x + 25y = 390 \qquad\qquad \underline{-5x - 25y = -390}$$
$$-20y = -280$$
$$y = 14$$

In equation 1, let $y = 14$.

$$x + 14 = 22$$
$$x = 8$$

Answer: There are 8 nickels and 14 quarters.

13) Use the basic formula distance = time x rate.

Keet the actual and hypothetical situations clear.

Let x represent the time walked at 5 kph.

Let y represent the time walked at 3 kph.

Actual distances:

5x is the distance walked at 5 kph.

3y is the distance walked at 3 kph.

19 is the total distance.

Hypothetical distances if times had been reversed:

3x would have been the distance walked at 5 kph.

3y would have been the distance walked at 3 kph.

19 - 2 = 17 would have been the total.

Actual total: $5x + 3y = 19$

Hypothetical total: $3x + 3y= 17$

$$\begin{array}{ll} 5x + 3y = 19 & -15x - 9y = -57 \\ 3x + 5y = 17 & \underline{15x + 25y = 85} \\ & 16y = 28 \\ & y = 1\frac{3}{4} \end{array}$$

In equation 1, let $y = 1\frac{3}{4}$.

$$5x + 3\left(\frac{7}{4}\right) = 19$$

$$5x = \frac{55}{4}$$

$$x = \frac{11}{4} = 2\ 3/4$$

Answer: David walked 2 3/4 hours at 5 kph, 1 3/4 hours at 3 kph, a total of 4 1/2 hours.

17) Let x represent the number of 39¢ letters.

Let y represent the number of 22¢ letters.

0.39x is the cost of mailing all the 39¢ letters.

0.22y is the cost of mailing all the 10¢ letters.

Total number of letters: $x + y = 910$

Total cost: $0.39x + 0.22y = 261.40$

$$\begin{array}{ll} x + y = 910 & -22x - 22y = -20020 \\ 0.39x + 0.22y = 261.40 & \underline{39x + 22y = 26140} \\ & 17x = 6120 \\ & x = 360 \end{array}$$

17) continued

In equation 1, let x = 360

$$350 + y = 910$$
$$y = 550$$

Answer: The store sent 360 letters at 39¢ and 550 letters at 22¢.

21) Let a represent the number of milliliters of Solution A.

Let b represent the number of milliliters of Solution B.

0.42a is the milliliters of acid contributed to the final mixture by Solution A.

0.18b is the milliliters of acid contributed by Solution B.

Total amount of mixture: $a + b = 32$

Total amount of acid in mixture: $0.42a + 0.18b = 9.6$

$a + b = 32$	$-18a - 18b = -576$
$0.42a + 0.18b = 9.6$	$42a + 18b = 960$
	$24a = 384$
	$a = 16$

In equation 1, let a = 16.

$$(16) + b = 32$$
$$b = 16$$

Answer: The mixture must contain 16 milliliters of each solution.

25) This problem requires careful attention to the units of time in the formula: distance = rate x time.

Let x represent the miles traveled by car.

Let y represent the miles traveled by train.

Since the problem states the total time, the time traveled by car and by train must be represented algebraically. If D = rt, then $t = \frac{D}{r}$.

$\frac{x}{36}$ represents the car travel time in <u>hours</u>.

$\frac{y}{60}$ represents the train travel time.

The total time of 1 hr. 22 min. is $1\frac{22}{60}$ hours.

0.15x is the cost of the car travel.

0.06x is the cost of the train travel.

25) continued

Total time: $\frac{x}{36} + \frac{y}{60} = \frac{82}{60}$

Total cost: $0.15x + 0.06y = \$5.22$

$\frac{x}{36} + \frac{y}{60} = \frac{82}{60}$ (multiply by 360) $\quad 10x + 6y = 492$

$0.15x + 0.06y = 5.22$ $\quad \underline{-15x - 6y = -522}$

$$-5x = -30$$
$$x = 6$$

In equation 2, let $x = 6$.

$0.15(6) + 0.06y = 5.22$

$0.90 + 0.06y = 5.22$

$0.06y = 4.32$

$y = 72$

Answer: Juanita travels 6 miles by car and 72 miles by train.

29) An indirect way to solve this problem is to assume the clerk is correct and work until a contradiction appears. Suppose there is a unit price for each fruit.

Let x represent the unit price for oranges, in cents.

Let y represent the unit price for tangerines.

Change \$2.34 to 234 cents.

Cost of smaller bag: $2x + 4y = 77$

Cost of larger bag: $6x + 12y = 234$

$2x + 4y = 77$ $\quad -6x - 12y = -231$

$6x + 12y = 234$ $\quad \underline{6x + 12y = 234}$

$$0 \neq 3$$

This system is inconsistent, so the oranges and tangerines cannot have the same prices in the two bags.

33) Let x represent the number of skate boards made.

8x is the cost of making x skateboards, not including building maintenance.

Let c represent the total cost of manufacturing x skateboards.

16x is the revenue from selling x skateboards.

Total cost: $c = 8x + 200$

Profit: $520 = 16x - c$

33) continued

Substitute (8x + 200) for c in equation 2.

$520 = 16x - (8x + 200)$

$520 = 16x - 8x - 200$

$720 = 8x$

$90 = x$

Answer: 90 skateboards must be made and sold to obtain a profit of \$520.

37) Let x represent the number of calculators produced.

12x is the cost of producing x calculators, not including the daily fixed cost.

Let c represent the total cost of making x calculators.

20x is the revenue from selling x calculators.

Total cost: $c = 12x + 720$

Profit: $200 = 20x - c$

Substitute (12x + 720) for c in equation 2.

$200 = 20x = (12x + 720)$

$200 = 20x - 12x - 720$

$920 = 8x$

$115 = x$

If 115 calculators are made and sold daily, the profit will be \$200.

41) Let x, y, and z represent the angles, from smallest to largest.

$x + y + z = 180$

$x + y = z$

$2x = z - 10$

The addition method can be used to solve this system, but substitution is very easy in this case.

Solve the equation 3 for z in terms of x.

$2x + 10 = z$

In equation two, substitute (2x + 10) for z to obtain y in terms of x.

$x + y = (2x + 10)$

$y = x + 10$

In equation one, substitute (x + 10) for y and (2x + 10) for z to obtain an equation in x only.

41) continued

$$x + (x + 10) + (2x + 10) = 180$$
$$4x + 20 = 180$$
$$4x = 160$$
$$x = 40$$

In equation 3, let $x = 40$

$$2(40) = z - 10$$
$$90 = z$$

In equation 1, let $x = 40$ and $z = 90$.

$$(40) + y + (90) = 180$$
$$y = 50$$

Answer: the angles are 40°, 50°, and 90°.

45) Let a, b, and c represent the grams of foods A, B, and C respectively.

Total vitamins: $35a + 10b + 20c = 900$

Total minerals: $15a + 20b + 15c = 750$

Total fat: $10a + 10b + 5c = 350$

(Eq. 1) $35a + 10b + 20c = 900$ $\quad 70a + 20b + 40c = 1800$

(Eq. 2) $15a + 20b + 15c = 750$ $\quad -15a - 20b - 15c = -750$

(Eq. 4) $55a + 25c = 1050$

(Eq. 2) $15a + 20b + 15c = 750$ $\quad 15a + 20b + 15c = 750$

(Eq. 3) $10a + 10b + 5c = 350$ $\quad -20a - 20b - 10c = -700$

(Eq. 5) $-5a + 5c = 50$

(Eq. 4) $55a + 25c = 1050$ $\quad 11a + 5c = 210$

(Eq. 5) $-5a + 5c = 50$ $\quad 5a - 5c = -50$

$$16a = 160$$
$$a = 10$$

Divide equation 5 by 5 to get $-a + c = 10$, then let $a = 10$.

$$-(10) + c = 10$$
$$c = 20$$

In equation 3 let $a = 10$ and $c = 20$.

$$10(10) + 10b + 5(20) = 350$$
$$10b = 150$$
$$b = 15$$

Answer: The mixture should contain 10 grams of food A, 15 grams of food B, and 20 grams of food C.

Section 11.3 Solving Linear Systems Using Matrices

Note: There are many correct ways to transform a matrix to solve a system of equations. One solution is shown for each of the exercises below. You may find others that are easier for you. It is strongly recommended that you make a note of what you are doing in each step. This reduces errors and helps in tracking down mistakes.

1) $x + 5y = -9$
$4x - 3y = -13$

$$\left[\begin{array}{cc|c} 1 & 5 & -9 \\ 4 & -3 & -13 \end{array}\right]$$

$$\left[\begin{array}{cc|c} 1 & 5 & -9 \\ 0 & -23 & 23 \end{array}\right] \leftarrow -4(R1) + R2$$

$$\left[\begin{array}{cc|c} 1 & 5 & -9 \\ 0 & 1 & -1 \end{array}\right] \leftarrow -\frac{1}{23}(R2)$$

$x + 5y = -9$

$y = -1 \quad x + 5(-1) = -9$

$x = -4$

Solution: $(-4, -1)$

5) $4x - 5y = -2$
$16x + 2y = 3$

$$\left[\begin{array}{cc|c} 4 & -5 & -2 \\ 16 & 2 & 3 \end{array}\right]$$

$$\left[\begin{array}{cc|c} 4 & -5 & -2 \\ 0 & 22 & 11 \end{array}\right] \leftarrow -4(R1) + R2$$

$$\left[\begin{array}{cc|c} 1 & -\frac{5}{4} & -\frac{1}{2} \\ 0 & 1 & \frac{1}{2} \end{array}\right] \begin{array}{l} \leftarrow \frac{1}{4}(R1) \\ \leftarrow \frac{1}{22}(R2) \end{array}$$

$x - \frac{5}{4}y = -\frac{1}{2}$

$y = \frac{1}{2} \quad x - \frac{5}{4}\left(\frac{1}{2}\right) = -\frac{1}{2}$

$x - \frac{5}{8} = -\frac{1}{2}$

$x = \frac{1}{8}$

Solution: $\left(\frac{1}{8}, \frac{1}{2}\right)$

9) $30x + 45y = 60$
$4x + 6y = 8$

$$\left[\begin{array}{cc|c} 30 & 45 & 60 \\ 4 & 6 & 8 \end{array}\right]$$

$$\left[\begin{array}{cc|c} 2 & 3 & 4 \\ 2 & 3 & 4 \end{array}\right] \begin{array}{l} \leftarrow \frac{1}{15}(R1) \\ \leftarrow \frac{1}{2}(R2) \end{array}$$

$$\left[\begin{array}{cc|c} 2 & 3 & 4 \\ 0 & 0 & 0 \end{array}\right]$$

This system has an infinite number of solutions.

Let $y = c$, where c is any number.

$$2x + 3c = 4$$
$$2x = 4 - 3c$$
$$x = 2 - \frac{3}{2}c$$

Solution: $(2 - \frac{3}{2}c,\ c)$ for all c

13) $x - 2y + 3z = -2$
$-4x + 10y + 2z = -2$
$3x + y + 10z = 7$

$$\left[\begin{array}{ccc|c} 1 & -2 & 3 & -2 \\ -4 & 10 & 2 & -2 \\ 3 & 1 & 10 & 7 \end{array}\right]$$

$$\left[\begin{array}{ccc|c} 1 & -2 & 3 & -2 \\ -2 & 5 & 1 & -1 \\ 3 & 1 & 10 & 7 \end{array}\right] \leftarrow \frac{1}{2}(R2)$$

$$\left[\begin{array}{ccc|c} 1 & -2 & 3 & -2 \\ 0 & 1 & 7 & -5 \\ 3 & 1 & 10 & 7 \end{array}\right] \leftarrow 2(R1) + R2$$

$$\left[\begin{array}{ccc|c} 1 & -2 & 3 & -2 \\ 0 & 1 & 7 & -5 \\ 0 & 7 & 1 & 13 \end{array}\right] \leftarrow -3(R1) + R3$$

$$\left[\begin{array}{ccc|c} 1 & -2 & 3 & -2 \\ 0 & 1 & 7 & -5 \\ 0 & 0 & -48 & 48 \end{array}\right] \leftarrow -7(R2) + R3$$

$$\left[\begin{array}{ccc|c} 1 & -2 & 3 & -2 \\ 0 & 1 & 7 & -5 \\ 0 & 0 & 0 & -1 \end{array}\right] \leftarrow -\frac{1}{48}(R3)$$

$$x - 2y + 3z = -2$$
$$y + 7z = -5$$
$$z = -1 \quad y + 7(-1) = -5$$
$$y = 2 \quad x - 2(2) + 3(-1) = -2$$
$$x - 4 - 3 = -2$$
$$x = 5$$

Solution: (5, 2, -1)

17) $x - 2z = 5$

$3y + 4z = -2$

$-2x + 3y + 8z = 4$

$$\left[\begin{array}{ccc|c} 1 & 0 & -2 & 5 \\ 0 & 3 & 4 & -2 \\ -2 & 3 & 8 & 4 \end{array}\right]$$

$$\left[\begin{array}{ccc|c} 1 & 0 & -2 & 5 \\ 0 & 3 & 4 & -2 \\ 0 & 3 & 4 & 14 \end{array}\right] \begin{array}{l} \\ \\ \leftarrow 2(R1) + R3 \end{array}$$

$$\left[\begin{array}{ccc|c} 1 & 0 & -2 & 5 \\ 0 & 3 & 4 & -2 \\ 0 & 0 & 0 & 16 \end{array}\right] \begin{array}{l} \\ \\ \leftarrow -1(R2) + R3 \end{array}$$

No solutions.

21) $w - x + 2y + 2z = 0$

$2w - y - 3z = 0$

$4x - 3y + z = -2$

$-3w + 2x + 4z = 1$

$$\left[\begin{array}{cccc|c} 1 & -1 & 2 & 2 & 0 \\ 2 & 0 & -1 & -3 & 0 \\ 0 & 4 & -3 & 1 & -2 \\ -3 & 2 & 0 & 4 & 1 \end{array}\right]$$

$$\left[\begin{array}{cccc|c} 1 & -1 & 2 & 2 & 0 \\ 0 & 2 & -5 & -7 & 0 \\ 0 & 4 & -3 & 1 & -2 \\ -3 & 2 & 0 & 4 & 1 \end{array}\right] \begin{array}{l} \\ \leftarrow -2(R1) + R2 \\ \\ \\ \end{array}$$

$$\left[\begin{array}{cccc|c} 1 & -1 & 2 & 2 & 0 \\ 0 & 2 & -5 & -7 & 0 \\ 0 & 0 & 7 & 15 & -2 \\ 0 & -1 & 6 & 10 & 1 \end{array}\right] \begin{array}{l} \\ \\ \leftarrow -2(R2) + R3 \\ \leftarrow 3(R1) + R4 \end{array}$$

$$\left[\begin{array}{cccc|c} 1 & -1 & 2 & 2 & 0 \\ 0 & 1 & -6 & -10 & -1 \\ 0 & 0 & 7 & 15 & -2 \\ 0 & 2 & -5 & -7 & 0 \end{array}\right] \begin{array}{l} \\ \leftarrow -R4 \\ \\ \leftarrow R2 \end{array}$$

$$\left[\begin{array}{cccc|c} 0 & -1 & 2 & 2 & 0 \\ 0 & 1 & -6 & -10 & -1 \\ 0 & 0 & 7 & 15 & -2 \\ 0 & 0 & 7 & 13 & 2 \end{array}\right] \begin{array}{l} \\ \\ \\ \leftarrow -2(R2) + R4 \end{array}$$

$$\left[\begin{array}{cccc|c} 1 & -1 & 2 & 2 & 0 \\ 0 & 1 & -6 & -10 & -1 \\ 0 & 0 & 7 & 15 & -2 \\ 0 & 0 & 0 & -2 & 4 \end{array}\right] \begin{array}{l} \\ \\ \\ \leftarrow -(R3) + R4 \end{array}$$

21) continued

$$\left[\begin{array}{cccc|c} 1 & -1 & 2 & 2 & 0 \\ 0 & 1 & -6 & -10 & -1 \\ 0 & 0 & 1 & \frac{15}{7} & -\frac{2}{7} \\ 0 & 0 & 0 & 1 & -2 \end{array}\right] \begin{array}{l} \\ \\ \leftarrow \frac{1}{15}\ (R3) \\ \leftarrow -\frac{1}{2}\ (R4) \end{array}$$

$w - x + 2y + 2z = 0$

$x - 6y - 10z = -1$

$y + \frac{15}{7}z = -\frac{2}{7}$

$z = -2 \qquad y + \frac{15}{7}(-2) = -\frac{2}{7}$

$y = 4 \qquad x - 6(4) - 10(-2) = -1$

$x = 3$

$w - (3) + 2(4) + 2(-2) = 0$

$w = -1$

Solution: (-1, 3, 4, -2)

Section 11.4 Matrix Algebra

1) $\begin{bmatrix} 5 & -1 \\ -3 & 4 \end{bmatrix} + \begin{bmatrix} 1 & 2 \\ 3 & 4 \end{bmatrix} = \begin{bmatrix} (5+1) & (-1+2) \\ (-3+3) & (4+4) \end{bmatrix} = \begin{bmatrix} 6 & 1 \\ 0 & 8 \end{bmatrix}$

5) $\begin{bmatrix} 1 & -3 & 2 \\ -5 & 2 & 0 \\ 0 & -1 & 3 \end{bmatrix} + 2\begin{bmatrix} 1 & 1 & 1 \\ -1 & 1 & 1 \\ -1 & -1 & 1 \end{bmatrix} = \begin{bmatrix} 1 & -3 & 2 \\ -5 & 2 & 0 \\ 0 & -1 & 3 \end{bmatrix} + \begin{bmatrix} 2 & 2 & 2 \\ -2 & 2 & 2 \\ -2 & -2 & 2 \end{bmatrix}$

$= \begin{bmatrix} 3 & -1 & 4 \\ -7 & 4 & 2 \\ -2 & -3 & 5 \end{bmatrix}$

9) $\begin{bmatrix} 5 & -1 \\ -3 & 4 \end{bmatrix}\begin{bmatrix} 1 & 2 \\ 3 & 4 \end{bmatrix} = \begin{bmatrix} (5-3) & (10-4) \\ (-3+12) & (-6+16) \end{bmatrix} = \begin{bmatrix} 2 & 6 \\ 9 & 10 \end{bmatrix}$

13) $\begin{bmatrix} 1 & -4 & 0 \\ 2 & -3 & 5 \end{bmatrix} \begin{bmatrix} 1 & -3 & 2 \\ -5 & 2 & 0 \\ 0 & -1 & 3 \end{bmatrix} = \begin{bmatrix} (1+20+0) & (-3-8+0) & (2+0+0) \\ (2+15+0) & (-6-6-5) & (4+0+15) \end{bmatrix}$

$= \begin{bmatrix} 21 & -11 & 2 \\ 17 & -17 & 19 \end{bmatrix}$

17) $\begin{bmatrix} 5 & 2 & -3 & 0 \\ 1 & 0 & 3 & -1 \\ 4 & -2 & 0 & 1 \end{bmatrix} \begin{bmatrix} 1 & 0 \\ 0 & -1 \\ 1 & 1 \\ 0 & 0 \end{bmatrix} = \begin{bmatrix} (5+0-3+0) & (0-2-3+0) \\ (1+0+3+0) & (0+0+3+0) \\ (4+0+0+0) & (0+2+0+0) \end{bmatrix}$

$= \begin{bmatrix} 2 & -5 \\ 4 & 3 \\ 4 & 2 \end{bmatrix}$

21) $\begin{bmatrix} 1 & 1 & 1 \\ -1 & 1 & 1 \\ -1 & 1 & 1 \end{bmatrix} \begin{bmatrix} 0 \\ 0 \\ 0 \end{bmatrix} = \begin{bmatrix} 0 \\ 0 \\ 0 \end{bmatrix}$

25) $\begin{bmatrix} 0 & 6 \\ 5 & -4 \end{bmatrix} + \begin{bmatrix} 3 & 7 \\ -1 & -2 \end{bmatrix} = \begin{bmatrix} (0+3) & (6+7) \\ (5-1) & (-4-2) \end{bmatrix} = \begin{bmatrix} 3 & 13 \\ 4 & -6 \end{bmatrix}$

$\begin{bmatrix} 3 & 7 \\ -1 & -2 \end{bmatrix} + \begin{bmatrix} 0 & 6 \\ 5 & -4 \end{bmatrix} = \begin{bmatrix} (3+0) & (7+6) \\ (-1+5) & (-2-4) \end{bmatrix} = \begin{bmatrix} 3 & 13 \\ 4 & -6 \end{bmatrix}$

29) $\left(\begin{bmatrix} 2 & 1 \\ -4 & 1 \end{bmatrix} + \begin{bmatrix} 0 & 6 \\ 5 & -4 \end{bmatrix} \right) \begin{bmatrix} 3 & 7 \\ -1 & -2 \end{bmatrix} = \begin{bmatrix} 2 & 7 \\ 1 & -3 \end{bmatrix} \begin{bmatrix} 3 & 7 \\ -1 & -2 \end{bmatrix} = \begin{bmatrix} -1 & 0 \\ 6 & 13 \end{bmatrix}$

$\begin{bmatrix} 2 & 1 \\ -4 & 1 \end{bmatrix} \begin{bmatrix} 3 & 7 \\ -1 & -2 \end{bmatrix} + \begin{bmatrix} 0 & 6 \\ 5 & -4 \end{bmatrix} \begin{bmatrix} 3 & 7 \\ -1 & -2 \end{bmatrix} = \begin{bmatrix} 5 & 12 \\ -13 & -30 \end{bmatrix} + \begin{bmatrix} -6 & -12 \\ 19 & 43 \end{bmatrix}$

$= \begin{bmatrix} -1 & 0 \\ 6 & 13 \end{bmatrix}$

33) a) $\begin{bmatrix} 0 & 1 & 0 \\ 1 & 0 & 0 \\ 0 & 0 & 1 \end{bmatrix} \begin{bmatrix} a & b & c \\ d & e & f \\ g & h & i \end{bmatrix} = \begin{bmatrix} d & e & f \\ a & b & c \\ g & h & i \end{bmatrix}$

33) continued

b) Matrix E interchanged rows 1 and 2.

$$F = \begin{bmatrix} 0 & 0 & 1 \\ 0 & 1 & 0 \\ 1 & 0 & 0 \end{bmatrix} \begin{matrix} \leftarrow \text{This will bring row 3 to row 1.} \\ \leftarrow \text{This will leave row 2 in place.} \\ \leftarrow \text{This will bring row 1 to row 3.} \end{matrix}$$

37) a) $$\begin{bmatrix} x & 0 & 0 \\ 0 & y & 0 \\ 0 & 0 & z \end{bmatrix}\begin{bmatrix} x & 0 & 0 \\ 0 & y & 0 \\ 0 & 0 & z \end{bmatrix} = \begin{bmatrix} x^2 & 0 & 0 \\ 0 & y^2 & 0 \\ 0 & 0 & z^2 \end{bmatrix}$$

b) $$\begin{bmatrix} x^2 & 0 & 0 \\ 0 & y^2 & 0 \\ 0 & 0 & z^2 \end{bmatrix}\begin{bmatrix} x & 0 & 0 \\ 0 & y & 0 \\ 0 & 0 & z \end{bmatrix} = \begin{bmatrix} x^3 & 0 & 0 \\ 0 & y^3 & 0 \\ 0 & 0 & z^3 \end{bmatrix}$$

c) $$\begin{bmatrix} x^n & 0 & 0 \\ 0 & y^n & 0 \\ 0 & 0 & z^n \end{bmatrix}$$

Section 11.5 Inverses

1) $$\left[\begin{array}{cc|cc} 4 & -1 & 1 & 0 \\ 2 & 0 & 0 & 1 \end{array}\right]$$

$$\left[\begin{array}{cc|cc} 2 & 0 & 0 & 1 \\ 4 & -1 & 1 & 0 \end{array}\right] \begin{matrix} \leftarrow R2 \\ \leftarrow R1 \end{matrix}$$

$$\left[\begin{array}{cc|cc} 1 & 0 & 0 & \frac{1}{2} \\ 4 & -1 & 1 & 0 \end{array}\right] \begin{matrix} \leftarrow \frac{1}{2}(R1) \\ \end{matrix}$$

$$\left[\begin{array}{cc|cc} 1 & 0 & 0 & \frac{1}{2} \\ 0 & -1 & 1 & -2 \end{array}\right] \begin{matrix} \\ \leftarrow -4(R1) + R2 \end{matrix}$$

$$\left[\begin{array}{cc|cc} 1 & 0 & 0 & \frac{1}{2} \\ 0 & 1 & -1 & 2 \end{array}\right]$$

1) continued

$$A^{-1} = \begin{bmatrix} 0 & \frac{1}{2} \\ -1 & 2 \end{bmatrix}$$

$$AA^{-1} = \begin{bmatrix} 4 & -1 \\ 2 & 0 \end{bmatrix}\begin{bmatrix} 0 & \frac{1}{2} \\ -1 & 2 \end{bmatrix} = \begin{bmatrix} (0 + 1) & (2 - 2) \\ (0 + 0) & (1 + 0) \end{bmatrix} = \begin{bmatrix} 1 & 0 \\ 0 & 1 \end{bmatrix} = I_2$$

$$A^{-1}A = \begin{bmatrix} 0 & \frac{1}{2} \\ -1 & 2 \end{bmatrix}\begin{bmatrix} 4 & -1 \\ 2 & 0 \end{bmatrix} = \begin{bmatrix} (0 + 1) & (0 + 0) \\ (-4 + 4) & (1 + 0) \end{bmatrix} = \begin{bmatrix} 1 & 0 \\ 0 & 1 \end{bmatrix} = I_2$$

5) $$\left[\begin{array}{cc|cc} 2 & -5 & 1 & 0 \\ -3 & 4 & 0 & 1 \end{array}\right]$$

$$\left[\begin{array}{cc|cc} -1 & -1 & 1 & 1 \\ -3 & 4 & 0 & 1 \end{array}\right] \leftarrow R2 + R1$$

$$\left[\begin{array}{cc|cc} 1 & 1 & -1 & -1 \\ -3 & 4 & 0 & 1 \end{array}\right] \leftarrow -(R1)$$

$$\left[\begin{array}{cc|cc} 1 & 1 & -1 & -1 \\ 0 & 7 & -3 & -2 \end{array}\right] \leftarrow 3(R1) + R2$$

$$\left[\begin{array}{cc|cc} 1 & 1 & -1 & -1 \\ 0 & 1 & -\frac{3}{7} & -\frac{2}{7} \end{array}\right] \leftarrow \frac{1}{7}(R2)$$

$$\left[\begin{array}{cc|cc} 1 & 0 & -\frac{4}{7} & -\frac{5}{7} \\ 0 & 1 & -\frac{3}{7} & -\frac{2}{7} \end{array}\right] \leftarrow -(R2) + R1$$

$$A^{-1} = \begin{bmatrix} -\frac{4}{7} & -\frac{5}{7} \\ -\frac{3}{7} & -\frac{2}{7} \end{bmatrix}$$

$$AA^{-1} = \begin{bmatrix} 2 & -5 \\ -3 & 4 \end{bmatrix}\begin{bmatrix} -\frac{4}{7} & -\frac{5}{7} \\ -\frac{3}{7} & -\frac{2}{7} \end{bmatrix} = \begin{bmatrix} -\frac{8}{7} + \frac{15}{7} & -\frac{10}{7} + \frac{10}{7} \\ \frac{12}{7} - \frac{12}{7} & \frac{15}{7} - \frac{8}{7} \end{bmatrix} = \begin{bmatrix} 1 & 0 \\ 0 & 1 \end{bmatrix} = I_2$$

5) continued

$$A^{-1}A = \begin{bmatrix} -\frac{4}{7} & -\frac{5}{7} \\ -\frac{3}{7} & -\frac{2}{7} \end{bmatrix} \begin{bmatrix} 2 & -5 \\ -3 & 4 \end{bmatrix} = \begin{bmatrix} -\frac{8}{7} + \frac{15}{7} & \frac{20}{7} - \frac{20}{7} \\ -\frac{6}{7} + \frac{6}{7} & \frac{15}{7} - \frac{8}{7} \end{bmatrix} = \begin{bmatrix} 1 & 0 \\ 0 & 1 \end{bmatrix} = I_2$$

9) $$\left[\begin{array}{cc|cc} 0 & a & 1 & 0 \\ b & 0 & 0 & 1 \end{array}\right]$$

$$\left[\begin{array}{cc|cc} b & 0 & 0 & 1 \\ 0 & a & 1 & 0 \end{array}\right] \begin{array}{l} \leftarrow R2 \\ \leftarrow R1 \end{array}$$

$$\left[\begin{array}{cc|cc} 1 & 0 & 0 & \frac{1}{b} \\ 0 & 1 & \frac{1}{a} & 0 \end{array}\right] \begin{array}{l} \leftarrow \frac{1}{b}(R1) \\ \leftarrow \frac{1}{a}(R2) \end{array}$$

$$A^{-1} = \begin{bmatrix} 0 & \frac{1}{b} \\ \frac{1}{a} & 0 \end{bmatrix}$$

$$AA^{-1} = \begin{bmatrix} 0 & a \\ b & 0 \end{bmatrix} \begin{bmatrix} 0 & \frac{1}{b} \\ \frac{1}{a} & 0 \end{bmatrix} = \begin{bmatrix} (0 + 1) & (0 + 0) \\ (0 + 0) & (1 + 0) \end{bmatrix} = \begin{bmatrix} 1 & 0 \\ 0 & 1 \end{bmatrix} = I_2$$

$$A^{-1}A = \left[\begin{array}{cc|cc} 0 & \frac{1}{b} & 0 & a \\ \frac{1}{a} & 0 & b & 0 \end{array}\right] = \begin{bmatrix} (0 + 1) & (0 + 0) \\ (0 + 0) & (0 + 1) \end{bmatrix} = \begin{bmatrix} 1 & 0 \\ 0 & 1 \end{bmatrix} = I_2$$

13) $$\left[\begin{array}{ccc|ccc} 1 & 0 & 2 & 1 & 0 & 0 \\ 2 & -1 & 0 & 0 & 1 & 0 \\ 0 & 3 & 4 & 0 & 0 & 1 \end{array}\right]$$

$$\left[\begin{array}{ccc|ccc} 1 & 0 & 2 & 1 & 0 & 0 \\ 0 & -1 & -4 & -2 & 1 & 0 \\ 0 & 3 & 4 & 0 & 0 & 1 \end{array}\right] \begin{array}{l} \\ \leftarrow -2(R1) + R2 \\ \\ \end{array}$$

$$\left[\begin{array}{ccc|ccc} 1 & 0 & 2 & 1 & 0 & 0 \\ 0 & -1 & -4 & -2 & 1 & 0 \\ 0 & 0 & -8 & -6 & 3 & 1 \end{array}\right] \begin{array}{l} \\ \\ \leftarrow 3(R2) + R3 \end{array}$$

13) continued

$$\left[\begin{array}{ccc|ccc} 1 & 0 & 2 & 1 & 0 & 0 \\ 0 & 1 & 4 & 2 & -1 & 0 \\ 0 & 0 & 1 & \frac{3}{4} & -\frac{3}{8} & -\frac{1}{8} \end{array}\right] \begin{array}{l} \\ \leftarrow -(R2) \\ \leftarrow -\frac{1}{8}(R3) \end{array}$$

$$\left[\begin{array}{ccc|ccc} 1 & 0 & 2 & 1 & 0 & 0 \\ 0 & 1 & 0 & -1 & \frac{1}{2} & \frac{1}{2} \\ 0 & 0 & 1 & \frac{3}{4} & -\frac{3}{8} & -\frac{1}{8} \end{array}\right] \begin{array}{l} \\ \leftarrow -4(R3) + R2 \\ \\ \end{array}$$

$$\left[\begin{array}{ccc|ccc} 1 & 0 & 0 & -\frac{1}{2} & \frac{3}{4} & \frac{1}{4} \\ 0 & 1 & 0 & -1 & \frac{1}{2} & \frac{1}{2} \\ 0 & 0 & 1 & \frac{3}{4} & -\frac{3}{8} & -\frac{1}{8} \end{array}\right] \begin{array}{l} \leftarrow -2(R3) + R1 \\ \\ \\ \end{array}$$

$$A^{-1} = \begin{bmatrix} -\frac{1}{2} & \frac{3}{4} & \frac{1}{4} \\ -1 & \frac{1}{2} & \frac{1}{2} \\ \frac{3}{4} & -\frac{3}{8} & -\frac{1}{8} \end{bmatrix}$$

$$AA^{-1} = \begin{bmatrix} 1 & 0 & 2 \\ 2 & -1 & 0 \\ 0 & 3 & 4 \end{bmatrix} \begin{bmatrix} -\frac{1}{2} & \frac{3}{4} & \frac{1}{4} \\ -1 & \frac{1}{2} & \frac{1}{2} \\ \frac{3}{4} & -\frac{3}{8} & -\frac{1}{8} \end{bmatrix} = \begin{bmatrix} -\frac{1}{2}+0+\frac{3}{2} & \frac{3}{4}+0-\frac{3}{4} & \frac{1}{4}+0-\frac{1}{4} \\ (-1+1+0) & \frac{3}{2}-\frac{1}{2}+0 & \frac{1}{2}-\frac{1}{2} = 0 \\ (0-3+3) & 0+\frac{3}{2}-\frac{3}{2} & 0+\frac{3}{2}-\frac{1}{2} \end{bmatrix}$$

$$= \begin{bmatrix} 1 & 0 & 0 \\ 0 & 1 & 0 \\ 0 & 0 & 1 \end{bmatrix} = I_3$$

$$A^{-1}A = \begin{bmatrix} -\frac{1}{2} & \frac{3}{4} & \frac{1}{4} \\ -1 & \frac{1}{2} & \frac{1}{2} \\ \frac{3}{4} & -\frac{3}{8} & -\frac{1}{8} \end{bmatrix} \begin{bmatrix} 1 & 0 & 2 \\ 2 & -1 & 0 \\ 0 & 3 & 4 \end{bmatrix}$$

13) continued

$$= \begin{bmatrix} -\frac{1}{2}+\frac{3}{2}+0 & 0-\frac{3}{4}-\frac{3}{4} & (-1+0+1) \\ (-1+1+0) & 0-\frac{1}{2}+\frac{3}{2} & (-2+0+2) \\ \frac{3}{4}-\frac{3}{4}+0 & 0+\frac{3}{8}+\frac{3}{8} & \frac{3}{2}+0-\frac{1}{2} \end{bmatrix} = \begin{bmatrix} 1 & 0 & 0 \\ 0 & 1 & 0 \\ 0 & 0 & 1 \end{bmatrix} = I_3$$

17) $$\left[\begin{array}{ccc|ccc} -11 & 2 & 2 & 1 & 0 & 0 \\ -4 & 0 & 1 & 0 & 1 & 0 \\ 6 & -1 & -1 & 0 & 0 & 1 \end{array}\right]$$

$$\left[\begin{array}{ccc|ccc} 1 & 0 & 0 & 1 & 0 & 2 \\ -4 & 0 & 1 & 0 & 1 & 0 \\ 6 & -1 & -1 & 0 & 0 & 1 \end{array}\right] \begin{array}{l} \leftarrow 2(R3)+R1 \\ \\ \\ \end{array}$$

$$\left[\begin{array}{ccc|ccc} 1 & 0 & 0 & 1 & 0 & 2 \\ 6 & -1 & -1 & 0 & 0 & 1 \\ -4 & 0 & 1 & 0 & 1 & 0 \end{array}\right] \begin{array}{l} \\ \leftarrow R3 \\ \leftarrow R2 \end{array}$$

$$\left[\begin{array}{ccc|ccc} 1 & 0 & 0 & 1 & 0 & 2 \\ 0 & -1 & -1 & -6 & 0 & -11 \\ 0 & 0 & 1 & 4 & 1 & 8 \end{array}\right] \begin{array}{l} \\ \leftarrow -6(R1)+R2 \\ \leftarrow 4(R1)+R3 \end{array}$$

$$\left[\begin{array}{ccc|ccc} 1 & 0 & 0 & 1 & 0 & 2 \\ 0 & 1 & 1 & 6 & 0 & 11 \\ 0 & 0 & 1 & 4 & 1 & 8 \end{array}\right] \begin{array}{l} \\ \leftarrow -(R2) \\ \\ \end{array}$$

$$\left[\begin{array}{ccc|ccc} 1 & 0 & 0 & 1 & 0 & 2 \\ 0 & 1 & 0 & 2 & -1 & 3 \\ 0 & 0 & 1 & 4 & 1 & 8 \end{array}\right] \begin{array}{l} \\ \leftarrow -(R3)+R2 \\ \\ \end{array}$$

$$A^{-1} = \begin{bmatrix} 1 & 0 & 2 \\ 2 & -1 & 3 \\ 4 & 1 & 8 \end{bmatrix}$$

$$AA^{-1} = \begin{bmatrix} -11 & 2 & 2 \\ -4 & 0 & 1 \\ 6 & -1 & -1 \end{bmatrix} \begin{bmatrix} 1 & 0 & 2 \\ 2 & -1 & 3 \\ 4 & 1 & 8 \end{bmatrix} = \begin{bmatrix} (-11+4+8) & (0-2+2) & (-22+6+16) \\ (-4+0+4) & (0+0+1) & (-8+0+8) \\ (6-2-4) & (0+1-1) & (12-3-8) \end{bmatrix}$$

17) continued

$$= \begin{bmatrix} 1 & 0 & 0 \\ 0 & 1 & 0 \\ 0 & 0 & 1 \end{bmatrix} = I_3$$

$$A^{-1}A = \begin{bmatrix} 1 & 0 & 2 \\ 2 & -1 & 3 \\ 4 & 1 & 8 \end{bmatrix} \begin{bmatrix} -11 & 2 & 2 \\ -4 & 0 & 1 \\ 6 & -1 & -1 \end{bmatrix} = \begin{bmatrix} (-11+0+12) & (2+0-2) & (2+0-2) \\ (-22+4+18) & (4+0-3) & (4-1-3) \\ (-44-4+48) & (8+0-8) & (8+1-8) \end{bmatrix}$$

$$= \begin{bmatrix} 1 & 0 & 0 \\ 0 & 1 & 0 \\ 0 & 0 & 1 \end{bmatrix} = I_3$$

21) $Ax = C$

$$\begin{bmatrix} 3 & 1 \\ 2 & -2 \end{bmatrix} \begin{bmatrix} x \\ y \end{bmatrix} = \begin{bmatrix} 9 \\ 14 \end{bmatrix}$$

$3x + y = 9$

$2x - 2y = 14$

25) $\frac{1}{3}x + \frac{2}{3}y = -8$

$\frac{2}{3}x + \frac{1}{3}y = 5$

$$A = \begin{bmatrix} \frac{1}{3} & \frac{2}{3} \\ \frac{2}{3} & \frac{1}{3} \end{bmatrix} \qquad X = \begin{bmatrix} x \\ y \end{bmatrix} \qquad C = \begin{bmatrix} -8 \\ 5 \end{bmatrix}$$

From exercise 7, $A^{-1} = \begin{bmatrix} -1 & 2 \\ 2 & -1 \end{bmatrix}$

$X = A^{-1}C$

$$\begin{bmatrix} x \\ y \end{bmatrix} = \begin{bmatrix} -1 & 2 \\ 2 & -1 \end{bmatrix} \begin{bmatrix} -8 \\ 5 \end{bmatrix}$$

$$\begin{bmatrix} x \\ y \end{bmatrix} = \begin{bmatrix} (8 + 10) \\ (-16 - 5) \end{bmatrix} = \begin{bmatrix} 18 \\ -21 \end{bmatrix}$$

$x = 18 \qquad y = -21$

Solution: $(18, -21)$

29) $x + y - z = 1$

$x - y - z = 2$

$-x - y - z = 3$

$$A = \begin{bmatrix} 1 & 1 & -1 \\ 1 & -1 & -1 \\ -1 & -1 & -1 \end{bmatrix} \quad X = \begin{bmatrix} x \\ y \\ z \end{bmatrix} \quad C = \begin{bmatrix} 1 \\ 2 \\ 3 \end{bmatrix}$$

From exercise 14,

$$A^{-1} = \begin{bmatrix} 0 & \frac{1}{2} & -\frac{1}{2} \\ \frac{1}{2} & -\frac{1}{2} & 0 \\ -\frac{1}{2} & 0 & -\frac{1}{2} \end{bmatrix}$$

$$X = A^{-1}C$$

$$\begin{bmatrix} x \\ y \\ z \end{bmatrix} = \begin{bmatrix} 0 & \frac{1}{2} & -\frac{1}{2} \\ \frac{1}{2} & -\frac{1}{2} & 0 \\ -\frac{1}{2} & 0 & -\frac{1}{2} \end{bmatrix} \begin{bmatrix} 1 \\ 2 \\ 3 \end{bmatrix}$$

$$\begin{bmatrix} x \\ y \\ z \end{bmatrix} = \begin{bmatrix} 0 + 1 - \frac{3}{2} \\ \frac{1}{2} - 1 + 0 \\ -\frac{1}{2} + 0 - \frac{3}{2} \end{bmatrix} = \begin{bmatrix} -\frac{1}{2} \\ -\frac{1}{2} \\ -2 \end{bmatrix}$$

$x = -\frac{1}{2} \qquad y = -\frac{1}{2} \qquad z = -2$

Solution: $\left(-\frac{1}{2}, -\frac{1}{2}, -2\right)$

33) $w + x + 2z = 2$

$-w + 2y - z = -6$

$2x - 2z = 0$

$2w + 5z = 8$

$$A = \begin{bmatrix} 1 & 1 & 0 & 2 \\ -1 & 0 & 2 & -1 \\ 0 & 2 & 0 & -2 \\ 2 & 0 & 0 & 5 \end{bmatrix} \qquad X = \begin{bmatrix} w \\ x \\ y \\ z \end{bmatrix} \qquad C = \begin{bmatrix} 2 \\ -6 \\ 0 \\ 8 \end{bmatrix}$$

33) continued

From exercise 19,

$$A^{-1} = \begin{bmatrix} -5 & 0 & \frac{5}{2} & 3 \\ 2 & 0 & -\frac{1}{2} & -1 \\ -\frac{3}{2} & \frac{1}{2} & \frac{3}{4} & 1 \\ 2 & 0 & -1 & -1 \end{bmatrix}$$

$X = A^{-1}C$

$$\begin{bmatrix} w \\ x \\ y \\ z \end{bmatrix} = \begin{bmatrix} -5 & 0 & \frac{5}{2} & 3 \\ 2 & 0 & -\frac{1}{2} & -1 \\ -\frac{3}{2} & \frac{1}{2} & \frac{3}{4} & 1 \\ 2 & 0 & -1 & -1 \end{bmatrix} \begin{bmatrix} 2 \\ -6 \\ 0 \\ 8 \end{bmatrix}$$

$$\begin{bmatrix} w \\ x \\ y \\ z \end{bmatrix} = \begin{bmatrix} (-10 + 0 + 0 + 24) \\ (4 + 0 + 0 - 8) \\ (-3 - 3 + 0 + 8) \\ (4 + 0 + 0 - 8) \end{bmatrix} = \begin{bmatrix} 14 \\ -4 \\ 2 \\ -4 \end{bmatrix}$$

$w = 14$ $x = -4$ $y = 2$ $z = -4$

Solution: $(14, -4, 2, -4)$

37) a) $AB = I$

$C(AB) = CI$

$(CA)B = C$

$IB = C$

$B = C$

b) The inverse of a matrix is unique.

Section 11.6 Determinants and a General Solution

1) $\begin{vmatrix} 5 & -1 \\ -3 & 4 \end{vmatrix} = (5)(4) - (-3)(-1) = 20 - 3 = 17$

5) $\begin{vmatrix} 10 & 5 \\ 6 & -3 \end{vmatrix} = (10)(-3) - (6)(5) = -30 - 30 = -60$

9) $3x + 9y = 15$ $\qquad$ $x + 3y = 5$

$6x + 12y = 18$ $\qquad$ $x + 2y = 3$

$$x = \frac{\begin{vmatrix} 5 & 3 \\ 3 & 2 \end{vmatrix}}{\begin{vmatrix} 1 & 3 \\ 1 & 2 \end{vmatrix}} = \frac{10 - 9}{2 - 3} = \frac{1}{-1} = -1$$

$$y = \frac{\begin{vmatrix} 1 & 5 \\ 1 & 3 \end{vmatrix}}{\begin{vmatrix} 1 & 3 \\ 1 & 2 \end{vmatrix}} = \frac{3 - 5}{2 - 3} = \frac{-2}{-1} = 2$$

Solution: (-1, 2)

13) $5x + 2y = 3$

$2x + 3y = -1$

$$x = \frac{\begin{vmatrix} 3 & 2 \\ -1 & 3 \end{vmatrix}}{\begin{vmatrix} 5 & 2 \\ 2 & 3 \end{vmatrix}} = \frac{9 - (-2)}{15 - 4} = \frac{11}{11} = 1$$

$$y = \frac{\begin{vmatrix} 5 & 3 \\ 2 & -1 \end{vmatrix}}{\begin{vmatrix} 5 & 2 \\ 2 & 3 \end{vmatrix}} = \frac{-5 - 6}{15 - 4} = -\frac{11}{11} = -1$$

Solution: (1, -1)

17) $3x + y = 20$ $\qquad$ $3x + y = 20$

$y = x$ $\qquad$ $-x + y = 0$

$$x = \frac{\begin{vmatrix} 20 & 1 \\ 0 & 1 \end{vmatrix}}{\begin{vmatrix} 3 & 1 \\ -1 & 1 \end{vmatrix}} = \frac{20 - 0}{3 - (-1)} = \frac{20}{4} = 5$$

$$y = \frac{\begin{vmatrix} 3 & 20 \\ -1 & 0 \end{vmatrix}}{\begin{vmatrix} 3 & 1 \\ -1 & 1 \end{vmatrix}} = \frac{0 - (-20)}{3 - (-1)} = \frac{20}{4} = 5$$

Solution: (5, 5)

21) $9x - 12 = 4y$ $\qquad$ $9x - 4y = 12$

$3x + 2y = 3$ $\qquad$ $3x + 2y = 3$

$$x = \frac{\begin{vmatrix} 12 & -4 \\ 3 & 2 \end{vmatrix}}{\begin{vmatrix} 9 & -4 \\ 3 & 2 \end{vmatrix}} = \frac{24 - (-12)}{18 - (-12)} = \frac{36}{30} = \frac{6}{5}$$

$$y = \frac{\begin{vmatrix} 9 & 12 \\ 3 & 3 \end{vmatrix}}{\begin{vmatrix} 9 & -4 \\ 3 & 2 \end{vmatrix}} = \frac{27 - 36}{18 - (-12)} = \frac{-9}{30} = -\frac{3}{10}$$

Solution: $\left(\frac{6}{5}, -\frac{3}{10}\right)$

25) $16x - 4y = 20$

$12x - 3y = 15$

$$\begin{vmatrix} 16 & -4 \\ 12 & -3 \end{vmatrix} = -48 - (-48) = 0$$

$16x - 4y = 20$ $\qquad$ $4x - y = 5$

$12x - 3y = 15$ $\qquad$ $-4x + y = -5$

$0 = 0$

Dependent.

29) $\begin{vmatrix} x & 2 \\ 5 & 3 \end{vmatrix} = 8$

$3x - 10 = 8$

$3x = 18$

$x = 6$

33) $\begin{vmatrix} x & y \\ 2 & 4 \end{vmatrix} = 5$ $\qquad$ $\begin{vmatrix} 1 & y \\ -1 & x \end{vmatrix} = -\frac{1}{2}$

$4x - 2y = 5$ $\qquad$ $x + y = -\frac{1}{2}$

$4x - 2y = 5$ $\qquad$ $4x - 2y = 5$

$x + y = -\frac{1}{2}$ $\qquad$ $-4x - 4y = 2$

$-6y = 7$

$y = -\frac{7}{6}$

33) continued

In equation 2, let $y = -\frac{7}{6}$.

$$x + \left(-\frac{7}{6}\right) = -\frac{1}{2}$$

$$x = \frac{4}{6} = \frac{2}{3}$$

Solution: $\left(\frac{2}{3}, -\frac{7}{6}\right)$

37) In the determinant $\begin{vmatrix} a_1 & b_1 \\ a_2 & b_2 \end{vmatrix}$ let $b_1 = ka_1$ and $b_2 = ka_2$ where k is any number.

$$\begin{vmatrix} a_1 & ka_1 \\ a_2 & ka_2 \end{vmatrix} = \begin{vmatrix} a_1 & a_2 \\ ka_1 & ka_2 \end{vmatrix}$$ As shown in exercise 35.

$$= 0$$ As shown in exercise 36.

41) Begin with the general form of the determinant $\begin{vmatrix} a_1 & b_1 \\ a_2 & b_2 \end{vmatrix}$.

Multiply each element of column 2 by k and add to column 1.

$$\begin{vmatrix} a_1 + kb_1 & b_1 \\ a_2 + kb_2 & b_2 \end{vmatrix} = (a_1 + kb_1)b_2 - (a_2 + kb_2)b_1$$

$$= a_1b_2 + kb_1b_2 - a_2b_1 - kb_1b_2 = a_1b_2 - a_2b_1$$

$$= \begin{vmatrix} a_1 & b_1 \\ a_2 & b_2 \end{vmatrix}$$

Section 11.7 Third-Order Determinants

1) a) $$\begin{vmatrix} 6 & -2 & -1 \\ 0 & -9 & 4 \\ -3 & 5 & 1 \end{vmatrix} = 6\begin{vmatrix} -9 & 4 \\ 5 & 1 \end{vmatrix} - 0\begin{vmatrix} -2 & -1 \\ 5 & 1 \end{vmatrix} + (-3)\begin{vmatrix} -2 & -1 \\ -9 & 4 \end{vmatrix}$$

$$= 6(-29) - 0 - 3(-17) = -123$$

1) continued

b) $\begin{vmatrix} 6 & -2 & -1 \\ 0 & -9 & 4 \\ -3 & 5 & 1 \end{vmatrix} = -3\begin{vmatrix} -2 & -1 \\ -9 & 4 \end{vmatrix} -5\begin{vmatrix} 6 & 1 \\ 0 & 4 \end{vmatrix} +1\begin{vmatrix} 6 & -2 \\ 0 & -9 \end{vmatrix}$

$$= -3(-17) - 5(24) + 1(-54) = -123$$

c) $\begin{vmatrix} 6 & -2 & -1 \\ 0 & -9 & 4 \\ -3 & 5 & 1 \end{vmatrix} = \left|\begin{matrix} 6 & -2 & -1 \\ 0 & -9 & 4 \\ -3 & 5 & 1 \end{matrix}\right|\begin{matrix} 6 & -2 \\ 0 & -9 \\ -3 & 5 \end{matrix}$

$$= (-54) + 24 + 0 - (-27) - 120 - 0 = -123$$

5) Expanding along a row or column containing a zero will shorten the work. Expanding along the third column:

$$\begin{vmatrix} 1 & -3 & 2 \\ -5 & 2 & 0 \\ 4 & -1 & 3 \end{vmatrix} = 2\begin{vmatrix} -5 & 2 \\ 4 & -1 \end{vmatrix} - 0 + 3\begin{vmatrix} 1 & -3 \\ -5 & 2 \end{vmatrix} = 2(-3) + 3(-13) = -45$$

9) $\begin{vmatrix} -1 & x & -1 \\ x & -3 & 0 \\ -3 & 5 & -1 \end{vmatrix} = 0 = -1\begin{vmatrix} x & -3 \\ -3 & 5 \end{vmatrix} + 0 + (-1)\begin{vmatrix} -1 & x \\ x & -3 \end{vmatrix}$

$$0 = -(5x - 9) - (3 - x^2) = -5x + 9 - 3 + x^2$$

$$= x^2 + 5x + 6 = (x - 2)(x - 3)$$

$$x - 2 = 0 \quad \text{or} \quad x - 3 = 0$$

$$x = 2 \qquad\qquad x = 3$$

13) $D_x = \begin{vmatrix} 2 & -1 & 4 \\ 0 & 3 & -7 \\ 12 & -4 & 4 \end{vmatrix} = 2\begin{vmatrix} 3 & -7 \\ -4 & 4 \end{vmatrix} - 0 + 12\begin{vmatrix} -1 & 4 \\ 3 & -7 \end{vmatrix}$

$$= 2(-16) + 12(-5) = -92$$

17) $D = \begin{vmatrix} 1 & 2 & 3 \\ 3 & -1 & 0 \\ -4 & 0 & 1 \end{vmatrix} = 3\begin{vmatrix} 3 & -1 \\ -4 & 0 \end{vmatrix} - 0 + 1\begin{vmatrix} 1 & 2 \\ 3 & -1 \end{vmatrix}$

$$= 3(-4) + 1(-7) = -19$$

17) continued

$$D_x = \begin{vmatrix} 5 & 2 & 3 \\ -3 & -1 & 0 \\ 6 & 0 & 1 \end{vmatrix} = 3\begin{vmatrix} -3 & -1 \\ 6 & 0 \end{vmatrix} - 0 + 1\begin{vmatrix} 5 & 2 \\ -3 & -1 \end{vmatrix}$$

$$= 3(6) + 1(1) = 19$$

$$D_y = \begin{vmatrix} 1 & 5 & 3 \\ 3 & -3 & 0 \\ -4 & 6 & 1 \end{vmatrix} = 3\begin{vmatrix} 3 & -3 \\ -4 & 6 \end{vmatrix} - 0 + 1\begin{vmatrix} 1 & 5 \\ 3 & -3 \end{vmatrix}$$

$$= 3(6) + 1(-18) = 0$$

$$D_z = \begin{vmatrix} 1 & 2 & 5 \\ 3 & -1 & -3 \\ -4 & 0 & 6 \end{vmatrix} = - 4\begin{vmatrix} 2 & 5 \\ -1 & -3 \end{vmatrix} - 0 + 6\begin{vmatrix} 1 & 2 \\ 3 & -1 \end{vmatrix}$$

$$= -4(-1) + 6(-7) = -38$$

$$x = \frac{D_x}{D} = \frac{19}{-19} = -1 \qquad y = \frac{D_y}{D} = \frac{0}{-19} = 0 \qquad z = \frac{D_z}{D} = \frac{-38}{-19} = 2$$

Solution: (-1, 0, 2)

21)

$$D = \begin{vmatrix} 6 & 3 & -4 \\ \frac{3}{2} & 1 & -4 \\ 3 & -1 & 8 \end{vmatrix} = 6\begin{vmatrix} 1 & -4 \\ -1 & 8 \end{vmatrix} - \frac{3}{2}\begin{vmatrix} 3 & -4 \\ -1 & 8 \end{vmatrix} + 3\begin{vmatrix} 3 & -4 \\ 1 & -4 \end{vmatrix}$$

$$= 6(4) - \frac{3}{2}(20) + 3(-8) = -30$$

$$D_x = \begin{vmatrix} 5 & 3 & -4 \\ 0 & 1 & -4 \\ 5 & -1 & 8 \end{vmatrix} = 5\begin{vmatrix} 1 & -4 \\ -1 & 8 \end{vmatrix} - 0 + 5\begin{vmatrix} 3 & -4 \\ 1 & -4 \end{vmatrix}$$

$$= 5(4) + 5(-8) = -20$$

$$D_y = \begin{vmatrix} 6 & 5 & -4 \\ \frac{3}{2} & 0 & -4 \\ 3 & 5 & 8 \end{vmatrix} = -5\begin{vmatrix} \frac{3}{2} & -4 \\ 3 & 8 \end{vmatrix} + 0 - 5\begin{vmatrix} 6 & -4 \\ \frac{3}{2} & -4 \end{vmatrix}$$

$$= -5(24) - 5(-18) = -30$$

21) continued

$$D_z = \begin{vmatrix} 6 & 3 & 5 \\ \frac{3}{2} & 1 & 0 \\ 3 & -1 & 5 \end{vmatrix} = 5\begin{vmatrix} \frac{3}{2} & 1 \\ 3 & -1 \end{vmatrix} - 0 + 5\begin{vmatrix} 6 & 3 \\ \frac{3}{2} & 1 \end{vmatrix}$$

$$= 5\left(-\frac{9}{2}\right) + 5\left(\frac{3}{2}\right) = -15$$

$$x = \frac{D_x}{D} = \frac{-20}{-30} = \frac{2}{3} \qquad y = \frac{D_y}{D} = \frac{-30}{-30} = 1 \qquad z = \frac{D_z}{D} = \frac{-15}{-30} = \frac{1}{2}$$

Solution: $\left(\frac{2}{3}, 1, \frac{1}{2}\right)$

25) $$\begin{vmatrix} c_1 & b_1 & a_1 \\ c_2 & b_2 & a_2 \\ c_3 & b_3 & a_3 \end{vmatrix} = c_1\begin{vmatrix} b_2 & a_2 \\ b_3 & a_3 \end{vmatrix} - c_2\begin{vmatrix} b_1 & a_1 \\ b_3 & a_3 \end{vmatrix} + c_3\begin{vmatrix} b_1 & a_1 \\ b_2 & a_2 \end{vmatrix}$$

$$= c_1b_2a_3 - c_1b_3a_2 - c_2b_1a_3 + c_2b_3a_1 + c_3b_1a_2 - c_3b_2a_1$$

$$= a_3b_2c_1 - a_2b_3c_1 - a_3b_1c_2 + a_1b_3c_2 + a_2b_1c_3 - a_1b_2c_3$$

$$= -a_1b_2c_3 + a_1b_3c_2 + a_2b_1c_3 - a_2b_3c_1 - a_3b_1c_2 + a_3b_2c_1$$

$$= -(a_1b_2c_3 - a_1b_3c_2 - a_2b_1c_3 + a_2b_3c_1 + a_3b_1c_2 - a_3b_2c_1)$$

$$= -\left(a_1\begin{vmatrix} b_2 & c_2 \\ b_3 & c_3 \end{vmatrix} - a_2\begin{vmatrix} b_1 & c_1 \\ b_3 & c_3 \end{vmatrix} + a_3\begin{vmatrix} b_1 & c_1 \\ b_2 & c_2 \end{vmatrix}\right)$$

$$= -\begin{vmatrix} a_1 & b_1 & c_1 \\ a_2 & b_2 & c_2 \\ a_3 & b_3 & c_3 \end{vmatrix}$$

29) $\begin{vmatrix} 7 & 2 & -14 \\ 0 & 6 & 0 \\ -3 & 1 & 6 \end{vmatrix} = 0$ because the third column is -2 times the first column.

33) $\begin{vmatrix} 5 & -4 & 3 \\ -6 & 6 & 2 \\ -7 & 3 & 4 \end{vmatrix} = \begin{vmatrix} 1 & -4 & 3 \\ 0 & 6 & 2 \\ -4 & 3 & 4 \end{vmatrix}$ ← c2 + c1

$= \begin{vmatrix} 1 & -4 & 3 \\ 0 & 6 & 2 \\ 0 & -13 & 16 \end{vmatrix}$ ← 4(R1) + R3

$= 1\begin{vmatrix} 6 & 2 \\ -13 & 16 \end{vmatrix}$ Expansion along column 1.

$= 2\begin{vmatrix} 3 & 1 \\ -13 & 16 \end{vmatrix} = 2(61) = 122$

37) $2\begin{vmatrix} 0 & 5 & -3 \\ 2 & -4 & 6 \\ 3 & 0 & 1 \end{vmatrix} - (-1)\begin{vmatrix} 1 & 5 & -3 \\ 0 & -4 & 6 \\ -5 & 0 & 1 \end{vmatrix} + 3\begin{vmatrix} 1 & 0 & -3 \\ 0 & 2 & 6 \\ -5 & 3 & 1 \end{vmatrix} + 0$

Find the value of each term.

$2\begin{vmatrix} 0 & 5 & -3 \\ 2 & -4 & 6 \\ 3 & 0 & 1 \end{vmatrix} = 2\left(0 - 2\begin{vmatrix} 5 & -3 \\ 0 & 1 \end{vmatrix} + 3\begin{vmatrix} 5 & -3 \\ -4 & 6 \end{vmatrix}\right)$

$= 2(-2(5) + 3(18)) = 2(44) = 88$

$1\begin{vmatrix} 1 & 5 & -3 \\ 0 & -4 & 6 \\ -5 & 0 & 1 \end{vmatrix} = 1\left(1\begin{vmatrix} -4 & 6 \\ 0 & 1 \end{vmatrix} - 0 - 5\begin{vmatrix} 5 & -3 \\ -4 & 6 \end{vmatrix}\right)$

$= 1(1(-4) - 5(18)) = -94$

$3\begin{vmatrix} 1 & 0 & -3 \\ 0 & 2 & 6 \\ -5 & 3 & 1 \end{vmatrix} = 3\left(1\begin{vmatrix} 2 & 6 \\ 3 & 1 \end{vmatrix} - 0 - 5\begin{vmatrix} 0 & -3 \\ 2 & 6 \end{vmatrix}\right)$

$= 3(1(-16) - 0 - 5(6)) = 3(-46) = -138$

Add these three values:

88 - 94 - 138 = -144

Section 11.8 Systems of Linear Inequalities

1) $y \geq x - 2$

First graph $y = x - 2$. Then shade the region with <u>greater or equal</u> y-values, the points on or above the line.

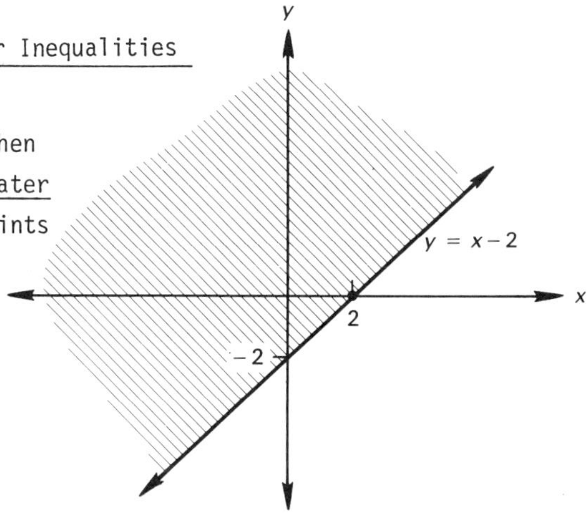

5) $x + y \leq 4$

$y \leq -x + 4$

First graph $y = -x + 4$. Then shade the region with <u>smaller or equal</u> y-values, the points on and below the line.

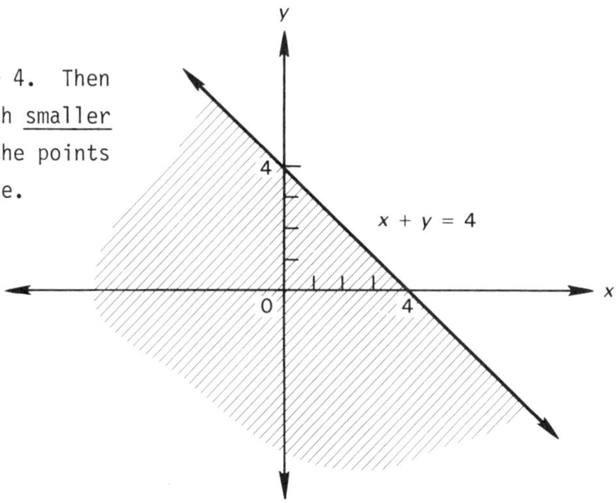

9) $|x| < 2$

$-2 < x < 2$

First graph $x = 2$ and $x = -2$. The graph is the region with x-values between -2 and 2 and y-values any number.

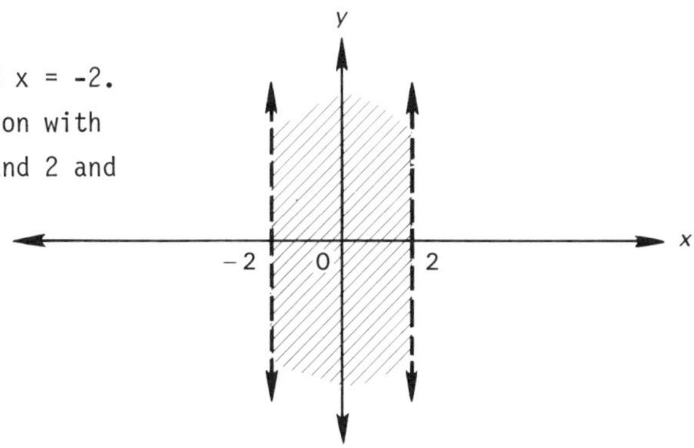

13) $y \geq x + 2$

$y \leq -x + 1$

First graph $y = x + 2$ with a broken line and shade the region above the line with vertical strokes. Next graph $y = -x + 1$ with a broken line and shade the region below the line with horizontal strokes. The solution consists of the points in the double-shaded region and the parts of the line making the boundary of this region.

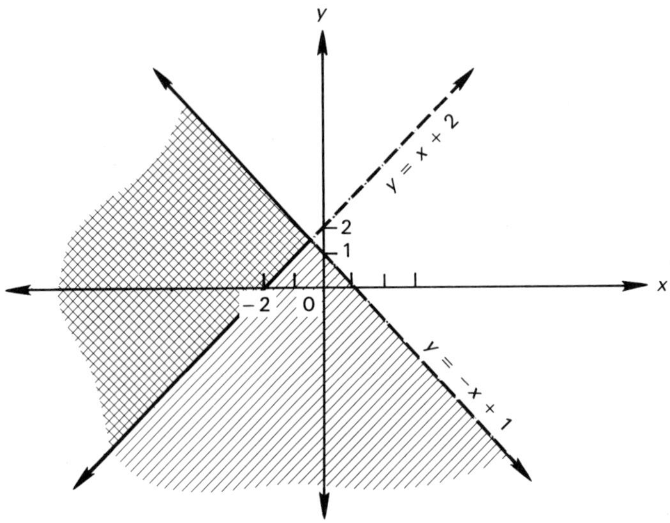

17) $3x + y < 6$

$y < -3x + 6$

$x > 1$

First graph $y = -3x + 6$ with a broken line and shade the region below the line with vertical strokes. Next graph $x = 1$ with a broken line and shade the region to the right with horizontal strokes. The solution consists of all points in the double-shaded region.

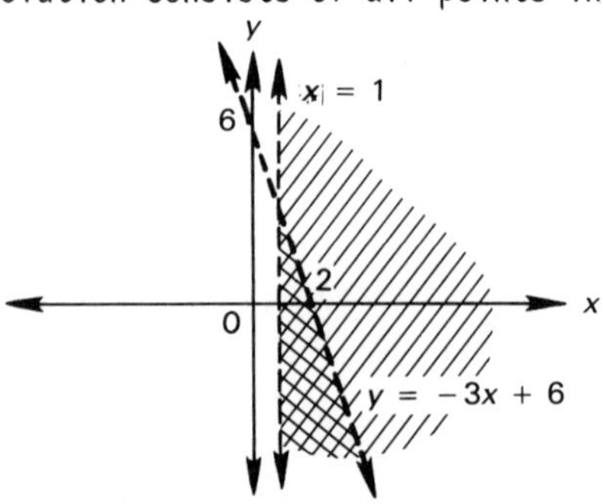

21) $x + 2y \le 10$ $\qquad$ $3x + 2y \ge 18$ $\qquad$ $y \ge 0$

$y \le -\frac{1}{2}x + 5$ $\qquad$ $y \ge -\frac{3}{2}x + 9$ $\qquad$ $y \ge 0$

The third inequality restricts the solution to points above or on the x-axis

Graph $y = -\frac{1}{2}x + 5$ with a broken line and shade the region below this line and above the x-axis with vertical strokes.

Graph $y = -\frac{3}{2}x + 9$ with a broken line and shade the region above this line and above the x-axis with horizontal strokes.

The solution consists of the points in the double-shaded region, including the boundaries.

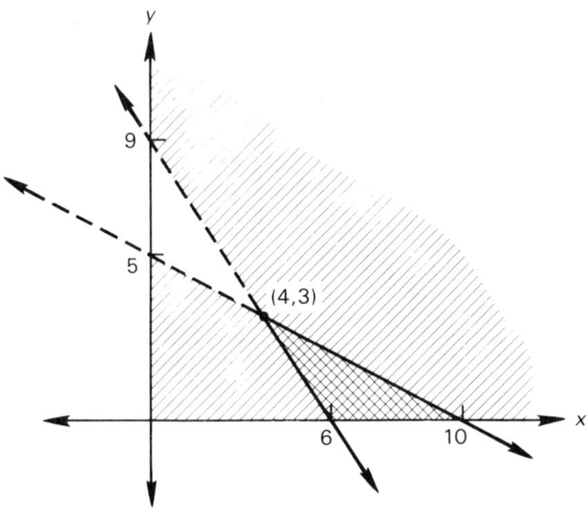

25) $x - y \ge -1$ $\qquad$ $2x + y \le 7$ $\qquad$ $4x - y \le 11$ $\qquad$ $x \ge 0$ $\qquad$ $y \ge 0$

$y \le x + 1$ $\qquad$ $y \le -2x + 7$ $\qquad$ $y \ge 4x - 11$ $\qquad$ $x \ge 0$ $\qquad$ $y \ge 0$

The fourth and fifth inequalities restrict the solution to the first quadrant, including the non-negative parts of the two axes.

Graph $y = x + 1$ and shade the region under the line and in the first quadrant with vertical strokes.

Graph $y = -2x + 7$ and shade the region under the line and in the first quadrant with horizontal strokes.

25) continued

Graph $y = 4x - 11$ and shade the region above the line and in the first quadrant with diagonal strokes.

The solution consists of the triple-shaded region and its boundaries.

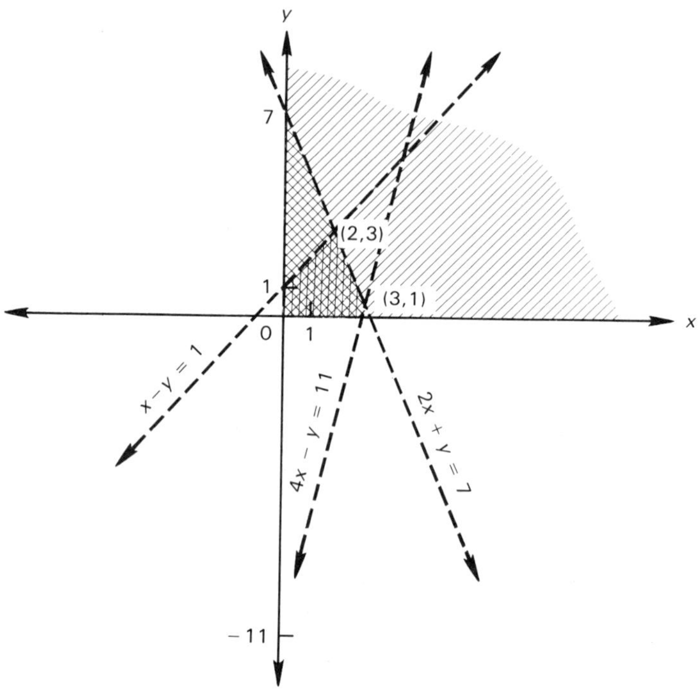

Section 11.9 | Linear Programming

1) $x + 3y \leq 6$ $\quad x \geq 0 \quad$ $y \geq 0$

$y \leq -\frac{1}{3}x + 2$

a) The graph consists of the region in the first quadrant under the line $y = -\frac{1}{3}x + 2$, including the boundaries of the region.

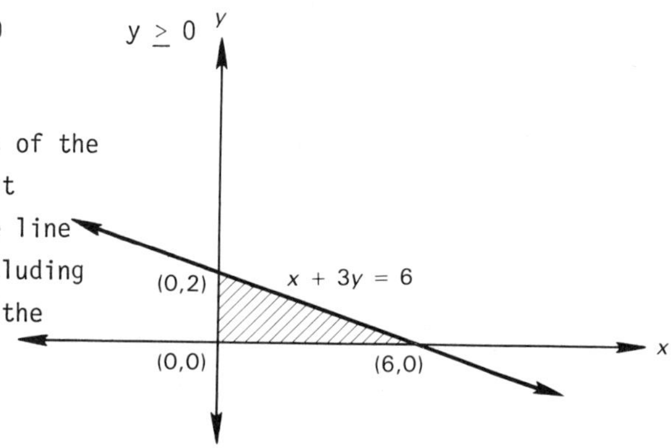

1) continued

b) The vertices are the intercepts (0, 2) and (6, 0) and the origin.

c)

Vertex	$p = x + y$
(0, 0)	$p = 0 + 0 = 0$
(0, 2)	$p = 0 + 2 = 2$
(6, 0)	$p = 6 + 0 = 6$

Maximum 6

Minimum 0

d)

Vertex	$q = 6x + 10y$
(0, 0)	$q = 6 \cdot 0 + 10 \cdot 0 = 0$
(0, 2)	$q = 6 \cdot 0 + 10 \cdot 20 = 20$
(6, 0)	$q = 6 \cdot 6 + 10 \cdot 0 = 36$

Maximum: 36

Minimum: 0

e)

Vertex	$r = 2x + 9y$
(0, 0)	$r = 2 \cdot 0 + 9 \cdot 0 = 0$
(0, 2)	$r = 2 \cdot 0 + 9 \cdot 2 = 18$
(6, 0)	$r = 2 \cdot 6 + 9 \cdot 0 = 12$

Maximum: 18

Minimum: 0

5)

$y - x \leq 3$	$y \leq x + 3$
$x + 4y \leq 22$	$y \leq -\frac{1}{4}x + \frac{11}{2}$
$2x + y \leq 16$	$y \leq -2x + 16$
$x - 3y \leq 1$	$y \geq \frac{1}{3}x - \frac{1}{3}$
$x + 4y \geq 8$	$y \geq -\frac{1}{4}x + 2$
$x \geq 0$	$x \geq 0$

a)

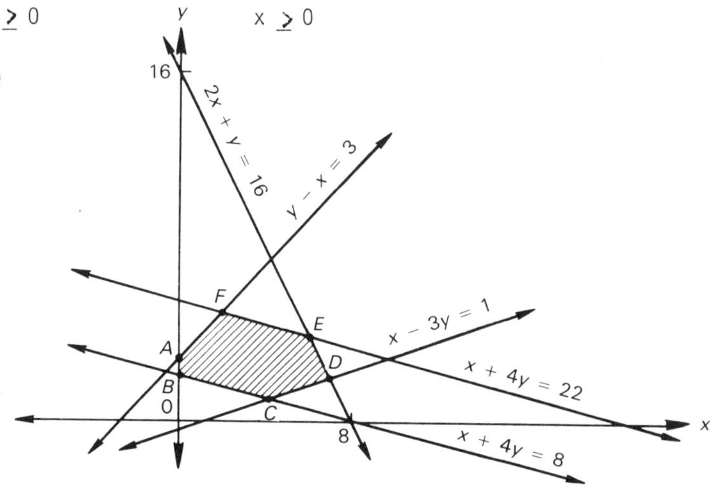

5) continued

b) The vertices are the points satisfying these pairs of equations:

A. $y = x + 3$

$x = 0$

$(0, 3)$

B. $y = -\frac{1}{4}x + 2$

$x = 0$

$(0, 2)$

C. $y = -\frac{1}{4}x + 2$

$y = \frac{1}{3}x - \frac{1}{3}$

$$\begin{aligned} 4y &= -x + 8 \\ \underline{3y} &\underline{= x - 1} \\ 7y &= 7 \\ y &= 1 \end{aligned}$$

$$\begin{aligned} (1) &= \frac{1}{3}x - \frac{1}{3} \\ \frac{4}{3} &= \frac{1}{3}x \\ 4 &= x \end{aligned}$$

$(4, 1)$

D. $y = \frac{1}{3}x - \frac{1}{3}$

$y = -2x + 16$

$$\begin{aligned} 6y &= 2x - 2 \\ \underline{y} &\underline{= -2x + 16} \\ 7y &= 14 \\ y &= 2 \end{aligned}$$

$$\begin{aligned} (2) &= -2x + 16 \\ -14 &= -2x \\ 7 &= x \end{aligned}$$

$(7, 2)$

E. $y = -\frac{1}{4}x + \frac{11}{2}$

$y = -2x + 16$

$$\begin{aligned} -8y &= 2x - 44 \\ \underline{y} &\underline{= -2x + 16} \\ -7y &= -28 \\ y &= 4 \end{aligned}$$

$$\begin{aligned} (4) &= -2x + 16 \\ -12 &= -2x \\ 6 &= x \end{aligned}$$

$(6, 4)$

5) continued

b) continued

F. $y = x + 3$

$y = -\frac{1}{4}x + \frac{11}{2}$

$$\begin{aligned} y &= x + 3 \\ 4y &= -x + 22 \\ \hline 5y &= 25 \\ y &= 5 \end{aligned}$$

$(5) = x + 3$

$2 = x$

$(2, 5)$

c)

Vertex	$p = x + y$
(0, 3)	$p = 0 + 3 = 3$
(0, 2)	$p = 0 + 2 = 2$
(4, 1)	$p = 4 + 1 = 5$
(7, 2)	$p = 7 + 2 = 9$
(6, 4)	$p = 6 + 4 = 10$
(2, 5)	$p = 2 + 5 = 7$

Maximum: 10

Minimum: 2

d)

Vertex	$q = 6x + 10y$
(0, 3)	$q = 6 \cdot 0 + 10 \cdot 3 = 30$
(0, 2)	$q = 6 \cdot 0 + 10 \cdot 2 = 20$
(4, 1)	$q = 6 \cdot 4 + 10 \cdot 1 = 34$
(7, 2)	$q = 6 \cdot 7 + 10 \cdot 2 = 62$
(6, 4)	$q = 6 \cdot 6 + 10 \cdot 4 = 76$
(2, 5)	$q = 6 \cdot 2 + 10 \cdot 5 = 72$

Maximum: 76

Minimum: 20

e)

Vertex	$r = 2x + 9y$
(0, 3)	$r = 2 \cdot 0 + 9 \cdot 3 = 27$
(0, 2)	$r = 2 \cdot 0 + 9 \cdot 2 = 18$
(4, 1)	$r = 2 \cdot 4 + 9 \cdot 1 = 17$
(7, 2)	$r = 2 \cdot 7 + 9 \cdot 2 = 32$
(6, 4)	$r = 2 \cdot 6 + 9 \cdot 4 = 48$
(2, 5)	$r = 2 \cdot 2 + 9 \cdot 5 = 49$

Maximum: 49

Minimum: 17

9) a) $x \geq 0$ and $y \geq 0$ because a negative number of either model is impossible.

$\frac{3}{2}x + y \leq 12$ is the time machine M_1 works in a day.

$\frac{3}{4}x + \frac{3}{2}y \leq 12$ is the time machine M_2 works in a day.

$\frac{4}{3}x + \frac{4}{3}y \leq 12$ is the time machine M_3 works in a day.

9) continued

b) 5x is the profit on all model I's made

8y is the profit on the model II's

$P = 5x + 8y$

c) Solve inequalities 3, 4, 5 for y:

$\frac{3}{2}x + y \leq 12$

$y \leq -\frac{3}{2}x + 12$

$\frac{3}{4}x + \frac{3}{2}y \leq 12$

$\frac{3}{2}y \leq -\frac{3}{4}x + 12$

$y \leq -\frac{1}{2}x + 8$

$\frac{4}{3}x + \frac{4}{3}y \leq 12$

$y \leq -x + 9$

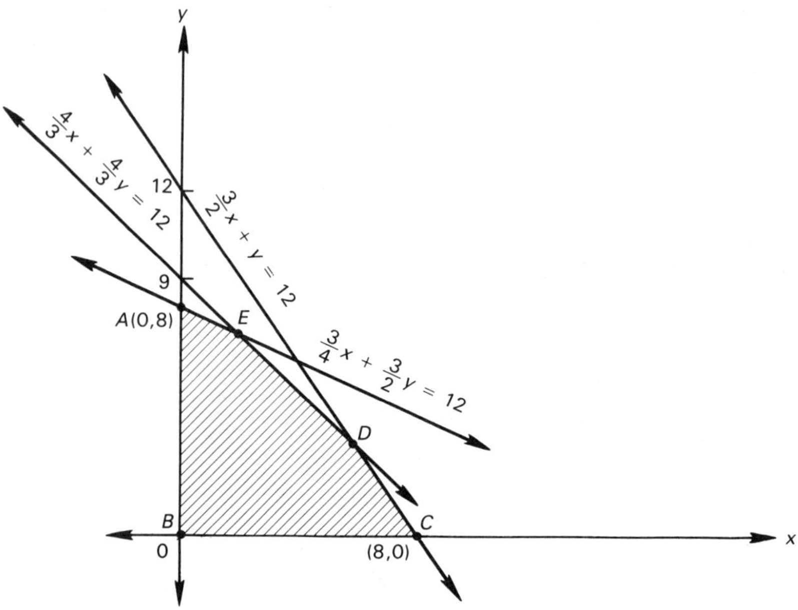

The vertices are the points satisfying these pairs of equations:

A. $y = -\frac{1}{2}x + 8$

$x = 0$

(0, 8)

B. $x = 0$

$y = 0$

(0, 0)

C. $y = -\frac{3}{2}x + 12$

$y = 0$

(8, 0)

9) continued

c) continued

D. $y = -\frac{3}{2}x + 12$

$y = -x + 9$

$$\begin{array}{rcl} -\frac{2}{3}y & = & x - 8 \\ \underline{y} & \underline{=} & \underline{-x + 9} \\ \frac{1}{3}y & = & 1 \\ y & = & 3 \end{array}$$

$(3) = -x + 9$

$6 = x$

$(6, 3)$

E. $y = -x + 9$

$y = -\frac{1}{2}x + 8$

$$\begin{array}{rcl} y & = & -x + 9 \\ \underline{-2y} & \underline{=} & \underline{x - 16} \\ -y & = & -7 \\ y & = & 7 \end{array}$$

$(7) = -x + 9$

$2 = x$

$(2, 7)$

d)

Vertex	$p = 5x + 8y$
(0, 8)	$p = 5 \cdot 0 + 8 \cdot 8 = 64$
(0, 0)	$p = 5 \cdot 0 + 8 \cdot 0 = 0$
(8, 0)	$p = 5 \cdot 8 + 8 \cdot 0 = 40$
(6, 3)	$p = 5 \cdot 6 + 8 \cdot 3 = 54$
(2, 7)	$p = 5 \cdot 2 + 8 \cdot 7 = 66$

The maximum profit is $66 when 2 model I's and 7 model II's are made.

e)

Vertex	$p = 8x + 5y$
(0, 8)	$p = 8 \cdot 0 + 5 \cdot 8 = 40$
(0, 0)	$p = 8 \cdot 0 + 5 \cdot 0 = 0$
(8, 0)	$p = 8 \cdot 8 + 5 \cdot 0 = 64$
(6, 3)	$p = 8 \cdot 6 + 5 \cdot 3 = 63$
(2, 7)	$p = 8 \cdot 2 + 5 \cdot 7 = 51$

The maximum profit is $64 when 8 model I's and no model II's are made.

Sample Test Questions for Chapter 11

Part I

1. $2x - y = 7$
$-3x + 2y = -11$

$$\begin{array}{r} 4x - 2y = 14 \\ \underline{-3x + 2y = -11} \\ x \quad = 3 \end{array}$$

In equation 1, let $x = 3$.

$$\begin{aligned} 2(3) - y &= 7 \\ -y &= 6 \\ y &= -1 \end{aligned}$$

Solution: $(3, -1)$

2. $2x + 3y - z = -1$
$3x - y + 2z = 5$
$-3x + 4y - 2z = 1$

(Eq. 1) $2x + 3y - z = -1$
(Eq. 2) $3x - y + 2z = 5$

$$\begin{array}{r} 4x + 6y - 2z = -2 \\ \underline{3x - y + 2z = 5} \\ \text{(Eq. 4) } 7x + 5y \quad = 3 \end{array}$$

(Eq. 1) $2x + 3y - z = -1$
(Eq. 3) $-3x + 4y - 2z = 1$

$$\begin{array}{r} -4x - 6y + 2z = 2 \\ \underline{-3x + 4y - 2z = 1} \\ \text{(Eq. 5) } -7x - 2y \quad = 3 \end{array}$$

$$\begin{array}{r} \text{(Eq. 4) } 7x + 5y = 3 \\ \text{(Eq. 5) } \underline{-7x - 2y = 3} \\ 3y = 6 \\ y = 2 \end{array}$$

In equation 4, let $y = 2$.

$$\begin{aligned} 7x + 5(2) &= 3 \\ 7x &= -7 \\ x &= -1 \end{aligned}$$

In equation 1, let $x = -1$ and $y = 2$.

$$\begin{aligned} 2(-1) + 3(2) - z &= -1 \\ -z &= -5 \\ z &= 5 \end{aligned}$$

Solution: $(-1, 2, 5)$

3. $4x - 5y = 12$
$-2x + \frac{5}{2}y = -6$

$$\begin{array}{r} 4x - 5y = 12 \\ \underline{-4x + 5y = -12} \\ 0 = 0 \end{array}$$

dependent

4. $7x + 4y = 1$ 　　　　 $7x + 4y = 1$

$-3x + 2y = -19$ 　　　　 $6x - 4y = 38$

$13x = 39$ 　 consistent

$x = 3$

In equation 1, let $x = 3$.

$$7(3) + 4y = 1$$
$$4y = -20$$
$$y = -5$$

Solution: $(3, -5)$

5. $-6x + 4y = -24$ (multiply by $\frac{3}{2}$) $-9x + 6y = -36$

$9x - 6y = 14$ 　　　　 $9x - 6y = 14$

$0 \neq -22$

inconsistent

6. Let x represent the number of 22¢ letters.

Let y represent the number of 39¢ letters.

22x is the cost of mailing all the 22¢ letters.

39y is the cost of mailing all the 39¢ letters.

Total number: $x + y = 40$

Total cost (change \$11.35 to cents): $22x + 39y = 1135$

$x + y = 40$ 　　　　 $-22x - 22y = -880$

$22x + 39y = 1135$ 　　　　 $22x + 39y = 1135$

$17y = 225$

$y = 15$

In equation 1, let $y = 15$.

$$x + (15) = 40$$
$$x = 25$$

The store mailed 25 letters at 22¢ and 15 letters at 39¢.

7. Let x represent the rate of the Porsche in kph.

Let y represent the rate of the Mercedes.

Time for both cars is $\frac{1}{6}$ hours. Using distance = rate x time, represent the distance each car travels.

$\frac{1}{6}x$ is the distance the Porsche travels.

$\frac{1}{6}y$ is the distance the Mercedes travels.

7. continued

Relationship of two rates: $y = \frac{4}{5}x$

Total distance $\frac{1}{6}x + \frac{1}{6}y = 60$

In the second equation, substitute $\frac{4}{5}x$ for y.

$$\frac{1}{6}x + \frac{1}{6}\left(\frac{4}{5}x\right) = 60$$

$$\frac{1}{6}x + \frac{4}{30}x = 60$$

$$5x + 4x = 1800$$

$$9x = 1800$$

$$x = 200$$

The Porsche is traveling 200 kph.

8. Let x represent the number of items made in a day.

At the break-even point the daily cost and daily revenue are the same. Represent this amount with y.

Daily cost: $y = 2.50x + 450$

Daily revenue: $y = 3.10x$

Set the right sides equal and solve for x.

$$2.50x + 450 = 3.10x$$

$$450 = 0.60x$$

$$750 = x$$

In equation 2, let x = 7.

$$y = 3.10(750) = \$2325$$

The break-even point is (750, 2325).

9. Let x, y, z represent the number of ounces of foods I, II, III respectively.

Total of ingredient A: $25x + 35y + 40z = 550$

Total of ingredient B: $20x + 15y + 20z = 300$

Total of ingredient C: $15x + 25y + 20z = 350$

(Eq. 1) $25x + 35y + 40z = 550$

(Eq. 2) $20x + 15y + 20z = 300$

$$\begin{array}{r} 25x + 35y + 40z = 550 \\ -40x - 30y - 40z = -600 \\ \hline -15x + 5y \qquad = -50 \end{array}$$

(Eq. 4) $3x - y = 10$

9. continued

$$\begin{array}{lrcl} \text{(Eq. 2)} & 20x + 15y + 20z &=& 300 \\ -\text{(Eq. 3)} & \underline{-15x - 25y - 20z} &\underline{=}& \underline{-350} \\ & 5x - 10y &=& -50 \end{array}$$

(Eq. 5) $x - 2y = -10$

(Eq. 4) $3x - y = 10$ $\qquad 6x - 2y = 20$

(Eq. 5) $x - 2y = -10$ $\qquad \underline{-x + 2y = 10}$

$$5x = 30$$

$$x = 6$$

In equation 5, let x = 6.

$$(6) - 2y = -10$$
$$-2y = -16$$
$$y = 8$$

Divide equation 3 by 5 and let x = 6 and y = 8.

$$3x + 5y + 4z = 70$$
$$3(6) + 5(8) + 4z = 70$$
$$18 + 40 + 4z = 70$$
$$4z = 12$$
$$z = 3$$

The mixture should contain 6 oz. of food I, 8 oz. of food II, and 3 oz. of food III.

10. $2x + 5y = 4$

$4x - 3y = -18$

$$\left[\begin{array}{cc|c} 2 & 5 & 4 \\ 4 & -3 & -18 \end{array}\right]$$

$$\left[\begin{array}{cc|c} 2 & 5 & 4 \\ 0 & -13 & -26 \end{array}\right] \leftarrow -2(R1) + R2$$

$$\left[\begin{array}{cc|c} 2 & 5 & 4 \\ 0 & 1 & 2 \end{array}\right] \leftarrow -\frac{1}{13}(R2)$$

$$\left[\begin{array}{cc|c} 2 & 0 & -6 \\ 0 & 1 & 2 \end{array}\right] \leftarrow -3(R2) + R1$$

$$\left[\begin{array}{cc|c} 1 & 0 & -3 \\ 0 & 1 & 2 \end{array}\right] \leftarrow \frac{1}{2} R1$$

x = -3

y = 2

Solution: (-3, 2)

11. $-6x + 3y = 9$
$10x - 5y = -12$

$$\left[\begin{array}{cc|c} -6 & 3 & 9 \\ 10 & -5 & -12 \end{array}\right]$$

$$\left[\begin{array}{cc|c} 2 & -1 & -3 \\ 2 & -1 & -\frac{12}{5} \end{array}\right] \begin{array}{l} \leftarrow -\frac{1}{3}R1 \\ \leftarrow \frac{1}{5}R2 \end{array}$$

$$\left[\begin{array}{cc|c} 2 & -1 & -3 \\ 0 & 0 & \frac{3}{5} \end{array}\right] \begin{array}{l} \\ \leftarrow -R1 + R2 \end{array}$$

No solutions.

12. $2x - y + 2z = -3$
$x + 4y - 3z = 18$
$-4x + 2y - 3z = 0$

$$\left[\begin{array}{ccc|c} 2 & -1 & 2 & -3 \\ 1 & 4 & -3 & 18 \\ -4 & 2 & -3 & 0 \end{array}\right]$$

$$\left[\begin{array}{ccc|c} 1 & -5 & 5 & -21 \\ 1 & 4 & -3 & 18 \\ -4 & 2 & -3 & 0 \end{array}\right] \begin{array}{l} \leftarrow -(R2) + R1 \\ \\ \\ \end{array}$$

$$\left[\begin{array}{ccc|c} 1 & -5 & 5 & -21 \\ 0 & 9 & -8 & 39 \\ 0 & -18 & 17 & -84 \end{array}\right] \begin{array}{l} \\ \leftarrow -(R1) + R2 \\ \leftarrow 4(R1) + R3 \end{array}$$

$$\left[\begin{array}{ccc|c} 1 & -5 & 5 & -21 \\ 0 & 9 & -8 & 39 \\ 0 & 0 & 1 & -6 \end{array}\right] \begin{array}{l} \\ \\ \leftarrow 2(R2) + R3 \end{array}$$

$$\left[\begin{array}{ccc|c} 1 & -5 & 5 & -21 \\ 0 & 9 & 0 & -9 \\ 0 & 0 & 1 & -6 \end{array}\right] \begin{array}{l} \\ \leftarrow 8(R3) + R2 \\ \\ \end{array}$$

$$\left[\begin{array}{ccc|c} 1 & -5 & 5 & -21 \\ 0 & 1 & 0 & -1 \\ 0 & 0 & 1 & -6 \end{array}\right] \begin{array}{l} \\ \leftarrow \frac{1}{9}R2 \\ \\ \end{array}$$

$$\left[\begin{array}{ccc|c} 1 & 0 & 5 & -26 \\ 0 & 1 & 0 & -1 \\ 0 & 0 & 1 & -6 \end{array}\right] \begin{array}{l} \leftarrow 5(R2) + R1 \\ \\ \\ \end{array}$$

$$\left[\begin{array}{ccc|c} 1 & 0 & 0 & 4 \\ 0 & 1 & 0 & -1 \\ 0 & 0 & 1 & -6 \end{array}\right] \begin{array}{l} \leftarrow -5(R3) + R1 \\ \\ \\ \end{array}$$

$x = 4 \quad y = -1 \quad z = -6$

Solution: $(4, -1, -6)$

13. $x + y + 2z = 3$

$3x + 2z = 2$

$-x + 2y + 2z = 4$

$$\left[\begin{array}{ccc|c} 1 & 1 & 2 & 3 \\ 3 & 0 & 2 & 2 \\ -1 & 2 & 2 & 4 \end{array}\right]$$

$$\left[\begin{array}{ccc|c} 1 & 1 & 2 & 3 \\ 0 & -3 & -4 & -7 \\ 0 & 3 & 4 & 7 \end{array}\right] \begin{array}{l} \\ \leftarrow 3(R1) + R2 \\ \leftarrow R1 + R2 \end{array}$$

$$\left[\begin{array}{ccc|c} 1 & 1 & 2 & 3 \\ 0 & -3 & -4 & -7 \\ 0 & 0 & 0 & 0 \end{array}\right] \begin{array}{l} \\ \\ \leftarrow R2 + R3 \end{array}$$

$x + y + 2z = 3$

$-3y - 4z = -7$

Let $z = k$, any number.

In equation 2, let $z = k$.

$$-3y - 4k = -7$$

$$3y = -4k + 7$$

$$y = -\frac{4}{3}k + \frac{7}{3}$$

In equation 1, let $z = k$ and $y = -\frac{4}{3}k + \frac{7}{3}$.

$$x + \left(-\frac{4}{3}k + \frac{7}{3}\right) + 2k = 3$$

$$x + \frac{2}{3}k + \frac{7}{3} = 3$$

$$x = -\frac{2}{3}k + \frac{2}{3}$$

Solution: $\left(-\frac{2}{3}k + \frac{2}{3},\ -\frac{4}{3}k + \frac{7}{3},\ k\right)$ for k any real number.

14. $2x + 5y = 4$

$4x - 3y = -18$

$$\begin{bmatrix} 2 & 5 \\ 4 & -3 \end{bmatrix} \begin{bmatrix} x \\ y \end{bmatrix} = \begin{bmatrix} 4 \\ -18 \end{bmatrix}$$

15. $2x - y + 2z = -3$

$x + 4y - 3z = 18$

$-4x + 2y - 3z = 0$

$$\begin{bmatrix} 2 & -1 & 2 \\ 1 & 4 & -3 \\ -4 & 2 & -3 \end{bmatrix} \begin{bmatrix} x \\ y \\ z \end{bmatrix} = \begin{bmatrix} -3 \\ 18 \\ 0 \end{bmatrix}$$

16. $\begin{bmatrix} 1 & 0 & -1 \\ 0 & -1 & 1 \\ -1 & 0 & 11 \end{bmatrix} + \begin{bmatrix} 0 & -1 & 1 \\ 6 & 2 & 3 \\ -5 & -3 & -8 \end{bmatrix} = \begin{bmatrix} 1 & -1 & 0 \\ 6 & 1 & 4 \\ -6 & -3 & 3 \end{bmatrix}$

17. Does not exist because the matrices are of different dimensions.

18. $5\begin{bmatrix} 3 & 0 \\ 1 & 1 \end{bmatrix} -2\begin{bmatrix} -5 & 7 \\ 4 & -9 \end{bmatrix} = \begin{bmatrix} 15 & 0 \\ 5 & 5 \end{bmatrix} + \begin{bmatrix} 10 & -14 \\ -8 & 18 \end{bmatrix} = \begin{bmatrix} 25 & -14 \\ -3 & 23 \end{bmatrix}$

19. $\begin{bmatrix} 2 & 0 & 3 \\ -1 & 4 & 9 \end{bmatrix} \begin{bmatrix} 1 & 0 & -1 \\ 0 & -1 & 1 \\ -1 & 0 & 11 \end{bmatrix} = \begin{bmatrix} (2 - 3) \; 0 \; (-2 + 33) \\ (-1 - 9) \; -4 \; (1 + 4 + 99) \end{bmatrix}$

$$= \begin{bmatrix} -1 & 0 & 31 \\ -10 & -4 & 104 \end{bmatrix}$$

20. $\begin{bmatrix} 7 & 8 & 9 \end{bmatrix} \begin{bmatrix} 0 \\ -1 \\ 0 \end{bmatrix} = \begin{bmatrix} -8 \end{bmatrix}$

21. $\begin{bmatrix} 2 & 0 & 3 \\ -1 & 4 & 9 \end{bmatrix} \begin{bmatrix} -1 & -1 \\ 2 & 2 \\ -3 & -3 \end{bmatrix} = \begin{bmatrix} (-2 - 9) & (-2 - 9) \\ (1 + 8 - 27) & (1 + 8 - 27) \end{bmatrix} = \begin{bmatrix} -11 & -11 \\ -18 & -18 \end{bmatrix}$

22. A(EF) Does not exist because (EF) will be 2 x 2 while A is 2 x 3.

23. $\begin{bmatrix} 3 & 0 \\ 1 & 1 \end{bmatrix} \begin{bmatrix} 3 & 0 \\ 1 & 1 \end{bmatrix} \begin{bmatrix} 3 & 0 \\ 1 & 1 \end{bmatrix} = \begin{bmatrix} 3 & 0 \\ 1 & 1 \end{bmatrix} \begin{bmatrix} (9) & (0) \\ (3 + 1) & (1) \end{bmatrix} = \begin{bmatrix} 3 & 0 \\ 1 & 1 \end{bmatrix} \begin{bmatrix} 9 & 0 \\ 4 & 1 \end{bmatrix}$

$$= \begin{bmatrix} 27 & 0 \\ (9 + 4) & 1 \end{bmatrix} = \begin{bmatrix} 27 & 0 \\ 13 & 1 \end{bmatrix}$$

24. $\begin{bmatrix} -1 \\ 1 \\ 2 \end{bmatrix} \begin{bmatrix} 2 & -2 & 4 \end{bmatrix} = \begin{bmatrix} -2 & 2 & -4 \\ 2 & -2 & 4 \\ 4 & -4 & 8 \end{bmatrix}$

25. CBDA

DA will be 3 x 3 because D has 3 rows and A 3 columns.

B(DA) will be 3 x 3 because B has 3 rows and (DA) 3 columns.

C(BDA) will be 4 x 3 because C has 4 rows and (BDA) 3 columns.

26. Does not exist.

27. $(|E|)(|F|) = \left(\begin{vmatrix} 3 & 0 \\ 1 & 1 \end{vmatrix}\right)\left(\begin{vmatrix} -5 & 7 \\ 4 & -9 \end{vmatrix}\right) = (3)(17) = 51$

28. $\begin{vmatrix} 0 & -1 & 1 \\ 6 & 2 & 3 \\ -5 & -3 & -8 \end{vmatrix} = 0 - (-1)\begin{vmatrix} 6 & 3 \\ -5 & -8 \end{vmatrix} + (1)\begin{vmatrix} 6 & 2 \\ -5 & -3 \end{vmatrix} = (-33) + (-8) = -41$

29. $(A + \frac{1}{2}B)(C - 2D) = AC + \frac{1}{2}BC - BD - 2AD$

$$AC = [11, 7, 2]\begin{bmatrix} -5 \\ -2 \\ 4 \end{bmatrix} = [\,-55 - 14 + 8\,] = [-61]$$

$$\frac{1}{2}BC = \frac{1}{2}[-2, 6, -2]\begin{bmatrix} -5 \\ -2 \\ 4 \end{bmatrix} = \frac{1}{2}[\,10 - 12 - 8\,] = \frac{1}{2}[-10] = [-5]$$

$$BD = [-2, 6, -2]\begin{bmatrix} -8 \\ 5 \\ 1 \end{bmatrix} = [\,16 + 30 - 2\,] = [44]$$

$$2AD = 2[11, 7, 2]\begin{bmatrix} -8 \\ 5 \\ 1 \end{bmatrix} = 2[\,-88 + 35 + 2\,] = 2[-51] = [-102]$$

$$AC + \frac{1}{2}BC - BD - 2AD = [-61] + [-5] - [44] - [-102] = [-8]$$

30. $\left[\begin{array}{cc|cc} 3 & 1 & 1 & 0 \\ 2 & -2 & 0 & 1 \end{array}\right]$

$\left[\begin{array}{cc|cc} 1 & 3 & 1 & -1 \\ 2 & -2 & 0 & 1 \end{array}\right] \leftarrow -(R2) + R1$

$\left[\begin{array}{cc|cc} 1 & 3 & 1 & -1 \\ 0 & -8 & -2 & 3 \end{array}\right] \leftarrow -2(R1) + R2$

$\left[\begin{array}{cc|cc} 1 & 3 & 1 & -1 \\ 0 & 1 & \frac{1}{4} & -\frac{3}{8} \end{array}\right] \leftarrow -\frac{1}{8} R2$

$\left[\begin{array}{cc|cc} 1 & 0 & \frac{1}{4} & \frac{1}{8} \\ 0 & 1 & \frac{1}{4} & -\frac{3}{8} \end{array}\right]$

$A^{-1} = \begin{bmatrix} \frac{1}{4} & \frac{1}{8} \\ \frac{1}{4} & -\frac{3}{8} \end{bmatrix}$

31. $3x + y = 9$
$2x - 2y = 14$

$A = \begin{bmatrix} 3 & 1 \\ 2 & -2 \end{bmatrix} \quad X = \begin{bmatrix} x \\ y \end{bmatrix} \quad C = \begin{bmatrix} 9 \\ 14 \end{bmatrix}$

$A^{-1} = \begin{bmatrix} \frac{1}{4} & \frac{1}{8} \\ \frac{1}{4} & -\frac{3}{8} \end{bmatrix}$

$X = A^{-1}C$

$\begin{bmatrix} x \\ y \end{bmatrix} = \begin{bmatrix} \frac{1}{4} & \frac{1}{8} \\ \frac{1}{4} & -\frac{3}{8} \end{bmatrix} \begin{bmatrix} 9 \\ 14 \end{bmatrix}$

$\begin{bmatrix} x \\ y \end{bmatrix} = \begin{bmatrix} (\frac{9}{4} + \frac{14}{8}) \\ (\frac{9}{4} - \frac{42}{8}) \end{bmatrix} = \begin{bmatrix} 4 \\ -3 \end{bmatrix}$

$x = 4 \qquad y = -3$

Solution: (4, -3)

32. $\left[\begin{array}{ccc|ccc} 1 & 0 & -2 & 1 & 0 & 0 \\ 0 & 1 & 0 & 0 & 1 & 0 \\ 3 & 2 & 0 & 0 & 0 & 1 \end{array}\right]$

$\left[\begin{array}{ccc|ccc} 1 & 0 & -2 & 1 & 0 & 0 \\ 0 & 1 & 0 & 0 & 1 & 0 \\ 0 & 2 & 6 & -3 & 0 & 1 \end{array}\right]$

$\left[\begin{array}{ccc|ccc} 1 & 0 & -2 & 1 & 0 & 0 \\ 0 & 1 & 0 & 0 & 1 & 0 \\ 0 & 0 & 6 & -3 & -2 & 1 \end{array}\right]$

$\left[\begin{array}{ccc|ccc} 1 & 0 & -2 & 1 & 0 & 0 \\ 0 & 1 & 0 & 0 & 1 & 0 \\ 0 & 0 & 1 & -\frac{1}{2} & -\frac{1}{3} & \frac{1}{6} \end{array}\right]$

$\left[\begin{array}{ccc|ccc} 1 & 0 & 0 & 0 & -\frac{2}{3} & \frac{1}{3} \\ 0 & 1 & 0 & 0 & 1 & 0 \\ 0 & 0 & 1 & -\frac{1}{2} & -\frac{1}{3} & \frac{1}{6} \end{array}\right]$

$A^{-1} = \begin{bmatrix} 0 & -\frac{2}{3} & \frac{1}{3} \\ 0 & 1 & 0 \\ -\frac{1}{2} & -\frac{1}{3} & \frac{1}{6} \end{bmatrix}$

33. $-11x + 2y + 2z = 3$

$-4x + z = -1$

$6x - y - z = -2$

$$A = \begin{bmatrix} -11 & 2 & 2 \\ -4 & 0 & 1 \\ 6 & -1 & -1 \end{bmatrix} \qquad X = \begin{bmatrix} x \\ y \\ z \end{bmatrix} \qquad C = \begin{bmatrix} 3 \\ -1 \\ -2 \end{bmatrix}$$

$$A^{-1} = \begin{bmatrix} 1 & 0 & 2 \\ 2 & -1 & 3 \\ 4 & 1 & 8 \end{bmatrix}$$

$X = A^{-1}C$

$$\begin{bmatrix} x \\ y \\ z \end{bmatrix} = \begin{bmatrix} 1 & 0 & 2 \\ 2 & -1 & 3 \\ 4 & 1 & 8 \end{bmatrix} \begin{bmatrix} 3 \\ -1 \\ -2 \end{bmatrix} = \begin{bmatrix} (3 - 4) \\ (6 + 1 - 6) \\ (12 - 1 - 16) \end{bmatrix} = \begin{bmatrix} -1 \\ 1 \\ -5 \end{bmatrix}$$

$x = -1$ $y = 1$ $z = -5$

Solution: $(-1, 1, -5)$

34. $\begin{vmatrix} 4a & b \\ -a & 3b \end{vmatrix} = 12ab - (-ab) = 13ab$

35. $\begin{vmatrix} -2 & 4 & 3 \\ 6 & -1 & -4 \\ 5 & 2 & 0 \end{vmatrix} \begin{matrix} -2 & 4 \\ 6 & -1 \\ 5 & 2 \end{matrix} = 0 - 80 + 36 - (-15) - 16 - 0 = -45$

36. $4x + 2y = 15$

$3x + 2y = -8$

$D = \begin{vmatrix} 4 & -2 \\ 3 & 2 \end{vmatrix} = 14$ $\qquad D_x = \begin{vmatrix} 15 & -2 \\ -8 & 2 \end{vmatrix} = 14$ $\qquad D_y = \begin{vmatrix} 4 & 15 \\ 3 & -8 \end{vmatrix} = -77$

$x = \dfrac{D_x}{D} = \dfrac{14}{14} = 1$ $\qquad y = \dfrac{D_y}{D} = -\dfrac{77}{14} = -\dfrac{11}{2}$

Solution: $(1, -\frac{11}{2})$

37. $2x + y - z = -3$

$x - 2y + z = 8$

$3x - y - 2z = -1$

$$D = \begin{vmatrix} 2 & 1 & -1 \\ 1 & -2 & 1 \\ 3 & -1 & -2 \end{vmatrix} = 2\begin{vmatrix} -2 & 1 \\ -1 & -2 \end{vmatrix} - 1\begin{vmatrix} 1 & 1 \\ 3 & -2 \end{vmatrix} + (-1)\begin{vmatrix} 1 & -2 \\ 3 & -1 \end{vmatrix}$$

$$= 2(5) - (-5) - (5) = 10$$

$$D_x = \begin{vmatrix} -3 & 1 & -1 \\ 8 & -2 & 1 \\ -1 & -1 & -2 \end{vmatrix} = -1\begin{vmatrix} 1 & -1 \\ -2 & 1 \end{vmatrix} - (-1)\begin{vmatrix} -3 & -1 \\ 8 & 1 \end{vmatrix} - 2\begin{vmatrix} -3 & 1 \\ 8 & -2 \end{vmatrix}$$

$$= -(-1) + (5) - 2(-2) = 10$$

$$D_y = \begin{vmatrix} 2 & -3 & -1 \\ 1 & 8 & 1 \\ 3 & -1 & -2 \end{vmatrix} = -1\begin{vmatrix} 1 & 8 \\ 3 & -1 \end{vmatrix} - 1\begin{vmatrix} 2 & -3 \\ 3 & -1 \end{vmatrix} - 2\begin{vmatrix} 2 & -3 \\ 1 & 8 \end{vmatrix}$$

$$= -(-25) - (7) - 2(19) = -20$$

$$D_z = \begin{vmatrix} 2 & 1 & -3 \\ 1 & -2 & 8 \\ 3 & -1 & -1 \end{vmatrix} = 2\begin{vmatrix} -2 & 8 \\ -1 & -1 \end{vmatrix} - 1\begin{vmatrix} 1 & -3 \\ -1 & -1 \end{vmatrix} + 3\begin{vmatrix} 1 & -3 \\ -2 & 8 \end{vmatrix}$$

$$= 2(10) - (-4) + 3(2) = 30$$

$$x = \frac{D_x}{D} = \frac{10}{10} = 1 \qquad y = \frac{D_y}{D} = -\frac{20}{10} = -2 \qquad z = \frac{D_z}{D} = \frac{30}{10} = 3$$

Solution: (1, -2, 3)

38. $x + y - 2 \leq 0$ $\qquad$ $2x - y + 2 \geq 0$

$y \leq -x + 2$ $\qquad$ $y \leq 2x + 2$

First graph $y = -x + 2$ with a broken line and shade the region below the line with vertical strokes.

Next graph $y = 2x + 2$ with a broken line and shade the area below the line with horizontal strokes.

The solution is the double-shaded region and its boundary lines.

38. continued

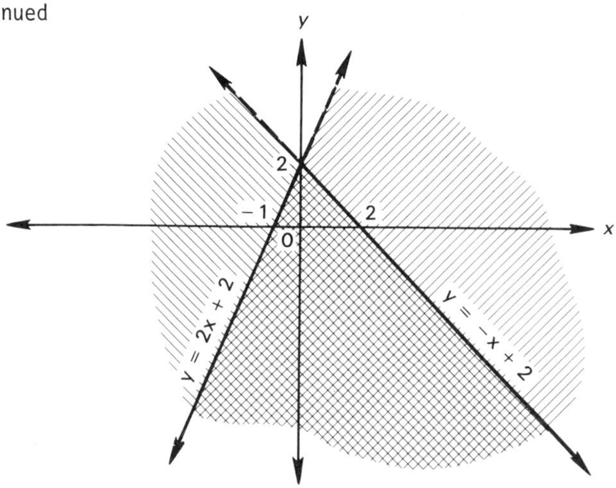

39. $x \geq 0$ $y \geq 0$ $x + 2y \leq 8$ $x + y \leq 5$ $2x + y \leq 8$

$y \leq -\frac{1}{2}x + 4$ $y \leq -x + 5$ $y \leq -2x + 8$

The first and second inequalities restrict the solution to the first quadrant, including the non-negative parts of the axes.

Graph $y = -\frac{1}{2}x + 4$ and shade the region under the line and in the first quadrant with vertical strokes.

Graph $y = -x + 5$ and shade the region under the line and in the first quadrant with horizontal strokes.

Graph $y = -2x + 8$ and shade the region under the line and in the first quadrant with diagonal strokes.

The solution is the triple-shaded region and its boundaries.

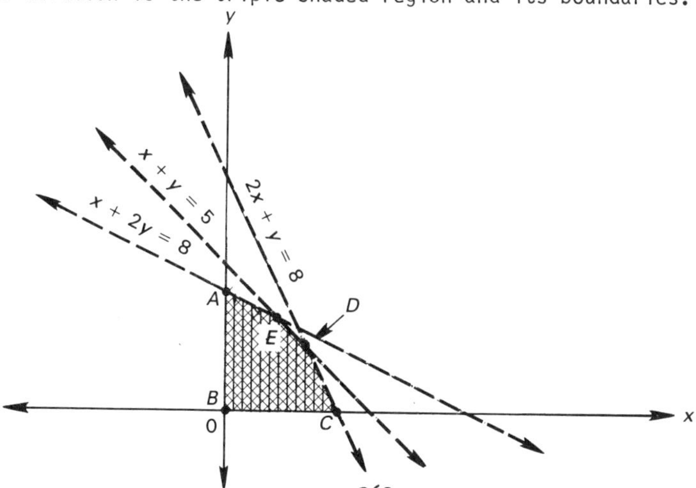

40. The maximum of $k = 3x + 2y$ will occur at one of the vertices of the figure in exercise 39.

The vertices occur at the points satisfying these pairs of equations.

A. $y = -\frac{1}{2}x + 4$

$x = 0$

$(0, 4)$

B. $x = 0$

$y = 0$

$(0, 0)$

C. $y = -2x + 8$

$y = 0$

$x = 4$

$(4, 0)$

D. $y = -2x + 8$

$y = -x + 5$

$$\begin{array}{r} -y = +2x - 8 \\ \underline{2y = -2x + 10} \\ y = 2 \end{array}$$

$(2) = -x + 5$

$x = 3$

$(3, 2)$

E. $y = -\frac{1}{2}x + 4$

$y = -x + 5$

$$\begin{array}{r} -2y = x - 8 \\ \underline{y = -x + 5} \\ -y = -3 \\ y = 3 \end{array}$$

$(3) = -x + 5$

$2 = x$

$(2, 3)$

Vertex	$k = 3x + 2y$
(0, 4)	$k = 3 \cdot 0 + 2 \cdot 4 = 8$
(0, 0)	$k = 3 \cdot 0 + 2 \cdot 0 = 0$
(4, 0)	$k = 3 \cdot 4 + 2 \cdot 0 = 12$
(3, 2)	$k = 3 \cdot 3 + 2 \cdot 2 = 13$
(2, 3)	$k = 3 \cdot 2 + 2 \cdot 3 = 12$

Maximum: 13

Sample Test Questions for Chapter 11

Part II

1. (c)

$$2x - y = 4 \qquad\qquad 2x - y = 4$$
$$x + 3y = 7 \qquad\qquad -2x - 6y = -14$$
$$-7y = -10$$
$$y = \frac{10}{7}$$

In equation 1, let $y = \frac{10}{7}$.

$$2x - \frac{10}{7} = 4$$
$$2x = \frac{38}{7}$$
$$x = \frac{14}{7}$$

Solution: $(\frac{14}{7}, \frac{10}{7})$

2. (a)

$$3x - 2y = 4 \qquad\qquad 3x - 2y = 4$$
$$6x - 4y = -5 \qquad\qquad -3x + 2y = \frac{5}{2}$$
$$0 = \frac{13}{2}$$

3. (d)

Let x represent the number of dolls made and sold.

Let y represent the total cost and total revenue at the break-even point.

$$y = 3x + 150$$
$$y = 9x$$
$$9x = 3x + 150$$
$$6x = 150$$
$$x = 25$$
$$y = 9(25) = 225$$
$$(25, 225)$$

4. (d)

5. (c)

6. (a)

$$\begin{bmatrix} -2 & 1 & 5 \\ 3 & 0 & 2 \end{bmatrix} \begin{bmatrix} 4 & -2 \\ 3 & 4 \\ 0 & -1 \end{bmatrix} = \begin{bmatrix} (-8 + 3) & (4 + 4 - 5) \\ (12) & (-6 - 2) \end{bmatrix} = \begin{bmatrix} -5 & 3 \\ 12 & -8 \end{bmatrix}$$

7. (b)

8. (d)

$$\left[\begin{array}{cc|cc} -1 & 2 & 1 & 0 \\ 2 & -3 & 0 & 1 \end{array}\right]$$

$$\left[\begin{array}{cc|cc} -1 & 2 & 1 & 0 \\ 0 & 1 & 2 & 1 \end{array}\right] \begin{array}{l} \\ \leftarrow 2(R1) + R2 \end{array}$$

$$\left[\begin{array}{cc|cc} 1 & -2 & -1 & 0 \\ 0 & 1 & 2 & 1 \end{array}\right] \begin{array}{l} \leftarrow -R1 \\ \\ \end{array}$$

$$\left[\begin{array}{cc|cc} 1 & 0 & 3 & 2 \\ 0 & 1 & 2 & 1 \end{array}\right] \begin{array}{l} \leftarrow 2(R2) + R1 \\ \\ \end{array}$$

9. (c)

Actual distances:

8x for first part

10y for second part

15 total

Distances had rates been reversed:

10x for first part

8y for second part

15 + 1.5 = 16.5 total

$8x + 10y = 15$

$10x + 8y = 16.5$

10. (d)

$$\left[\begin{array}{ccc|c} 2 & -3 & 1 & 11 \\ 3 & 1 & -1 & 9 \\ -1 & 3 & -2 & -1 \end{array}\right]$$

$$\left[\begin{array}{ccc|c} 1 & 0 & -1 & 10 \\ 3 & 1 & -1 & 9 \\ -1 & 3 & -2 & -1 \end{array}\right] \begin{array}{l} \leftarrow R3 + R1 \\ \\ \\ \end{array}$$

$$\left[\begin{array}{ccc|c} 1 & 0 & -1 & 10 \\ 0 & 10 & -7 & 6 \\ 0 & 3 & -3 & 9 \end{array}\right] \begin{array}{l} \\ \leftarrow 3(R3) + R2 \\ \leftarrow R1 + R3 \end{array}$$

$$\left[\begin{array}{ccc|c} 1 & 0 & -1 & 10 \\ 0 & 1 & 2 & -21 \\ 0 & 1 & -1 & 3 \end{array}\right] \begin{array}{l} \\ \leftarrow -3(R3) + R2 \\ \leftarrow \frac{1}{3}(R3) \end{array}$$

$$\left[\begin{array}{ccc|c} 1 & 0 & -1 & 10 \\ 0 & 1 & 2 & -21 \\ 0 & 0 & -3 & 24 \end{array}\right] \begin{array}{l} \\ \\ \leftarrow -(R2) + R3 \end{array}$$

$$\left[\begin{array}{ccc|c} 1 & 0 & -1 & 10 \\ 0 & 1 & 2 & -21 \\ 0 & 0 & 1 & -8 \end{array}\right] \begin{array}{l} \\ \\ \leftarrow \frac{1}{3}(R3) \end{array}$$

11. (c)

$$\begin{vmatrix} 1 & -2 & 3 \\ -1 & 0 & 1 \\ -2 & 3 & -1 \end{vmatrix} = -(-1)\begin{vmatrix} -2 & 3 \\ 3 & -1 \end{vmatrix} + 0 - 1\begin{vmatrix} 1 & -2 \\ -2 & 3 \end{vmatrix}$$

$$= -7 - (-1) = -6$$

12. (a)

$2x + y \leq 6$ $\quad$ $x + y \geq 4$ $\quad$ $y \geq 0$ $\quad$ $x \geq 0$

$y \leq -2 + 6$ $\quad$ $y \geq -x + 4$

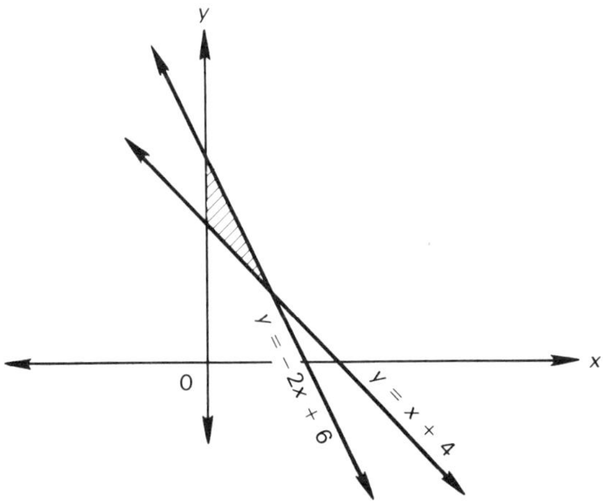

13. (e)

$$D_y = \begin{vmatrix} 2 & 5 & -3 \\ -1 & 1 & 1 \\ 3 & -3 & 2 \end{vmatrix}$$

14. (b)

$$\left[\begin{array}{ccc|ccc} 1 & 0 & 2 & 1 & 0 & 0 \\ 0 & 2 & -1 & 0 & 1 & 0 \\ 2 & 5 & 2 & 0 & 0 & 1 \end{array}\right]$$

$$\left[\begin{array}{ccc|ccc} 1 & 0 & 2 & 1 & 0 & 0 \\ 0 & 1 & -\frac{1}{2} & 0 & \frac{1}{2} & 0 \\ 0 & 5 & -2 & -2 & 0 & 1 \end{array}\right] \begin{array}{l} \\ \leftarrow \frac{1}{2}(R2) \\ \leftarrow -2(R1) + R3 \end{array}$$

14. continued

$$\left[\begin{array}{ccc|ccc} 1 & 0 & 2 & 1 & 0 & 0 \\ 0 & 1 & -\frac{1}{2} & 0 & \frac{1}{2} & 0 \\ 0 & 0 & \frac{1}{2} & -2 & -\frac{5}{2} & 1 \end{array}\right] \begin{array}{l} \\ \\ \leftarrow -5(R2) + R3 \end{array}$$

$$\left[\begin{array}{ccc|ccc} 1 & 0 & 2 & 1 & 0 & 0 \\ 0 & 1 & 0 & -2 & -2 & 1 \\ 0 & 0 & \frac{1}{2} & -2 & -\frac{5}{2} & 1 \end{array}\right] \begin{array}{l} \\ \leftarrow R3 + R2 \\ \\ \end{array}$$

$$\left[\begin{array}{ccc|ccc} 1 & 0 & 2 & 1 & 0 & -4 \\ 0 & 1 & 0 & -2 & -2 & 1 \\ 0 & 0 & 1 & -4 & -5 & 2 \end{array}\right] \begin{array}{l} \\ \\ \leftarrow 2(R3) \end{array}$$

$$\left[\begin{array}{ccc|ccc} 1 & 0 & 0 & 9 & 10 & -4 \\ 0 & 1 & 0 & -2 & -2 & 1 \\ 0 & 0 & 1 & -4 & -5 & 2 \end{array}\right] \begin{array}{l} \leftarrow -2(R3) + R1 \\ \\ \\ \end{array}$$

15. (c)

$x \geq 0$ $\quad$ $y \geq 0$ $\quad$ $x + 2y \leq 22$ $\quad$ $3x + 2y \leq 30$

$y \leq -\frac{1}{2}x + 11$ $\quad$ $y \leq -\frac{3}{2}x + 15$

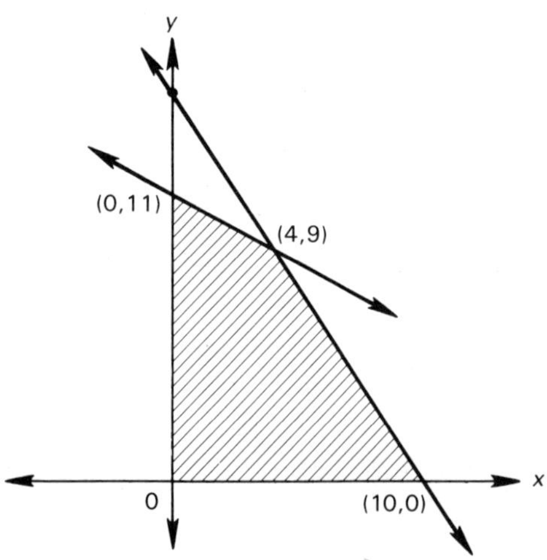

15. continued

The vertices are (0, 11), (0, 0), (10, 0) and the point of intersection of $y = -\frac{1}{2}x + 11$ and $y = -\frac{3}{2}x + 15$.

$$y = -\frac{1}{2}x + 11 \qquad\qquad y = -\frac{1}{2}x + 11$$

$$y = -\frac{3}{2}x + 15 \qquad\qquad \underline{-y = +\frac{3}{2}x - 15}$$

$$0 = x - 4$$

$$4 = x$$

$$y = -\frac{1}{2}(4) + 11$$

$$y = 9$$

The fourth vertex is (4, 9).

Vertex	$k = 4x + 3y$
(0, 11)	$k = 4 \cdot 0 + 3 \cdot 11 = 33$
(0, 0)	$k = 4 \cdot 0 + 3 \cdot 0 = 0$
(10, 0)	$k = 4 \cdot 10 + 3 \cdot 0 = 40$
(4, 9)	$k = 4 \cdot 4 + 3 \cdot 9 = 43$

CHAPTER 12

SEQUENCES AND SERIES

Section 12.1 Sequences

1) $s_n = 2n - 1$

$s_1 = 2 \cdot 1 - 1 = 1$

$s_2 = 2 \cdot 2 - 1 = 3$

$s_3 = 2 \cdot 3 - 1 = 5$

$s_4 = 2 \cdot 4 - 1 = 7$

$s_5 = 2 \cdot 5 - 1 = 9$

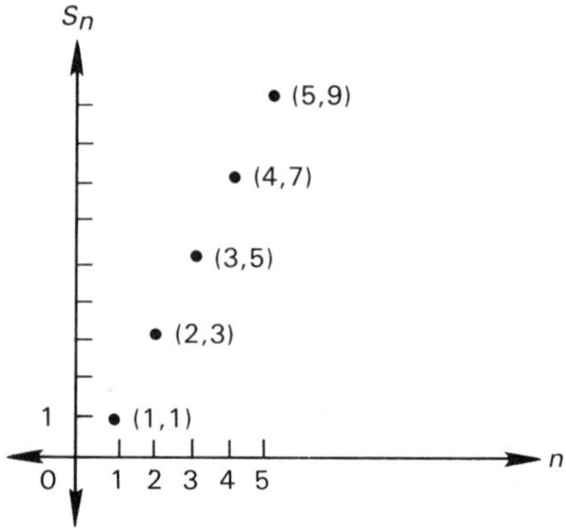

5) $s_i = 8\left(-\frac{1}{2}\right)^i$

$s_1 = 8\left(-\frac{1}{2}\right)^1 = 8\left(-\frac{1}{2}\right) = -4$

$s_2 = 8\left(-\frac{1}{2}\right)^2 = 8\left(\frac{1}{4}\right) = 2$

$s_3 = 8\left(-\frac{1}{2}\right)^3 = 8\left(-\frac{1}{8}\right) = -1$

$s_4 = 8\left(-\frac{1}{2}\right)^4 = 8\left(\frac{1}{16}\right) = \frac{1}{2}$

$s_5 = 8\left(-\frac{1}{2}\right)^5 = 8\left(-\frac{1}{32}\right) = -\frac{1}{4}$

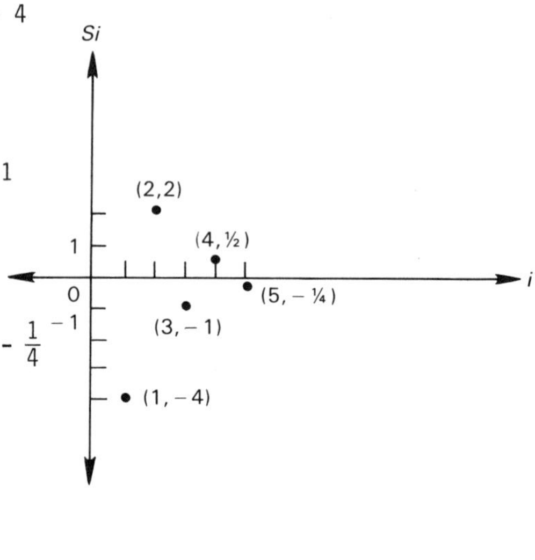

9) $s_j = 3\left(\frac{1}{10}\right)^j$

$s_1 = 3\left(\frac{1}{10}\right)^1 = 3\left(\frac{1}{10}\right) = \frac{3}{10}$

$s_2 = 3\left(\frac{1}{10}\right)^2 = 3\left(\frac{1}{100}\right) = \frac{3}{100}$

$s_3 = 3\left(\frac{1}{10}\right)^3 = 3\left(\frac{1}{1000}\right) = \frac{3}{1000}$

$s_4 = 3\left(\frac{1}{10}\right)^4 = 3\left(\frac{1}{10000}\right) = \frac{3}{10000}$

$\frac{3}{10}, \frac{3}{100}, \frac{3}{1000}, \frac{3}{10000}$

13) $s_n = \frac{1}{n} - \frac{1}{n+1}$

$s_1 = \frac{1}{1} - \frac{1}{1+1} = 1 - \frac{1}{2} = \frac{1}{2}$

$s_2 = \frac{1}{2} - \frac{1}{2+1} = \frac{1}{2} - \frac{1}{3} = \frac{1}{6}$

$s_3 = \frac{1}{3} - \frac{1}{3+1} = \frac{1}{3} - \frac{1}{4} = \frac{1}{12}$

$s_4 = \frac{1}{4} - \frac{1}{4+1} = \frac{1}{4} - \frac{1}{5} = \frac{1}{20}$

$\frac{1}{2}, \frac{1}{6}, \frac{1}{12}, \frac{1}{20}$

17) $s_n = -2 + (n - 1)3$

$s_1 = -2 + (1 - 1)3 = -2 + (0)3 = -2$

$s_2 = -2 + (2 - 1)3 = -2 + (1)3 = 1$

$s_3 = -2 + (3 - 1)3 = -2 + (2)3 = 4$

$s_4 = -2 + (4 - 1)3 = -2(3)3 = 7$

$-2, 1, 4, 7$

21) $s_n = \left(1 + \frac{1}{n}\right)^{n-1}$

$s_1 = \left(1 + \frac{1}{1}\right)^{1-1} = (2)^0 = 1$

$s_2 = \left(1 + \frac{1}{2}\right)^{2-1} = \left(\frac{3}{2}\right)^1 = \frac{3}{2}$

$s_3 = \left(1 + \frac{1}{3}\right)^{3-1} = \left(\frac{4}{3}\right)^2 = \frac{16}{9}$

$s_4 = \left(1 + \frac{1}{4}\right)^{4-1} = \left(\frac{5}{4}\right)^3 = \frac{125}{64}$

$1, \frac{3}{2}, \frac{16}{9}, \frac{125}{64}$

25) $s_k = \frac{k}{2^k}$

$s_1 = \frac{1}{2^1} = \frac{1}{2}$

$s_2 = \frac{2}{2^2} = \frac{2}{4} = \frac{1}{2}$

$s_3 = \frac{3}{2^3} = \frac{3}{8}$

$s_4 = \frac{4}{2^4} = \frac{4}{16} = \frac{1}{4}$

$\frac{1}{2}, \frac{1}{2}, \frac{3}{8}, \frac{1}{4}$

29) $s_n = 4$
$s_1 = 4$
$s_2 = 4$
$s_3 = 4$
$s_4 = 4$

4, 4, 4, 4

33) $s_k = 3(0.1)^{k-1}$

$s_7 = 3(0.1)^{7-1} = 3(0.1)^6 = 0.000003$

37) $s_i = (1 - i)^3$

$s_{12} = (1 - 12)^3 = (-11)^3 = -1331$

41) 5, 10, 15, 20, 25

$s_n = 5n$

45) $10 + 5 = 15$

$15 + 6 = 21$

$21 + 7 = 28$

15, 21, 28

49) For each n there are n factors in the numerator beginning with $2n + 1$, and there are n factors in the denominator beginning with $2n$.

$$s_1 = \frac{2 \cdot 1 + 1}{2 \cdot 1} = \frac{3}{2}$$

$$s_2 = \frac{(2 \cdot 2 + 1)(2 \cdot 2 - 1)}{(2 \cdot 2)(2 \cdot 2 - 2)} = \frac{5 \cdot 3}{4 \cdot 2} = \frac{15}{8}$$

$$s_3 = \frac{(2 \cdot 3 + 1)(2 \cdot 3 - 1)(2 \cdot 3 - 3)}{(2 \cdot 3)(2 \cdot 3 - 2)(2 \cdot 3 - 4)} = \frac{7 \cdot 5 \cdot 3}{6 \cdot 4 \cdot 2} = \frac{35}{16}$$

$$s_4 = \frac{(2 \cdot 4 + 1)(2 \cdot 4 - 1)(2 \cdot 4 - 3)(2 \cdot 4 - 5)}{(2 \cdot 4)(2 \cdot 4 - 2)(2 \cdot 4 - 4)(2 \cdot 4 - 6)} = \frac{9 \cdot 7 \cdot 5 \cdot 3}{8 \cdot 6 \cdot 4 \cdot 2}$$

$$= \frac{315}{128}$$

Section 12.2 Sums of Finite Sequences

1) $s_n = 3n$

$$s_1 + s_2 + s_3 + s_4 + s_5 = 3(1) + 3(2) + 3(3) + 3(4) + 3(5)$$

$$= 3 + 6 + 9 + 12 + 15 = 45$$

5) $s_k = \frac{3}{10^k}$

$$s_1 + s_2 + s_3 + s_4 + s_5 = \frac{3}{10} + \frac{3}{10^2} + \frac{3}{10^3} + \frac{3}{10^4} + \frac{3}{10^5}$$

$$= 0.3 + 0.03 + 0.003 + 0.0003 + 0.00003$$

$$= 0.33333$$

9) $$\sum_{k=1}^{20} 3 = \underbrace{3 + 3 + \cdots + 3}_{\text{20 addends}}$$

$$= 20 \cdot 3 = 60$$

13) $3 + \frac{3}{2} + 3\left(\frac{1}{2}\right)^{3-1} + 3\left(\frac{1}{2}\right)^{4-1} + 3\left(\frac{1}{2}\right)^{5-1} + 3\left(\frac{1}{2}\right)^{6-1} + 3\left(\frac{1}{2}\right)^{7-1}$

$= 3 + \frac{3}{2} + 3\left(\frac{1}{4}\right) + 3\left(\frac{1}{8}\right) + 3\left(\frac{1}{16}\right) + 3\left(\frac{1}{32}\right) + 3\left(\frac{1}{64}\right)$

$= \frac{3 \cdot 64 + 3 \cdot 32 + 3 \cdot 16 + 3 \cdot 8 + 3 \cdot 4 + 3 \cdot 2 + 3}{64} = \frac{381}{64}$

17) $\sum_{n=1}^{4} n^2 + \sum_{n=1}^{4} n = (1^2 + 2^2 + 3^2 + 4^2) + (1 + 2 + 3 + 4) = 30 + 10 = 40$

21) $\sum_{k=1}^{8} (-1)^k = (-1)^1 + (-1)^2 + (-1)^3 + (-1)^4 + (-1)^5 + (-1)^6 + (-1)^7$

$+ (-1)^8 = -1 + 1 - 1 + 1 - 1 + 1 - 1 + 1 = 0$

25) $\sum_{k=-3}^{3} \frac{1}{10^k} = \frac{1}{10^{-3}} + \frac{1}{10^{-2}} + \frac{1}{10^{-1}} + \frac{1}{10^0} + \frac{1}{10^1} + \frac{1}{10^2} + \frac{1}{10^3}$

$= 1000 + 100 + 10 + 1 + 0.1 + 0.01 + 0.001 = 1111.111$

29) $\sum_{n=1}^{3} \frac{n+1}{n} - \sum_{n=1}^{3} \frac{n}{n+1} = \left(\frac{1+1}{1} + \frac{2+1}{2} + \frac{3+1}{3}\right) -$

$\left(\frac{1}{1+1} + \frac{2}{2+1} + \frac{3}{3+1}\right) = 2 + \frac{3}{2} + \frac{4}{3} - \frac{1}{2} - \frac{2}{3} - \frac{3}{4}$

$= \frac{24 + 3 \cdot 6 + 4 \cdot 4 - 1 \cdot 6 - 2 \cdot 4 - 3 \cdot 3}{12} = \frac{35}{12}$

33) $5 + 10 + 15 + 20 + \cdots + 50 = 5 \cdot 1 + 5 \cdot 2 + 5 \cdot 3 + 5 \cdot 4 +$

$\cdots + 5 \cdot 10 = \sum_{n=1}^{10} 5n$

37) a) $\sum_{k=1}^{2} (2k - 1) = 2 \cdot 1 - 1 + 2 \cdot 2 - 1 = 1 + 3 = 4$

$\sum_{k=1}^{3} (2k - 1) = (2 \cdot 1 - 1 + 2 \cdot 2 - 1) + 2 \cdot 3 - 1 = 4 + 5 = 9$

$\sum_{k=1}^{4} (2k - 1) = (2 \cdot 1 - 1 + 2 \cdot 2 - 1 + 2 \cdot 3 - 1) + 2 \cdot 4 - 1$

$= 9 + 7 = 16$

37) continued

$$\sum_{k=1}^{5} (2k - 1) = (2 \cdot 1 - 1 + 2 \cdot 2 - 1 + 2 \cdot 3 - 1 + 2 \cdot 4 - 1) + 2 \cdot 5 - 1 = 16 + 9 = 25$$

$$\sum_{k=1}^{6} (2k - 1) = (2 \cdot 1 - 1 + 2 \cdot 2 - 1 + 2 \cdot 3 - 1 + 2 \cdot 4 - 1 + 2 \cdot 5 - 1) + 2 \cdot 6 - 1 = 25 + 11 = 36$$

b) Observe that when $n = 2$, the sum is 4 or 2^2. For $n = 3$, the sum is 9 or 3^2 and so on.

$$\sum_{k=1}^{n} (2k - 1) = n^2$$

41) $$\sum_{k=1}^{n} s_k + \sum_{k=1}^{n} t_k = (s_1 + s_2 + s_3 + \cdots + s_n) + (t_1 + t_2 + t_3 + \cdots + t_n)$$

$$= (s_1 + t_1) + (s_2 + t_2) + (s_3 + t_3) + \cdots + (s_n + t_n)$$

$$= \sum_{k=1}^{n} (s_k + t_k)$$

Section 12.3 Arithmetic Sequences and Series

1) 5, 7, 9

$s_n = dn + (a - d) = 2n + 1 - 2 = 2n - 1$

$$\sum_{k=1}^{20} (2n - 1) = \frac{n}{2}[2a + (n - 1)d] = \frac{20}{2}[2 \cdot 1 + 19 \cdot 2] = 10(40) = 400$$

5) $\frac{17}{2}, 9, \frac{19}{2}$

$s_n = dn + (a - d) = \frac{1}{2}n + \left(\frac{15}{2} - \frac{1}{2}\right) = \frac{n}{2} + 7$

$$\sum_{k=1}^{20} \left(\frac{n}{2} + 7\right) = \frac{n}{2}[2a + (n - 1)d] = \frac{20}{2}\left[2\left(\frac{15}{2}\right) + 19\left(\frac{1}{2}\right)\right] = 10\left(\frac{49}{2}\right) = 245$$

9) 150, 200, 250

$s_n = dn + (a - d) = 50n + (50 - 50) = 50n$

$\sum_{k=1}^{20} 50n = \frac{n}{2}[2a + (n - 1)d] = \frac{20}{2}[2 \cdot 50 + 19 \cdot 50] = 10(1050) = 10,500$

13) 5 + 10 + 15 + 20 + 25 + 30 + 35 + 40 + 45 + 50 + 55 + 60 + 65 = 455

$\frac{n}{2}[2a + (n - 1)d] = \frac{13}{2}[2 \cdot 5 + 12 \cdot 5] = \frac{13}{2}(70) = 455$

17) $S_{100} = \frac{100}{2}[2 \cdot 3 + 99 \cdot 3] = 50(303) = 15150$

21) $S_{100} = \frac{100}{2}\left[2\left(\frac{1}{7}\right) + 99 \cdot 5\right] = 50\left(\frac{3467}{7}\right) = \frac{173,350}{7}$

25) -8, 8, ⋯, 16n - 24

$a = -8, \quad d = 16$

$S_{28} = \frac{28}{2}[2(-8) + 27(16)] = 14(416) = 5824$

29) a) 1, 3, 5, ⋯, 2n - 1

$a = 1, \quad d = 2$

$S_{100} = \frac{100}{2}[2 \cdot 1 + 99 \cdot 2] = 10,000$

b) $S_n = \frac{n}{2}[2 + (n - 1)2] = \frac{2n}{2}(1 + n - 1) = n(n) = n^2$

33) $\sum_{k=1}^{30} (10k - 1)$

$s_k = (10k - 1) \qquad 10k = dk \qquad a = s_1 = 10 \cdot 1 - 1 = 9$

$10 = d$

$\sum_{k=1}^{30} (10k - 1) = \frac{30}{2}[2 \cdot 9 + 29 \cdot 10] = 15(308) = 4620$

37) $\sum_{k=1}^{n} 5k$

$s_k = 5k \qquad 5k = dk \qquad a = s_1 = 5 \cdot 1 = 5$

$5 = d$

$\sum_{k=1}^{n} 5k = \frac{n}{2}[2 \cdot 5 + (n - 1)5] = \frac{n}{2}[10 + 5n - 5] = \frac{n}{2}(5n + 5) = \frac{5n}{2}(n + 1)$

41) $-\frac{2}{3}, -\frac{1}{5}, \ldots$

$d = -\frac{1}{5} - \left(-\frac{2}{3}\right) = \frac{7}{15}$ $\qquad a = -\frac{2}{3}$

$s_{35} = -\frac{2}{3} + 34\left(\frac{7}{15}\right) = \frac{228}{15} = \frac{76}{5}$

45) $\sum_{n=6}^{20} (5n - 3)$

The formula $\frac{n}{2}[2a + (n - 1)d]$ represents a sum in which $a = s_1$, but in this case the first term is s_6. At least two strategies could be followed. It is possible to rewrite $\sum_{n=6}^{20} (5n - 3)$ as $\sum_{n=1}^{15} (5(n + 5) - 3)$ and apply the formula.

It is also possible to find $\sum_{n=1}^{20} (5n - 3)$ and subtract the first five terms, as follows:

$$\sum_{n=6}^{20} (5n - 3) = \sum_{n=1}^{20} (5n - 3) - \sum_{n=1}^{5} (5n - 3)$$

For both series:
$5n = dn$ $\quad a = s_1 =$
$5 = d$ $\quad 5 \cdot 1 - 3 = 2$

$$= \frac{20}{2}[2 \cdot 2 + 19 \cdot 5] - \frac{5}{2}[2 \cdot 2 + 4 \cdot 5]$$

$$= 10(99) - \frac{5}{2}(24) = 990 - 60 = 930$$

49) 3, u, v, 10

10 is the fourth term.

$s_4 = a + 3d = 3 + 3d = 10$

$3d = 7$

$d = \frac{7}{3}$

$u = 3 + \frac{7}{3} = \frac{16}{3}$

$v = \frac{16}{3} + \frac{7}{3} = \frac{23}{3}$

53) $s_k = -5k$

$s_1 = -5 \cdot 1 = -5$

$s_{80} = -5 \cdot 80 = -400$

$S_{80} = 80\left(\frac{-5 + (-400)}{2}\right)$

$= 80\left(-\frac{405}{2}\right)$

$= -16200$

Section 12.4 Geometric Sequences and Series

1) 16, 32, 64

$r = \frac{4}{2} = 2$

$s_n = ar^{n-1} = 2 \cdot 2^{n-1} = 2^n$

5) $\frac{1}{9},\ -\frac{1}{27},\ \frac{1}{81}$

$r = \frac{1}{-3} = -\frac{1}{3}$

$s_n = ar^{n-1} = -3\left(-\frac{1}{3}\right)^{n-1}$

9) $-\frac{16}{9},\ -\frac{32}{27},\ -\frac{64}{81}$

$r = \frac{-4}{-6} = \frac{2}{3}$

$s_n = ar^{n-1} = -6\left(\frac{2}{3}\right)^{n-1}$

13) $\sum_{n=1}^{6} 2^n = 2^1 + 2^2 + 2^3 + 2^4 + 2^5 + 2^6 = 2 + 4 + 8 + 16 + 32 + 64 = 126$

$\sum_{n=1}^{6} 2^n = \frac{a(1 - r^n)}{1 - r} = \frac{2(1 - 2^6)}{1 - 2} = \frac{2(-63)}{-1} = 126$

17) $\frac{1}{8},\ \frac{1}{4},\ \frac{1}{2},\ \ldots$

$r = \frac{\frac{1}{4}}{\frac{1}{8}} = 2$

$s_{14} = \left(\frac{1}{8}\right)(2)^{13} = \frac{8192}{8} = 1024$

21) $\sum_{k=1}^{10} 2^{k-1}$ $\qquad a = 2^0 = 1$

$r = 2$

$\sum_{k=1}^{10} 2^{k-1} = \frac{1(1 - 2^{10})}{1 - 2} = \frac{1(-1023)}{-1} = 1023$

25) $\sum_{k=1}^{5} 3^{k-4}$

$3^{k-4} = ar^{k-1}$

$3^{-3} \cdot 3^{k-1} = ar^{k-1}$

$\frac{1}{27} \cdot 3^{k-1} = ar^{k-1}$

$a = \frac{1}{27},\quad r = 3$

$\sum_{k=1}^{5} 3^{k-4} = \frac{3^{-3}(1 - 3^5)}{1 - 3} = \frac{1}{27}\left(\frac{-242}{-2}\right) = \frac{121}{27}$

29) $\sum_{k=1}^{8} 16\left(-\frac{1}{2}\right)^{k+2}$ $\qquad a = 16\left(-\frac{1}{2}\right)^{k+2} = 16\left(-\frac{1}{2}\right)^{3} = 16\left(-\frac{1}{8}\right) = -2$

$$r = \frac{\text{second term}}{\text{first term}} = \frac{16\left(-\frac{1}{2}\right)^{4}}{16\left(-\frac{1}{2}\right)^{1}} = -\frac{1}{2}$$

$$\sum_{k=1}^{8} 16\left(-\frac{1}{2}\right)^{k+2} = \frac{-2\left[1-\left(-\frac{1}{2}\right)^{8}\right]}{1-\left(-\frac{1}{2}\right)} = \frac{-2\left(\frac{255}{256}\right)}{\frac{3}{2}} = \frac{-4}{3}\cdot\frac{255}{256} = \frac{-85}{64}$$

33) The initial number is 1000 = a.

$r = 2$

$s_{10} = 1000(2^9) = 512,000$

$s_n = 1000(2^{n-1})$

37) The initial amount is 800 = a.
At the end of the first year, the investor has his principal plus 11% interest or 111% of his initial amount. $r = 1.11$.

a) The amount after 1 year is $800\ (1.11)^1$. Note that the exponent is the <u>same</u> as the year.

$s_n = 800\ (1.11)^n$

b) $s_5 = 800\ (1.11)^5$

$\log s_5 = \log 800 + \log (1.11)^5 = \log 800 + 5 \log 1.11$

$= 2.9031 + 5(.0453) = 2.9031 + 0.2265$

$\log s_5 = 3.1296$

$s_5 = 1.35 \times 10^3$

$s_5 = \$1350$

Section 12.5 Infinite Geometric Series

1) $2 + 1 + \frac{1}{2} + \cdots$

$a = 2 \qquad r = \frac{1}{2}$

$s_k = 2\left(\frac{1}{2}\right)^{k-1}$

$$\sum_{k=1}^{\infty} 2\left(\frac{1}{2}\right)^{k-1} = \frac{2}{1 - \frac{1}{2}} = \frac{2}{\frac{1}{2}} = 4$$

5) $1 - \frac{1}{2} + \frac{1}{4} - \cdots$

$a = 1 \qquad r = \frac{-\frac{1}{2}}{1} = -\frac{1}{2}$

$s_k = 1\left(-\frac{1}{2}\right)^{k-1}$

$$\sum_{k=1}^{\infty} 1\left(-\frac{1}{2}\right)^{k-1} = \frac{1}{1 - \left(-\frac{1}{2}\right)} = \frac{1}{\frac{3}{2}} = \frac{2}{3}$$

9) $-2 - \frac{1}{4} - \frac{1}{32} - \cdots$

$a = -2 \qquad r = \frac{-\frac{1}{4}}{-2} = \frac{1}{8}$

$s_k = -2\left(\frac{1}{8}\right)^{n}$

$$\sum_{k=1}^{\infty} -2\left(\frac{1}{8}\right)^{n} = \frac{-2}{1 - \frac{1}{8}} = \frac{-2}{\frac{7}{8}} = -\frac{16}{7}$$

13) $$\sum_{n=1}^{\infty} \left(-\frac{1}{3}\right)^{n-1} = \frac{1}{1 - \left(-\frac{1}{3}\right)} = \frac{1}{1 + \frac{1}{3}}$$

So the denominator of the right side, $\frac{1}{1 - \frac{1}{3}}$, is incorrect.

17) $\sum_{k=1}^{\infty} \left(\frac{1}{3}\right)^{k+1}$; $\left(\frac{1}{3}\right)^{k+1} = ar^{k-1}$

$$\left(\frac{1}{3}\right)^{2}\left(\frac{1}{3}\right)^{k-1} = ar^{k-1}$$

$$\frac{1}{9}\left(\frac{1}{3}\right)^{k-1} = ar^{k-1}$$

$$a = \frac{1}{9} , \qquad r = \frac{1}{3}$$

$$\sum_{k=1}^{\infty} \frac{1}{9}\left(\frac{1}{3}\right)^{k-1} = \frac{\frac{1}{9}}{1 - \frac{1}{3}} = \frac{\frac{1}{9}}{\frac{2}{3}} = \frac{3}{18} = \frac{1}{6}$$

21) $\sum_{k=1}^{\infty} 2(0.1)^{k-1}$ $\qquad a = 2 \qquad r = 0.1$

$$\sum_{k=1}^{\infty} 2(0.1)^{k-1} = \frac{2}{1 - 0.1} = \frac{2}{0.9} = \frac{20}{9}$$

25) $\sum_{k=1}^{\infty} (0.7)^{k-1}$ $\qquad a = (0.7)^{1-1} = 1 \qquad r = 0.7$

$$\sum_{k=1}^{\infty} 1(0.7)^{k-1} = \frac{1}{1 - 0.7} = \frac{1}{0.3} = \frac{10}{3}$$

29) $\sum_{k=1}^{\infty} 10\left(\frac{2}{3}\right)^{k-1}$ $\qquad a = 10 \qquad r = \frac{2}{3}$

$$\sum_{k=1}^{\infty} 10\left(\frac{2}{3}\right)^{k-1} = \frac{10}{1 - \frac{2}{3}} = \frac{10}{\frac{1}{3}} = 30$$

33) $\sum_{n=1}^{\infty} 7\left(-\frac{3}{4}\right)^{n-1}$ $\qquad a = 7 \qquad r = -\frac{3}{4}$

$$\sum_{n=1}^{\infty} 7\left(-\frac{3}{4}\right)^{n-1} = \frac{7}{1 - \left(-\frac{3}{4}\right)} = \frac{7}{\frac{7}{4}} = \frac{28}{7} = 4$$

37) $0.777\ldots = \frac{7}{10} + \frac{7}{100} + \frac{7}{1000} + \cdots + \frac{7}{10^n}$

$$s_k = ar^{k-1} = \frac{7}{10^k} = \frac{7}{10}\left(\frac{1}{10}\right)^{k-1}$$

$$\sum_{k=1}^{\infty} \frac{7}{10}\left(\frac{1}{10}\right)^{k-1} = \frac{\frac{7}{10}}{1 - \frac{1}{10}} = \frac{\frac{7}{10}}{\frac{9}{10}} = \frac{7}{9}$$

41) $0.0131313\ldots = \frac{13}{1000} + \frac{13}{100000} + \frac{13}{10000000} + \cdots + \frac{13}{10^{2n+1}}$

$$s_n = ar^{n-1} = \frac{13}{10^{2n+1}} = \frac{13}{10^3}\left(\frac{1}{10^2}\right)^{n-1}$$

$$\sum_{n=1}^{\infty} \frac{13}{10^3}\left(\frac{1}{10^2}\right)^{n-1} = \frac{\frac{13}{10^3}}{1 - \frac{1}{10^2}} = \frac{\frac{13}{1000}}{\frac{99}{100}} = \frac{13}{1000} \cdot \frac{100}{99} = \frac{13}{990}$$

45) a) Using $T = \frac{D}{R}$, the sequence of times will be

$$\frac{\frac{2}{3}}{\frac{1}{2}},\ \frac{\frac{2}{9}}{\frac{1}{2}},\ \frac{\frac{2}{27}}{\frac{1}{2}},\ \ldots,\ \frac{\frac{2}{3^n}}{\frac{1}{2}},\ \ldots = \frac{4}{3},\ \frac{4}{9},\ \frac{4}{27},\ \ldots,\ \frac{4}{3^n},\ \ldots$$

b) $s_n = ar^{n-1} = \frac{4}{3^n}$

$$\frac{4}{3} r^{n-1} = \frac{4}{3}\left(\frac{1}{3}\right)^{n-1}$$

$$\sum_{n=1}^{\infty} \frac{4}{3}\left(\frac{1}{3}\right)^{n-1} = \frac{\frac{4}{3}}{1 - \frac{1}{3}} = \frac{\frac{4}{3}}{\frac{2}{3}} = \frac{4}{2} = 2$$

49) The sequence of times for the second half mile is $\frac{2}{5}, \frac{4}{9}, \frac{40}{81}, \cdots$

$a = \frac{2}{5}$ $\qquad$ $r = \frac{\frac{4}{9}}{\frac{2}{5}} = \frac{20}{18} = \frac{10}{9}$

But $\left|\frac{10}{9}\right| > 1$, therefore this sequence does not have a finite sum and the horse will never be finish the race.

Section 12.6 Mathematical Induction

1) $S_n: 1 + 2 + 3 + \cdots + n = \frac{n(n+1)}{2}$

Part one, prove S_1 is true.

$$S_1: 1 = \frac{1(1+1)}{2} = 1 \qquad \text{TRUE}$$

Part two, prove If S_k, then S_{k+1}.

$$S_k: 1 + 2 + 3 + \cdots + k = \frac{k(k+1)}{2} \qquad \text{(Assume this is true)}$$

$$(k+1) = (k+1) \qquad \text{(Add the next term)}$$

$$1 + 2 + 3 + \cdots + k + (k+1) = \frac{k(k+1)}{2} + (k+1)$$

$$= \frac{k^2 + k + 2(k+1)}{2}$$

$$= \frac{k^2 + 3k + 2}{2}$$

$$= \frac{(k+1)(k+2)}{2} \qquad \text{(Show this is } S_{k+1}\text{)}$$

$$1 + 2 + 3 + \cdots + (k+1) = \frac{(k+1)[(k+1)+1]}{2}$$

Therefore S_{k+1} does follow from S_k.

Since S_1 is true and S_k implies S_{k+1}, the principle of mathematical induction makes S_n true for all integers $n \geq 1$.

Note: The preceding sentence is an appropriate statement to be put at the end of the remaining proofs. For the sake of brevity, however, it will not be repeated.

5) $S_n: \frac{5}{3} + \frac{4}{3} + 1 + \cdots + \left(-\frac{1}{3}n + 2\right) = \frac{n(11-n)}{6}$

Part one, prove S_1 is true.

$$S_1: \frac{5}{3} = \frac{(1)(11-1)}{6} = \frac{10}{6} = \frac{5}{3} \qquad \text{TRUE}$$

Part two, prove If S_k, then S_{k+1}.

$$S_R: \frac{5}{3} + \cdots + \left(-\frac{1}{3}k + 2\right) = \frac{k(11-k)}{6} \qquad \text{(Assume this is true.)}$$

5) continued

$$\left[-\frac{1}{3}(k+1)+2\right] = \left[-\frac{1}{3}(k+1)+2\right] \quad \text{(Add the next term.)}$$

$$\frac{5}{3}+\cdots+\left(-\frac{1}{3}k+2\right)+\left[-\frac{1}{3}(k+1)+2\right] = \frac{k(11-k)}{6}$$

$$+\left[-\frac{1}{3}(k+1)+2\right] = \frac{11k-k^2}{6}-\frac{(k+1)}{2}+2$$

$$= \frac{11k-k^2-2(k+1)+6(2)}{6}$$

$$= \frac{10+9k-k^2}{6} = \frac{(1+k)(10-k)}{6}$$

(Convert this into the form of the right side of S_{k+1})

Thus,

$$\frac{5}{3}+\cdots+\left[-\frac{1}{3}(k+1)+2\right] = \frac{(k+1)[11-(k+1)]}{6}$$

and S_{k+1} follows from S_k.

9) S_n: $-2-4-6-\cdots-2n = -n-n^2$

Part one, show S_1 is true.

S_1: $-2(1) = -1-(1)^2 = -2$ TRUE

Part two, show If S_k, then S_{k+1}.

S_k: $-2-4-6-\cdots-2k = -k-k^2$ (Assume this is true.)

$$-2(k+1) = -2(k+1) \quad \text{(Add the next term.)}$$

$$-2-4-6-\cdots-2k-2(k+1) = -k-k^2-2(k+1)$$

$$= -k-k^2-2k-2$$

$$= -k-1-k^2-2k-1$$

(convert this into the form of the right side of S_{k+1})

Thus

$$-2-4-6-\cdots-2(k+1) = -(k+1)-(k+1)^2$$

and S_{k+1} follows from S_k.

13) $S_n: 1^3 + 2^3 + 3^3 + \cdots + n^3 = \frac{n^2(n + 1)^2}{4}$

Part one, show S_1 is true.

$$S_1: 1^3 = \frac{1^2(1 + 1)^2}{4} = \frac{1^2(2)^2}{4} = \frac{4}{4} = 1 \qquad \text{TRUE}$$

Part two, show If S_k, then S_{k+1}.

$$S_k: 1^3 + 2^3 + 3^3 + \cdots + k^3 = \frac{k^2(k + 1)^2}{4} \quad \text{(Assume this is true.)}$$

$$(k + 1)^3 = (k + 1)^3 \quad \text{(Add the next term.)}$$

$$1^3 + 2^3 + 3^3 + \cdots + k^3 + (k + 1)^3 = \frac{k^2(k + 1)^2}{4} + (k + 1)^3$$

$$= \frac{k^2(k + 1)^2 + 4(k + 1)^3}{4} = \frac{k^2(k + 1)^2 + 4(k + 1)(k + 1)^2}{4}$$

$$= \frac{(k + 1)^2[k^2 + 4(k + 1)]}{4} = \frac{(k + 1)^2(k^2 + 4k + 4)}{4}$$

$$= \frac{(k + 1)^2(k + 2)^2}{4}$$

Thus,

$$1^3 + 2^3 + 3^3 + \cdots + (k + 1)^3 = \frac{(k + 1)^2[(k + 1) + 1]^2}{4}$$

and S_{k+1} follows from S_k.

17) $S_n: a^n < 1$ for $0 < a < 1$

Part one, prove S_1 is true.

$$S_1: a^1 < 1$$

$$a < 1 \qquad \text{TRUE}$$

Part two, prove If S_k, then S_{k+1}.

$$S_k: a^k < 1 \qquad \text{(Assume this is true.)}$$

$$a = a \qquad \text{(Multiply both sides by a.)}$$

$$a^{k+1} < a < 1$$

$$a^{k+1} < 1$$

Therefore S_{k+1} does follow from S_k.

21) $S_n: |a_0 + a_1 + \cdots + a_n| \le |a_0| + |a_1| + \cdots + |a_n|$

Part one, show S_1 is true.

S_1: $|a_0 + a_1| \le |a_0| + |a_1|$ TRUE

Part two, show If S_k, then S_{k+1}.

S_k: $|a_0 + a_1 + \cdots + a_k| \le |a_0| + |a_1| + \cdots + |a_k|$

(Assume this is true.)

Now begin with the left side of S_{k+1} and work toward the right side of S_{k+1}.

$$|a_0 + a_1 + \cdots + a_k + a_{k+1}| = |(a_0 + a_1 + \cdots + a_k) + a_{k+1}|$$

$$\le |a_0 + a_1 + \cdots + a_k| + |a_{k+1}| \quad \text{By } S_1$$

$$\le |a_0| + |a_1| + |a_2| + \cdots + |a_k| + |a_{k+1}| \quad \text{By } S_k$$

Thus,

$$|a_0 + a_1 + \cdots + a_{k+1}| \le |a_0| + |a_1| + |a_2| + \cdots + |a_{k+1}|$$

and S_{k+1} follows from S_k.

25) a) 1, 4, 16, 25

b) guess: Let S_2 be the statement $1 + 2 + 3 + \cdots + (n - 1) + n + (n - 1) + \cdots + 3 + 2 + 1 = n^2$.

c) Part one, show S_1 is true.

$S_1: 1 = 1$ TRUE

Part two, prove If S_k, then S_{k+1}.

Assume this is true:

$S_k: 1 + 2 + 3 + \cdots + (k - 1) + k + (k - 1) + \cdots + 3 + 2 + 1 = k^2$

25) continued

Add the two new "middle" numbers:

$$1 + 2 + 3 + \cdots + (k - 1) + k + \underline{(k + 1)} + \underline{k} + (k - 1) + \cdots + 3 + 2 + 1 = k^2 + (k + 1) + k = k^2 + 2k + 1$$

Thus,

$$1 + 2 + 3 + \cdots + k + (k + 1) + k + \cdots + 3 + 2 + 1 = (k + 1)^2$$

and S_{k+1} follows from S_k.

Sample Test Questions for Chapter 12

Part I

1. $s_n = \frac{n^2}{6 - 5n}$

$s_1 = \frac{(1)^2}{6 - 5(1)} = \frac{1}{1} = 1,$ $s_2 = \frac{(2)^2}{6 - 5(2)} = \frac{4}{-4} = -1,$

$s_3 = \frac{(3)^2}{6 - 5(3)} = \frac{9}{-9} = -1,$ $s_4 = \frac{4^2}{6 - 5(4)} = \frac{16}{-14} = -\frac{8}{7}$

2. $s_{10} = \frac{(10)^2}{6 - 5(10)} = \frac{100}{-44} = -\frac{25}{11}$

3. $s_n = (-1)^n + n$

$s_1 = (-1)^1 + 1 = 0,$ $s_2 = (-1)^2 + 2 = 3,$

$s_3 = (-1)^3 + 3 = 2,$ $s_4 = (-1)^4 + 4 = 5$

4. $s_{49} = -3 + (48)\left(\frac{1}{2}\right) = -3 + 24 = 21$

5. $S_{20} = \frac{20}{2}\left[2(-3) + 19\left(\frac{1}{2}\right)\right] = 10\left(\frac{7}{2}\right) = 35$

6. $-768, 192, -48 \cdots$

$r = \frac{192}{-768} = -\frac{1}{4}$

$12, -3, \frac{3}{4}$

7. $s_n = (-768)\left(-\frac{1}{4}\right)^{n-1}$

8. $\sum_{n=1}^{5} \frac{(-2)^n}{n} = \frac{(-2)^1}{1} + \frac{(-2)^2}{2} + \frac{(-2)^3}{3} + \frac{(-2)^4}{4} + \frac{(-2)^5}{5}$

$= -2 + 2 - \frac{8}{3} + 4 - \frac{32}{5} = \frac{-40 + 60 - 96}{15} = -\frac{76}{15}$

9. $\sum_{k=1}^{50} 7n$ This is an arithmetic series.

$a = 7 \cdot 1 = 7$ $d = 7$

$\sum_{k=1}^{50} \frac{50}{2}[2 \cdot 7 + 49 \cdot 7] = 25(357) = 8925$

10. $\sum_{k=1}^{4} 8\left(\frac{1}{2}\right)^k$ $\qquad a = 8\left(\frac{1}{2}\right)^1 = 4 \qquad r = \frac{1}{2}$

$$\sum_{k=1}^{4} 8\left(\frac{1}{2}\right)^k = \frac{4\left[1 - \left(\frac{1}{2}\right)^4\right]}{1 - \frac{1}{2}} = \frac{4\left(\frac{15}{16}\right)}{\frac{1}{2}} = \frac{15}{2}$$

11. $\sum_{j=1}^{101} (4j - 50)$ This is an arithmetic series.

$a = 4 \cdot 1 - 50 = 46 \qquad d = 4$

$$\sum_{j=1}^{101} (4j - 50) = \frac{101}{2}(2 \cdot 46 - 100(4)) = \frac{101}{2}(308) = 15554$$

12. $\sum_{k=1}^{8} 12\left(\frac{1}{2}\right)^{k-1}$ $\qquad a = 12\left(\frac{1}{2}\right)^{1-1} = 12 \qquad r = \frac{1}{2}$

$$\sum_{k=1}^{8} 12\left(\frac{1}{2}\right)^{k+1} = \frac{12\left[1 - \left(\frac{1}{2}\right)^8\right]}{1 - \frac{1}{2}} = \frac{12\left(\frac{255}{256}\right)}{\frac{1}{2}} = \frac{765}{32}$$

13. $\sum_{k=1}^{\infty} 8\left(\frac{3}{4}\right)^{k+1}$ $\qquad a = 8\left(\frac{3}{4}\right)^2 = \frac{9}{2} \qquad r = \frac{\text{second term}}{\text{first term}} = \frac{8\left(\frac{3}{4}\right)^3}{8\left(\frac{3}{4}\right)^2} = \frac{3}{4}$

$$\sum_{k=1}^{\infty} 8\left(\frac{3}{4}\right)^{k+1} = \frac{\frac{9}{2}}{1 - \frac{3}{4}} = \frac{9}{2} \cdot \frac{4}{1} = 18$$

14. $1 + \frac{3}{2} + \frac{9}{4} + \cdots$ $\qquad a = 1 \qquad r = \frac{\frac{3}{2}}{1} = \frac{3}{2}$

No finite sum because $\left|\frac{3}{2}\right| > 1$

15. $0.06 - 0.009 + 0.00135 - \cdots$ $\qquad a = 0.06 \qquad r = \frac{-0.009}{0.06} = -0.15$

$$\sum_{k=1}^{\infty} 0.06(-0.15)^{k-1} = \frac{0.06}{1 - (-0.15)} = \frac{0.06}{1.15} = \frac{6}{115}$$

16. $0.363636 = \frac{36}{100} + \frac{36}{10000} + \frac{36}{1000000} + \cdots$

$$= \frac{36}{10^2} + \frac{36}{10^4} + \frac{36}{10^6} + \cdots + \frac{36}{10^{2n}}$$

$a = \frac{36}{100} \qquad r = \frac{1}{10^2}$

16. continued

$$\sum_{k=1}^{\infty} \frac{36}{100}\left(\frac{1}{10^2}\right)^{k-1} = \frac{\frac{36}{100}}{1 - \frac{1}{10^2}} = \frac{36}{100} \cdot \frac{100}{99} = \frac{36}{99} = \frac{4}{11}$$

17. This is an arithmetic series with a = 10.00 and d = 0.10.

$$\sum_{k=1}^{52} [10 + (k - 1)(0.10)] \quad \cdot 10 + 51(0.10)) = 26(25.10) = \$652.60$$

18. This is an infinite geometric series with a = 24 and $r = \frac{1}{3}$.

$$\sum_{k=1}^{\infty} 24\left(\frac{1}{3}\right)^{k-1} = \frac{24}{1 - \frac{1}{3}} = 24 \cdot \frac{3}{2} = 36 \text{ feet.}$$

19. $S_n: 5 + 10 + 15 + \cdots + 5n = \frac{5n(n + 1)}{2}$

Part one; prove S_1 is true.

$S_1: 5 \cdot 1 = \frac{5(1)(1 + 1)}{2} = \frac{10}{2} = 5$ TRUE

Part two, prove If S_k, then S_{k+1}.

$S_k: 5 + 10 + 15 + \cdots + 5k = \frac{5k(k + 1)}{2}$

$5(k + 1) = 5(k + 1)$ (add the next term.)

$$5 + 10 + 15 + \cdots + 5k + 5(k + 1) = \frac{5k(k + 1)}{2} + 5(k + 1)$$

$$= \frac{5k(k + 1) + 10(k + 1)}{2}$$

$$= \frac{(k + 1)(5k + 10)}{2}$$

$$= \frac{5(k + 1)(k + 2)}{2}$$

Thus,

$$5 + 10 + 15 + \cdots + 5(k + 1) = \frac{5(k + 1)[(k + 1) + 1]}{2}$$

and S_{k+1} follows from S_k.

20. $S_n: \frac{1}{1 \cdot 3} + \frac{1}{3 \cdot 5} + \frac{1}{5 \cdot 7} + \cdots + \frac{1}{(2n-1)(2n+1)} = \frac{n}{2n+1}$

Part one, prove S_1 is true.

$$S_1: \frac{1}{1 \cdot 3} = \frac{1}{2(1)+1} = \frac{1}{3} \qquad \text{TRUE}$$

Part two, prove If S_k, then S_{k+1}.

$$S_k: \frac{1}{1 \cdot 3} + \frac{1}{3 \cdot 5} + \frac{1}{5 \cdot 7} + \cdots + \frac{1}{(2k-1)(2k+1)} = \frac{k}{2k+1}$$

$$\frac{1}{(2k+1)(2k+3)} = \frac{1}{(2k+1)(2k+3)} \quad \text{(add the next term)}$$

$$\frac{1}{1 \cdot 3} + \frac{1}{3 \cdot 5} + \frac{1}{5 \cdot 7} + \cdots + \frac{1}{(2k-1)(2k+1)} + \frac{1}{(2k+1)(2k+3)}$$

$$= \frac{k}{2k+1} + \frac{1}{(2k+1)(2k+3)} = \frac{k(2k+3)+1}{(2k+1)(2k+3)}$$

$$= \frac{2k^2+3k+1}{(2k+1)(2k+3)} = \frac{(2k+1)(k+1)}{(2k+1)(2k+3)} = \frac{k+1}{2k+3}$$

Thus,

$$\frac{1}{1 \cdot 3} + \frac{1}{3 \cdot 5} + \frac{1}{5 \cdot 7} + \cdots + \frac{1}{(2k+1)(2k+3)} = \frac{k+1}{2(k+1)+1}$$

and S_{k+1} follows from S_k.

Sample Test Questions for Chapter 12

Part II

1. (b)

$$s_{10} = \frac{10 + 2}{2(10)^2 + 3(10) - 2} = \frac{12}{228} = \frac{1}{19}$$

2. (a)

$s_1 = 1 + (-1)^1 = 0$
$s_2 = 1 + (-1)^2 = 2$
$s_3 = 1 + (-1)^3 = 0$
$s_4 = 1 + (-1)^4 = 2$

3. (c)

This is an arithmetic series with a = -2, d = -2.

$$\sum_{k=1}^{15} (-2k) = \frac{15}{2}[2(-2) + 14(-2)] = -240$$

4. (b)

$$\sum_{n=1}^{4} 3^{n-1}(n - 1) = s_1 + s_2 + s_3 + s_4$$

$s_1 = 3^{1-1}(1 - 1) = 0$

$s_2 = 3^{2-1}(2 - 1) = 3$

$s_3 = 3^{3-1}(3 - 1) = 18$

$s_4 = 3^{4-1}(4 - 1) = 81$

$s_1 + s_2 + s_3 + s_4 = 102$

5. (a)

$s_{50} = (-2) + 49(5) = 243$

6. (d)

$$S_{20} = \frac{20}{2}[2(2) + 19(-3)] = 10(-53) = -530$$

7. (d)

$$\sum_{k=1}^{50} (-2k + 3) \qquad a = -2(1) + 3 = 1 \qquad d = -2$$

7. continued

$$\sum_{k=1}^{50} (-2k + 3) = \frac{50}{2} [2(1) + 49(-2)] = 25(-96) = -2400$$

8. (c)

This is an arithmetic series with $a = 1$ and $d = \frac{1}{2}$.

$$S_{60} = \frac{60}{2}\left[2(1) + 59\left(\frac{1}{2}\right)\right] = 30\left(\frac{63}{2}\right) = 945$$

9. (c)

$$s_{100} = \left(-\frac{1}{2}\right)\left(\frac{1}{2}\right)^{99} = -\frac{1}{2^{100}}$$

10. (a)

$$\sum_{k=1}^{100} 3\left(\frac{1}{3}\right)^k \qquad 3\left(\frac{1}{3}\right)^k = ar^{k-1}$$

$$3\left(\frac{1}{3}\right)\left(\frac{1}{3}\right)^{k-1} = ar^{k-1}$$

$$1\left(\frac{1}{3}\right)^{k-1} = ar^{k-1}$$

$$a = 1 \qquad r = \frac{1}{3}$$

$$\sum_{k=1}^{100} \frac{1\left[1 - \left(\frac{1}{3}\right)^{100}\right]}{1 - \frac{1}{3}} = \frac{1 - \frac{1}{3^{100}}}{\frac{2}{3}} = \frac{3}{2}\left(1 - \frac{1}{3^{100}}\right)$$

11. (d)

This is a geometric series with $a = 512$ and $r = \frac{1}{2}$

$$\sum_{n=1}^{12} 512\left(\frac{1}{2}\right)^{n-1} = \frac{512\left[1 - \left(\frac{1}{2}\right)^{12}\right]}{1 - \frac{1}{2}} = 1024\left(\frac{4095}{4096}\right) = \$1023.75$$

12. (e)

-27, -9, -3, ...

$$a = -27 \qquad r = \frac{-9}{-27} = \frac{1}{3}$$

$$\sum_{k=1}^{\infty} (-27)\left(\frac{1}{3}\right)^{k-1} = \frac{-27}{1 - \frac{1}{3}} = \frac{-27}{\frac{2}{3}} = -\frac{81}{2}$$

13. (a)

$$\sum_{n=1}^{\infty} 100\left(\frac{7}{100}\right)^{n+1} \qquad 100\left(\frac{7}{100}\right)^{n+1} = ar^{n-1}$$

$$100\left(\frac{7}{100}\right)^{2}\left(\frac{7}{100}\right)^{n-1} = ar^{n-1}$$

$$\frac{49}{100}\left(\frac{7}{100}\right)^{n-1} = ar^{n-1}$$

$$a = \frac{49}{100} \qquad r = \frac{7}{100}$$

$$\sum_{n=1}^{\infty} \left(\frac{49}{100}\right)\left(\frac{7}{100}\right)^{n-1} = \frac{\frac{49}{100}}{1 - \frac{7}{100}} = \frac{49}{93}$$

14. (d)

$$\sum_{n=1}^{\infty} (-1)^n 2^n = \sum_{n=1}^{\infty} (-2)^n \qquad a = -2 \qquad r = -2$$

No finite sum because $|r| = |-2| > 1$

15. (c)

CHAPTER 13

PERMUTATIONS, COMBINATIONS, PROBABILITY

Section 13.1 Permutations

1) $\frac{7!}{6!} = \frac{7 \cdot 6!}{6!} = 7$

5) ${}_5P_4 = 5 \cdot 4 \cdot 3 \cdot 2 = 120$

9) Ways to choose a pitcher: 5
Ways to choose a catcher: 3
Ways to choose both: $5 \cdot 3 = 15$

13) Choices for the first letter: 4 (not <u>O</u>)
Choices for the second letter: 1 (Must be <u>O</u>)
Choices for the third letter: 3
Choices for all three: $4 \cdot 1 \cdot 3 = 12$

17) If <u>a</u> is the first letter, there are 4 choices for the second letter and 3 for the third, so $1 \cdot 4 \cdot 3 = 12$ words with "a" first.
For "a" second, $4 \cdot 1 \cdot 3 = 12$
For "a" third, $4 \cdot 3 \cdot 1 = 12$
In all: $12 + 12 + 12 = 36$

21) There are only 4 ways to choose the last digit. If it is chosen first, there are 8 choices for the second digit and 7 for the first.
$4 \cdot 7 \cdot 8 = 224$

25) ${}_5P_5 = 5! = 120$

29) Seat person A. There are 6 other people to arrange.
${}_6P_6 = 6! = 720.$

33) a) ${}_nP_1 = 10$

$$\frac{n!}{(n-1)!} = 10$$

$$\frac{n \cdot \cancel{(n-1)!}}{\cancel{(n-1)!}} = 10$$

$$n = 10$$

b) ${}_nP_2 = 12$

$$\frac{n!}{(n-2)!} = 12$$

$$\frac{n \cdot (n-1)\cancel{(n-2)!}}{\cancel{(n-2)!}} = 12$$

$$n(n-1) = 12$$

$$n^2 - n = 12$$

33) b) continued

$n^2 - n - 12 = 0$

$(n - 4)(n + 3) = 0$

$n - 4 = 0$ or $n + 3 = 0$

$n = 4$ $\quad n = -3$

Since n cannot be negative n = -3 is not a solution.

Section 13.2 Combinations

1) $_5C_2 = \frac{5!}{2!3!} = \frac{5 \cdot 4 \cdot 3!}{2 \cdot 1 \cdot 3!} = 10$ 5) $\binom{15}{15} = \frac{15!}{15!0!} = 1$

9) a) Order does not matter here.

$$_{20}C_4 = \frac{20!}{4!16!} = \frac{20 \cdot 19 \cdot 18 \cdot 17 \cdot 16!}{4 \cdot 3 \cdot 2 \cdot 1 \cdot 16!} = 4845$$

b) Order is essential here because we need to select the four officers out of 20 people by filling these 4 positions:

________ ________ ________ ________

President Vice-president Secretary Treasurer

There are 20 choices for president, 19 for vice-president, 18 for secretary, and 17 for treasurer. This is the permutation of 20 elements taken 4 at a time. Thus,

$$_{20}P_4 = \frac{20!}{4!} = 20 \cdot 19 \cdot 18 \cdot 17 = 116{,}280$$

13) Order does not matter here.

$$_{15}C_5 = \frac{15!}{5!10!} = \frac{15 \cdot 14 \cdot 13 \cdot 12 \cdot 11 \cdot 10!}{5 \cdot 4 \cdot 3 \cdot 2 \cdot 1 \cdot 10!} = 3003$$

17) Choosing the three women: $\binom{12}{3} = \frac{12!}{3!9!} = 220$

Choosing the four men: $\binom{10}{4} = \frac{10!}{4!6!} = 210$

Any group of the women may be paired with any group of the men, so the committee can be chosen in $\binom{12}{3}\binom{10}{4} = 46{,}200$ ways.

21) $_nC_4 = \frac{_nP_4}{4!} = \frac{1680}{4 \cdot 3 \cdot 2 \cdot 1} = 70$

25) For <u>each</u> suit, the possible 5-card hands are $\binom{13}{5} = \frac{13!}{5!8!} = 1287.$

Multiply by 4 because there are 4 suits. $4 \cdot \binom{13}{5} = 5148.$

29) a) $\binom{25}{5}\binom{19}{4} = \frac{25!}{5!20!} \cdot \frac{19!}{4!15!} = 53130 \cdot 3876 = 205{,}931{,}880$

b) $\binom{19}{5}\binom{25}{4} = \frac{19!}{5!14!} \cdot \frac{25!}{4!21!} = 11628 \cdot 12650 = 147{,}094{,}200$

33) $$\binom{n}{r-1} + \binom{n}{r} = \frac{n!}{(r-1)![n-(r-1)]!} + \frac{n!}{r!(n-r)!}$$

$$= \frac{rn!}{r(r-1)!(n-r+1)!} + \frac{n!(n-r+1)}{r!(n-r+1)(n-r)!}$$

$$= \frac{rn!}{r!(n-r+1)!} + \frac{n!(n-r+1)}{r!(n-r+1)!}$$

$$= \frac{rn! + n!(n-r+1)}{r!(n-r+1)!} = \frac{n!(r+n-r+1)}{r!(n-r+1)!}$$

$$= \frac{n!(n+1)}{r!(n+1-r)!} = \frac{(n+1)!}{r![(n+1)-r]!} = \binom{n+1}{r}$$

Section 13.3 The Binomial Expansion

1) $$(x+1)^5 = \binom{5}{0}x^5(1)^0 + \binom{5}{1}x^4(1)^1 + \binom{5}{2}x^3(1)^2 + \binom{5}{3}x^2(1)^3 + \binom{5}{4}x^1(1)^4 + \binom{5}{5}x^0(1)^5 = x^5 + 5x^4 + 10x^3 + 10x^2 + 5x + 1$$

5) $(a - b)^4 = \binom{4}{0} a^4b^0 - \binom{4}{1} a^3b^1 + \binom{4}{2} a^2b^2 - \binom{4}{3} ab^3 + \binom{4}{4} a^0b^4$

$= a^4 - 4a^3b + 6a^2b^2 - 4ab^3 + b^4$

9) $(a^2 + 1)^5 = \binom{5}{0} (a^2)^5(1)^0 + \binom{5}{1} (a^2)^4(1)^1 + \binom{5}{2} (a^2)^3(1)^2$

$+ \binom{5}{3} (a^2)^2(1)^3 + \binom{5}{4} (a^2)^1(1)^4 + \binom{5}{5} (a^2)^0(1)^4$

$= a^{10} + 5a^8 + 10a^6 + 10a^4 + 5a^2 + 1$

13) $\dfrac{(1 + h)^3 - 1}{h} = \dfrac{(1 + 3h + 3h^2 + h^3) - 1}{h} = \dfrac{3h + 3h^2 + h^3}{h} = 3 + 3h + h^2$

17) $\dfrac{2(x + h)^5 - 2x^5}{h}$

$= \dfrac{2\left[\binom{5}{0} x^5h^0 + \binom{5}{1} x^4h^1 + \binom{5}{2} x^3h^2 + \binom{5}{3} x^2h^3 + \binom{5}{4} x^1h^4 + \binom{5}{5} x^0h^5\right] - 2x^5}{h}$

$= \dfrac{2(x^5 + 5x^4h + 10x^3h^2 + 10x^2h^3 + 5xh^4 + h^5) - 2x^5}{h}$

$= \dfrac{2x^5 + 10x^4h + 20x^3h^2 + 20x^2h^3 + 10xh^4 + 2h^5 - 2x^5}{h}$

$= \dfrac{10x^4h + 20x^3h^2 + 20x^2h^3 + 10xh^4 + 2h^5}{h}$

$= 10x^4 + 20x^3h + 20x^2h^2 + 10xh^3 + 2h^4$

21) First five terms:

$\binom{20}{0} c^{20}h^0 + \binom{20}{1} c^{19}h^1 + \binom{20}{2} c^{18}h^2 + \binom{20}{3} c^{17}h^3 + \binom{20}{4} c^{16}h^4$

$= c^{20} + 20c^{19}h + 190c^{18}h^2 + 1140c^{17}h^3 + 4845c^{16}h^4 + \cdots$

$+ 4845c^4h^{16} + 1140c^3h^{17} + 190c^2h^{18} + 20ch^{19} + h^{20}$

25) <u>6th</u> 1 6 15 20 15 6 1

<u>7th</u> 1 7 21 35 35 21 7 1

<u>8th</u> 1 8 28 56 70 56 28 8 1

<u>9th</u> 1 9 36 84 126 126 84 36 9 1

<u>10th</u> 1 10 45 120 210 252 210 120 45 10 1

29) $r + 1 = 6,\quad r = 5 \quad \text{and} \quad n = 10$

$$\binom{n}{r} a^{n-r}b^r = \binom{10}{5} a^{10-5}(2b)^5 = 252a^5(32b^5) = 8064a^5b^5$$

33) Each term contains x^{8-r}, so for the term containing x^5, $8 - r = 5$, $r = 3$.

$$\binom{n}{r} a^{n-r}b^r = \binom{8}{3}(2x)^{8-3}(3y)^3 = 56(32x^5)(27y^3) = 48384x^5y^3$$

Section 13.4 Probability

1) The only result giving 3 heads is HHH.

$$P(3\text{ heads}) = \frac{\text{ways to get 3 heads}}{\text{total outcomes}} = \frac{1}{8}$$

5) "At most 1 heads" includes 1 heads and 0 heads.
The results with 1 heads: HTT, THT, HHT
The result with 0 heads: TTT

$$P(\text{at most 1 head}) = \frac{\text{ways to get 1 or 0 heads}}{\text{total outcomes}} = \frac{3 + 1}{8} = \frac{4}{8} = \frac{1}{2}$$

9) The results giving a sum of 7:

(1, 6), (2, 5), (3, 4), (4, 3), (5, 2), (6, 1)

$$P(\text{sum of } 7) = \frac{\text{ways to get } 7}{\text{total outcomes}} = \frac{6}{36} = \frac{1}{6}$$

13) Half of all the results give an odd sum.

$$P(\text{odd sum}) = \frac{\text{ways to get odd sum}}{\text{total outcomes}} = \frac{18}{36} = \frac{1}{2}$$

17) There is no way for both cards to be the ace of hearts. P(2 aces of hearts) = 0.

21) Because the card is replaced, the two drawings are independent events. On each drawing the chance of getting the ace of hearts is $\frac{1}{52}$.

$$P(\text{ace of hearts twice}) = \frac{1}{52} \cdot \frac{1}{52} = \frac{1}{2704}$$

25) Because the card is replaced, the two drawings are independent events.

$$P(\text{ace}) = \frac{\text{ways to get an ace}}{\text{total outcomes}} = \frac{4}{52} = \frac{1}{13}$$

$$P(\text{king}) = \frac{\text{ways to get a king}}{\text{total outcomes}} = \frac{4}{52} = \frac{1}{13}$$

$$P(\text{ace - king}) = \frac{1}{13} \cdot \frac{1}{13} = \frac{1}{169}$$

29) Number of ways to draw 2 reds: $\binom{8}{2}$

Number of ways to draw 1 green: $\binom{5}{1}$

Number of ways to draw 3 marbles: $\binom{13}{3}$

$$P(\text{2 red, 1 green}) = \frac{(\text{ways to pick 2 red}) \cdot (\text{ways to pick 1 green})}{\text{total outcomes}}$$

$$= \frac{\binom{8}{2}\binom{5}{1}}{\binom{13}{3}} = \frac{28 \cdot 5}{286} = \frac{140}{286} = \frac{70}{143}$$

33) a) Number of ways to pick 4 aces from 4: $\binom{4}{4}$

Number of ways to pick the fifth card: $\binom{48}{1}$

Total number of ways to pick 5 cards: $\binom{52}{5}$

$$P(4\text{ aces}) = \frac{\binom{4}{4}\binom{48}{1}}{\binom{52}{5}} = \frac{48}{2598960} = 0.0000185$$

b) Pick the face-value there will be 4 of: $\binom{13}{1}$

Pick the fifth card: $\binom{48}{1}$

Total number of ways to pick 5 cards: $\binom{52}{5}$

$$P(4\text{ of a kind}) = \frac{\binom{13}{1}\binom{48}{1}}{\binom{52}{5}} = \frac{624}{2598960} = 0.0002401$$

c) Pick the suit: $\binom{4}{1}$

Pick five cards from the suit: $\binom{13}{5}$

Total number of ways to pick five cards: $\binom{52}{5}$

$$P(\text{flush}) = \frac{\binom{4}{1}\binom{13}{5}}{\binom{52}{5}} = \frac{4 \cdot 1287}{2598960} = 0.0019808$$

37) $P(2 \text{ heads}) = \frac{1}{4}$

$P(1 \text{ head}) = \frac{2}{4} = \frac{1}{2}$

$P(0 \text{ heads}) = \frac{1}{4}$

$E = \frac{1}{4}(20) + \frac{1}{2}(10) + \frac{1}{4}(0) = 5 + 5 = \$10$

41) $P(7) = \frac{6}{36}$

$P(11) = \frac{2}{36}$

$P(7 \text{ or } 11) = \frac{8}{36}$

$P(\text{not } 7 \text{ or } 11) = 1 - \frac{8}{36} = \frac{28}{36}$

Odds <u>against</u> a 7 or 11: $\frac{28}{36} / \frac{8}{36} = \frac{7}{2}$; 7 to 2

Sample Test Questions for Chapter 13

Part I

1. a) $_{10}P_3 = \frac{10!}{(10 - 3)!} = \frac{10!}{7!} = 10 \cdot 9 \cdot 8 = 720$

   b) $_{10}C_3 = \frac{10!}{3!(10 - 3)!} = \frac{10!}{3!7!} = \frac{10 \cdot 9 \cdot 8}{3 \cdot 2 \cdot 1} = 120$

2. $_5P_5 = 5! = 5 \cdot 4 \cdot 3 \cdot 2 \cdot 1 = 120$

3. a) There are 9 choices for the hundreds digit and 10 choices for each of the other digits.

   $9 \cdot 10 \cdot 10 = 900$

   b) There are 9 choices for the hundreds digit.
   There are also 9 choices for the tens digit, because the one chosen for the hundreds cannot be used, but 0 is available.
   There are 8 choices for the units digit.

   $9 \cdot 9 \cdot 8 = 648$

4. Order makes no difference. $_{15}C_3 = \frac{15!}{3!12!} = 455$

5. Ways to choose the 2 boys: $\binom{12}{2} = \frac{12!}{10!2!} = 66$

   Ways to choose the 2 girls: $\binom{10}{2} = \frac{10!}{8!2!} = 45$

   Ways to choose all four: $66 \cdot 45 = 2970$

6. Each rearrangement is a new word, so order counts.

   $_6P_4 = \frac{6!}{(6 - 4)!} = \frac{6!}{2!} = 6 \cdot 5 \cdot 4 \cdot 3 = 360$

7. a) There are 2 ways to answer each question. $2^5 = 32$

   b) $P(100\%) = \frac{\text{number of ways to get all right}}{\text{total ways to answer test}} = \frac{1}{32}$

   c) $P(0\%) = \frac{\text{number of ways to get all wrong}}{\text{total ways to answer test}} = \frac{1}{32}$

8. $P(\text{even or} < 3) = P(\text{even}) + P(< 3) - P(\text{even and} < 3)$

$P(\text{even}) = \frac{3}{6}$

$P(< 3) = \frac{2}{6}$

$P(\text{even and} < 3) = \frac{1}{6}$

$P(\text{even or} < 3) = \frac{3}{6} + \frac{2}{6} - \frac{1}{6} = \frac{4}{6} = \frac{2}{3}$

9. a) There is only one outcome with a sum of 12, that is two 6's.

$P(12) = \frac{1}{36}$.

b) $P(\text{not } 12) = 1 - P(12) = 1 - \frac{1}{36} = \frac{35}{36}$

c) There are six outcomes with a sum less than 5:

(1, 1), (1, 2), (2, 1), (2, 2), (3, 1), (1, 3)

$P(\text{less than } 5) = \frac{6}{36} = \frac{1}{6}$

10. a) When the first card is replaced, the events are independent.

$P(2 \text{ aces or } 2 \text{ kings}) = P(2 \text{ aces}) + P(2 \text{ kings})$

$P(2 \text{ aces}) = \frac{4}{52} \cdot \frac{4}{52} = \frac{1}{13} \cdot \frac{1}{13} = \frac{1}{169}$

$P(2 \text{ kings}) = \frac{1}{13} \cdot \frac{1}{13} = \frac{1}{169}$

$P(2 \text{ aces or } 2 \text{ kings}) = \frac{1}{169} + \frac{1}{169} = \frac{2}{169}$

b) If the first card is not replaced, the events are no longer independent.

$P(2 \text{ aces}) = \frac{4}{52} \cdot \frac{3}{51} = \frac{12}{2652} = \frac{1}{221}$

$P(2 \text{ kings}) = \frac{1}{221}$

$P(2 \text{ aces or } 2 \text{ kings}) = \frac{1}{221} + \frac{1}{221} = \frac{2}{221}$

11. There are 12 picture cards in a deck.

Number of ways to get 4 picture cards: $\binom{12}{4}$

Number of choices for the fifth card: $\binom{40}{1}$

Total number of 5-card hands: $\binom{52}{5}$

$$P(\text{4 picture cards}) = \frac{\binom{12}{4}\binom{40}{1}}{\binom{52}{5}} = \frac{19800}{2598960} = 0.0076184$$

12. Ways to pick 4 red chips: $\binom{20}{4}$

Ways to pick 1 green chip: $\binom{15}{1}$

Total number of ways to pick 5 chips: $\binom{35}{5}$

$$P(\text{4 red}) = \frac{\binom{20}{4}\binom{15}{1}}{\binom{35}{5}} = \frac{72675}{324632} = 0.2238689$$

13. a) $P(HHHH) = \frac{1}{2^4} = \frac{1}{16}$ b) $P(TTTT) = \frac{1}{16}$

14. Number of ways to pick 2 odd integers: $\binom{5}{2}$

Total number of ways to pick 2 integers: $\binom{11}{2}$

$$P(\text{2 odds}) = \frac{\binom{5}{2}}{\binom{11}{2}} = \frac{10}{55} = \frac{2}{11}$$

15. The draw from each box is independent of the draw from the other.

P(red from A <u>and</u> red from B) = P(red from A) · P(red from B)

$$= \frac{3}{8} \cdot \frac{6}{10} = \frac{18}{80} = \frac{9}{40}$$

16. This event has 2 parts. First, a box is picked at random, then a chip is picked from that box.

$$P(\text{pick A and pick red}) = \frac{1}{2} \cdot \frac{3}{8} = \frac{3}{16}$$

$$P(\text{pick B and pick red}) = \frac{1}{2} \cdot \frac{6}{10} = \frac{6}{20} = \frac{3}{10}$$

$$P(\text{pick red}) = \frac{3}{16} + \frac{3}{10} = \frac{39}{80}$$

17. $$(x - 2y)^5 = \binom{5}{0} x^5(2y)^0 - \binom{5}{1} x^4(2y)^1 + \binom{5}{2} x^3(2y)^2 - \binom{5}{3} x^2(2y^3) + \binom{5}{4} x^1(2y)^4 - \binom{5}{5} x^0(2y)^5$$

$$= x^5 - 5x^4(2y) + 10x^3(4y^2) - 10x^2(8y^3) + 5x(16y^4) - 32y^5$$

$$= x^5 - 10x^4y + 40x^3y^2 - 80x^2y^3 + 80xy^4 - 32y^5$$

18. First four terms:

$$\binom{10}{0}\left(\frac{1}{2a}\right)^{10} b^0 - \binom{10}{1}\left(\frac{1}{2a}\right)^9 b^1 + \binom{10}{2}\left(\frac{2}{2a}\right)^8 b^2 - \binom{10}{3}\left(\frac{1}{2a}\right)^7 b^3$$

$$= \frac{1}{1024a^{10}} - \frac{10b}{512a^9} + \frac{45b^2}{256a^8} - \frac{120b^3}{128a^7}$$

$$= \frac{1}{1024a^{10}} - \frac{5b}{256a^9} + \frac{45b^2}{256a^8} - \frac{15b^3}{16a^7}$$

19. $r + 1 = 7, \quad r = 6 \quad$ and $\quad n = 11$

$$\binom{n}{r} a^{n-r}b^r = \binom{11}{6} (3a)^5b^6 = 462(243a^5)b^6 = 112{,}266a^5b^6$$

20. There are 17 terms in all, so the 9th term is the middle one.

$r + 1 = 9, \qquad r = 8 \qquad n = 16$

$$\binom{n}{r} a^{n-r}b^r = \binom{16}{8} (2x)^8(-y)^8 = 12870(256x^8)y^8 = 3{,}294{,}720x^8y^8$$

Sample Test Questions for Chapter 13

Part II

1. (c)

$9 \cdot 9 \cdot 8 \cdot 7 = 4536$

2. (d)

3. (c)

4. (b)

$\binom{5}{2}\binom{6}{3} = (10)(20) = 200$

5. (b)

$${}_nP_2 = \frac{n!}{(n-2)!} = 56$$

$$\frac{n(n-1)(n-2)!}{(n-2)!} = 56$$

$$n^2 - n - 56 = 0$$

$$(n-8)(n+7) = 0$$

$$n = 8 \quad \text{or} \quad n = -7$$

$n > 0$, therefore $n = 8$

6. (e)

I. true II. true

III. false

7. (d)

${}_6C_3 = 20$

8. (a)

$x^3 = x^r$

$3 = r \quad \binom{5}{3}(2)^2x^3 = 10(4)x^3 = 40x^3$

9. (b)

$r + 1 = 5$

$r = 4$

$$\binom{6}{4}(3a)^2(-b)^4 = 15(9a^2)b^4 = 135a^2b^4$$

10. (c)

There are seven terms, so the fourth is the middle term.

$r + 1 = 4$

$r = 3 \quad \binom{6}{3}x^3(-2)^3 = 20x^3(-8) = -160x^3$

11. (a)

12. (d)

There are 2 outcomes with all tosses the same, HHH and TTT. There are 8 outcomes in all, so 6 outcomes have the three tosses <u>not</u> the same.

P(not three the same) $= \frac{6}{8} = \frac{3}{4}$

13. (c)

14. (a)

These are independent events.

Ways to get non-spade: 39

Probability of non-spade on one draw: $\frac{39}{52} = \frac{3}{4}$

P(neither of two draws in a spade) $= \frac{3}{4} \cdot \frac{3}{4} = \frac{9}{16}$

15. (b)

There are three ways to get a sum of 10:

(4, 6), (6, 4), (5, 5).

P(sum of 10) $= \frac{3}{36} = \frac{1}{12}$